Gene Activity in Early Development

Second Edition

GENE ACTIVITY IN EARLY DEVELOPMENT

SECOND EDITION

Eric H. Davidson

California Institute of Technology

ACADEMIC PRESS   New York   San Francisco   London

A Subsidiary of Harcourt Brace Jovanovich, Publishers

ACADEMIC PRESS, INC.
111 Fifth Avenue, New York, New York 10003

United Kingdom Edition published by
ACADEMIC PRESS, INC. (LONDON) LTD.
24/28 Oval Road, London NW1

Library of Congress Cataloging in Publication Data

Davidson, Eric H   Date
    Gene activity in early development.

Bibliography: p.
    Includes index.
    1.   Developmental genetics.   I.   Title.
QH453.D38   1977   575.2$'$1   76-43320
ISBN 0−12−205160−2

To Jane Rigg
whose encouragement and intelligent
assistance made this project possible

Contents

4 Quantitative Aspects of Protein Synthesis in Early Embryos: The Role of Maternal Components

5 Transcription in Early Embryos

6 RNA Sequence Complexity and Structural Gene Transcription in Early Embryos

7   Localization of Morphogenetic Determinants in Egg Cytoplasm

8 Lampbrush Chromosomes and the Synthesis of Heterogeneous Nuclear and Messenger RNA's during Oogenesis

Preface

Knowledge of the molecular biology of early development derives from a tangled skein of measurements carried out on a number of diverse organisms. My ultimate objective in writing this second edition of "Gene Activity in Early Development" has been to review critically the many observations which are now available in order that a coherent view of at least some areas of this field might emerge. This is a nearly impossible endeavor, and at best can be only partially successful. In order to achieve a comprehensive picture, it has been necessary in many places to rely on my interpretations where direct knowledge is lacking and to choose between incompatible data. I have not shied away from this, for though I believe the book provides a documented review of certain areas of the literature, it is basically a work which is organized according to my own views of this subject. Many have of course changed since the first edition was written in 1967.

A major aim in this edition has been to develop the outlines of a quantitative treatment of some of the key classes of macromolecules in early embryos and oocytes. Thus I have devoted considerable space to estimates of RNA and protein synthesis rates, complexities, and amounts. Such information must underlie a molecular level resolution of the basic process with which development begins.

My hope is that this book will be useful to the friends, colleagues, and advanced students with whom I have spent so much time arguing the various subjects considered, and to others like them.

It is important and pleasurable for me to acknowledge the essential contributions of several of my colleagues and associates. The manuscript in its various drafts was reviewed critically and perceptively by Dr. Barbara R. Hough-Evans, Dr. William H. Klein, and Dr. Glenn A. Galau of our research group at Caltech, and I am particularly grateful for their detailed assistance. My partner, Dr. Roy J. Britten, encouraged me to carry out this project and suggested many important improvements. Professor Fotis Kafatos of Harvard University and Professor L. Dennis Smith of Purdue University each reviewed a major portion of the book, and Professor Gary Freeman of the University of Texas reviewed Chapter 7. I owe to these excellent scientists a large number of essential corrections, additions, and suggestions. I wish to extend my gratitude and thanks to these people and to the members of my research group who frequently assisted me in this project, and from whom my time and attention were often diverted. I would also like to thank Ms. Brooke Moyer who assisted with the cover design. This book is dedicated to Jane Rigg who transformed my imperfect drafts into a book and who so often remembered what I forgot.

Eric H. Davidson

1

Introduction: The Variable Gene Activity Theory of Cell Differentiation

The basic arguments leading to the proposal of transcription level regulation in animal cells are reviewed, and their history is briefly outlined. Nineteenth century cell biologists considered the possibility that differentiation can be accounted for by qualitative division of the genome during development. This view was rejected on the basis of classical experiments suggesting that the potentialities of embryonic cells are equivalent. A large amount of later evidence demonstrates genomic equivalence in differentiated cells within the same organism. The main forms of evidence include cases in which given cells or cell lineages are shown to carry out diverse functions successively ("trans-differentiation"), the observation that differentiated cells usually contain equal quantities of DNA and the same complements of DNA sequence, and proof that differentiated cell nuclei may contain all the genetic information necessary to program the development of a whole organism. Nor in general do differentiated cells which intensively express given genes contain extra copies of these genes. Current experiments show that only minor fractions of the genome are represented in the RNA of differentiated cells and that when various differentiated cells are compared, the transcribed regions constitute distinct, though overlapping, sets of DNA sequences. In addition, transcriptionally inactive DNA exists in all differentiated cells. Direct evidence for variable gene activity, i.e., transcriptional control, comes from measurements of specific messenger RNA accumulation. These show in general that when given messenger RNA's are present in the cytoplasmic polysomes, the structural genes from which they are derived are transcribed, while at other times or in other cell types, these genes are transcribed less often. The initial level of control is at the transcriptional, rather than post-transcriptional level. Thus, at least in some examples so far studied, structural gene sequences can be transcribed in chromatin only from cells in which the gene is being expressed, and sequences not represented in polysomal RNA are also undetectable in nuclear RNA. However, many levels of control are possible, and probably all are utilized to some extent. The molecular basis of transcription level regulation in animal cells is not understood, but its mechanism seems likely to depend on the way(s) in which DNA sequences are organized in the genome. Recent discoveries, showing that there exists an ordered pattern of interspersion of repetitive and nonrepetitive sequence in animal DNA, are briefly reviewed. At least some of the interspersed repetitive sequences probably play a role in structural gene function. The evidence for this is that structural genes are located in the immediate vicinity of interspersed repetitive sequences and that special subsets of repetitive sequences are contiguous to those structural genes expressed in a given state of differentiation. The view taken in this book is that transcription level regulation is the fundamental process underlying differentiation and development.

Two premises are required in arriving at the proposition that differentiation is a function of variable gene activity. The first of these is the well-understood relationship between the nucleotide sequence of the DNA in the genome and the amino acid sequence of the various proteins found in the cell. Since the structural and functional characteristics of the cell depend on its proteins, the cell requires the expression of genetic information specifying its proteins in order for these characteristics to materialize. Therefore, the differentiated state ultimately depends on the transcription of genomic information.

Early Evidence for the Informational Equivalence of Differentiated Cell Genomes

A second premise of the argument for the variable gene activity theory is that every living cell nucleus in a metazoan organism contains the same complete genome as was present in the zygote nucleus. The opposite view was proposed by Roux in 1883. Roux's idea was that differentiation of cell function results from the partition of qualitatively diverse genetic determinants into different cell nuclei. Thus, each cell would contain in its nucleus only those genes needed for the programming of its particular set of functional activities, so that developmental specialization would stem from the establishment of a mosaic of diverse partial genomes. Experiments designed specifically to test this point were carried out by Driesch (1892) and later by various other experimental embryologists (Morgan, 1927). In Driesch's experiments the normal pattern of distribution of cleavage stage nuclei into the diverse sectors of egg cytoplasm was transiently altered by forcing cleavage to occur under the pressure of a flat glass plate. When the plate was removed it was found that given nuclei had been partitioned into cells other than those normally inheriting them, but that normal development could still occur. Since nuclei normally assigned to endoderm cells could also direct the development of mesoderm, and vice versa, it was argued that these nuclei must contain the genes for mesoderm as well as those for endoderm properties. It follows that any cleavage-stage nucleus contains all the zygote genes.

The contemporaries of Driesch and his followers believed that the pressure plate experiments showed the theory of qualitative nuclear division to be incorrect (see, e.g., Wilson, 1925). However, it can be argued that these experiments demonstrate the genomic equality of nuclei only at a period of development which long precedes either the onset of cell differentiation or the onset of direct control over morphogenesis by the embryo nuclei. On the other hand, a variety of other observations suggest that even highly differentiated cells contain a complete genome equal to that contained in the zygote nucleus. It was recognized very early that the cells of an organism are normally equal in the number of distinct chromosomes which they possess. A significant early clue came from the study of dipteran polytene chromosomes, where chromosomal abnormalities associated with mutations affecting the structural characteristics of one tissue can be observed in the chromosomes of another tissue. An example was furnished by the *Bar* gene in *Drosophila*, which effects the morphogenesis of the eye. Bridges (1936) showed that a duplication in band 16A of the X chromosome is visible in the polytene chromosomes of salivary gland cells in flies bearing this mutation. Yet the salivary gland cells are evidently not responsible for the details of eye morphogenesis. Another early example was the *Notch* mutation in *Drosophila*, which in heterozygotes causes peripheral incisions and other morphological abnormalities in the wings. This phenotype was associated with a heterozygous deficiency in salivary chromosome band 3C7 (Demerec *et al.*, 1942). The nuclei of one differentiated cell type (the salivary gland) thus seem to bear genetic information required for the differentiated function of other kinds of cells, such as wing and eye forming cells.

Transdifferentiation

An interesting test of the idea that differentiated cells carry information normally expressed only in other cell types can be found in altered cell fate experiments, in which obviously differentiated cells are shown to change their specialized roles and to assume a new state of differentiation. This phenomenon is termed "transdifferentiation." For example, it was shown by Stone (1950) that in the regenerating newt eye neural retinal cells derive directly from cells which were formerly pigment cells. Changes in state of cellular differentiation also occur in the regeneration of the eye lens (reviewed by Yamada, 1967) and in other cases of regeneration, such as limb regeneration (for instance, see Namenwirth, 1974; reviewed by Hay, 1968). It has long been known that extensive changes in cell state also take place during regeneration in simple metazoa such as *Hydra* (e.g., Burnett *et al.*, 1973; Lowell and Burnett, 1973).

A great number of examples of transdifferentiation probably occur in the normal embryological development of higher animals, where cells performing a given specialized function at one stage later perform other functions. In developmental cases, however, it is often difficult to prove that the same cells or their lineal descendants are responsible for the new state of differentiation rather than clones descended from previously undifferentiated cell types. Several developmental examples have now been well described. A clear case is the transdifferentiation of larval silk gland cells in the moth. Selman and Kafatos (1974) have shown that in this animal the cuticular cells of the silk gland later redifferentiate into cells specialized for the secretion of comparatively huge volumes of $KHCO_3$ solution, which is used as a solvent for the hatching enzyme cocoonase. Another example from the same silk moth concerns cells of the labial gland. During the pupal stage these cells produce a thick cuticle, but as metamorphosis proceeds they synthesize and secrete cocoonase zymogen (Selman and Kafatos, 1975). A classic case of transdifferentiation claimed to occur many years ago by Maximow (1927) was the transformation of blood lymphocytes into phagocytic macrophages and then into collagen-secreting fibroblasts. Petrakis *et al.* (1961) studied this transformation, and showed that a culture of circulating mononuclear leukocytes sealed into a diffusion chamber is indeed able to give rise to a sheet of collageneous connective tissue fibroblasts after passing through an intermediate macrophage stage. The identity of the collagenous fibroblasts with their macrophage precursors was certified by their retention of India ink particles originally incorporated by the macrophages.

The occurrence of transdifferentiation in normal development, in regeneration, and in various other special experimental circumstances shows that differentiated cells contain genomic information other than that needed for their current specialized activities. However, it can be argued that each such case involves only a small fraction of the total genomic information possessed by the organism, since it concerns only a few functional traits. Such traits could be regarded as "closely related," *de facto*, since they belong to the repertoire of functions which are demonstrable in a single cell type. From a biochemical point of view this argument seems arbitrary, since the differences between a cell specialized for pigment synthesis and a neuron, between a leukocyte and a collagen-secreting fibroblast, or between a cuticle- and a salt-secreting cell would seem no less than those between a liver and a kidney cell. Nonetheless, it requires a considerable act of generalization to conclude that because transdifferentiation can occur, a differentiated cell nucleus actually contains the *whole* genome, and the case for this now rests to a large extent on other evidence.

DNA Constancy and Nuclear Transplantation

A critical element of evidence is the presence of twice the haploid amount of DNA in the nucleus of *every differentiated cell* (a few particular exceptions aside), except for the gametes, which contain half the somatic cell quantity. The constancy of DNA content among diploid cells was discovered by Boivin *et al*. (1948) and Mirsky and Ris (1949), and provided one of the major reasons for regarding DNA as the genetic material. Equality of DNA content among differentiated cell nuclei means that differentiation cannot in general be explained through the selective *loss* of massive fractions of unused genes from the nucleus, but this does not preclude the possibility that differentiation involves the inactivation of DNA coding for properties not manifest in a given cell type by means of chemical alterations in the genetic material. Furthermore, animal genomes are so large that the DNA of a large number of structural genes could be deleted without detectably affecting the total DNA content. It is now clear, however, that developmental alterations in the genomic DNA either do not occur or are reversible. This important conclusion rests to a large extent on nuclear transplantation experiments in which nuclei from differentiated cells are injected into mature eggs and are shown to possess the capacity to direct the complete course of development.

The most significant nuclear transplantation experiments relevant here are those carried out by Gurdon and his associates. Gurdon reported in 1962 that *Xenopus* tadpoles could be raised from enucleated eggs which had been injected with a nucleus derived from a differentiated tadpole intestinal cell. Some 24% of those nuclei able to promote cleavage were able to give rise to normal swimming tadpoles (Gurdon, 1962, 1963). Adult *Xenopus*, which were normal in all respects, including fertility, were subsequently raised from eggs injected with intestinal cell nuclei (Gurdon and Uehlinger, 1966). The intestinal cell nuclei therefore retained in a usable form all the genomic information needed for the ontogeny of circulatory, skeletal, sensory, endocrine, digestive systems, etc. This experimental *tour de force* was accomplished after a long succession of previous experiments with apparently less fortuitous material, experiments which had seemed to demonstrate that embryonic nuclei soon become irreversibly limited in their potentialities. However, in such difficult experiments positive results are by far the more significant. A number of factors are now known to affect the quantitative success of nuclear transplant experiments, including the medium in which the nuclei are transferred (Hennen, 1970) and the state of proliferation of the cells from which the donor nuclei are obtained (Kobel *et al.*, 1973). More recently cell nuclei derived from primary cultures of adult *Xenopus* kidney, lung, heart, testis, and skin have been injected into enucleated eggs, and all are able to give rise to normally differentiated swimming tadpoles (Laskey and Gurdon, 1970). Some adult frogs have also been derived from cultured epithelial cell nuclei (Gurdon and Laskey, 1970). Nuclear transplantation experiments which are essentially similar in import have now been performed with *Drosophila* eggs as well (Illmensee, 1972; Zalokar, 1973; Okada *et al.*, 1974c). Here the recipient egg nucleus or nuclei are not removed or destroyed as in the amphibian experiments, since both the donor and recipient nuclei can be marked genetically. It has been found that irrespective of site of origin (and hence the embryological fate) of donor blastoderm or preblastoderm nuclei, their descendants are capable of participating in all forms of adult differentiation, including the production of fertile gametes.

The nuclear transplantation experiments provide a powerful demonstration that differentiation need not involve irreversible changes in any significant part of the genome. Thus irreversible chemical changes in the DNA cannot be regarded as the underlying cause of distinction between active and inactive genes. The nuclear transplant experiments and those dealing with transdifferentiation lead to the view that *whatever the nature of the nuclear processes leading to differentiation, these processes are at least potentially reversible.*

Certain exceptions to the above generalization exist, some of which we shall have occasion to consider at more length in other contexts. In dipteran polytene chromosomes, DNA accumulates far beyond the 2C (i.e., twice haploid) value, and, on the other end of the scale, all the DNA is lost in certain terminally differentiated cell types such as mammalian erythrocytes or eye lens epithelium cells. In some tissues, e.g., mammalian liver, tetraploid cells occur at a regular low frequency. In most cases, however, the whole of the functional genome is lost or is duplicated, so far as is known, and these examples are therefore of little interest in interpreting the *differential* appearance of gene products. Differential DNA replication does occur in some instances. Thus, for example, satellite DNA is underreplicated during polytenization, but since satellite DNA is not transcribed, this is of little relevance here. Breuer and Pavan (1955) showed that additional DNA synthesis occurs at the sites of the large puffs in the polytene chromosomes of *Rhynchosciara*, though this is clearly not generally true for dipteran polytene chromosome puffs. Excess replication of ribosomal genes also occurs during oogenesis in many organisms. Cases in which a certain portion of the genome is lost rather than being specially replicated also exist, the most famous being that of chromosome diminution in *Ascaris* (see Chapter 7). These examples share a common characteristic: The specially replicated DNA is not passed on to cells of later generations. Thus, for example, cells bearing polytene chromosomes will never go on to divide. Similarly, the amplified ribosomal genes of the oocyte are lost at the termination of oogenesis and do not contribute to the ribosomal gene complement of the embryo (Brown and Blackler, 1972).

DNA Sequence Complements of Differentiated Cells Appear Identical

The first molecular level test of the concept of genomic equivalence is to be found in the 1964 DNA–DNA reassociation studies of McCarthy and Hoyer. The critical experiment is reproduced in Fig. 1.1, and it shows that DNA preparations extracted from mouse embryo, mouse brain, kidney, thymus, spleen, and liver are indistinguishable in their ability to compete with (labeled) mouse L cell DNA for complementary binding sites in mouse embryo DNA. McCarthy and Hoyer (1964) concluded from this and similar experiments that "all polynucleotide sequences in DNA are present in each somatic cell . . . (and that) all the sequences represented appear in the same relative proportions." This statement is manifestly true for all the DNA which participated in the reassociation reaction illustrated in Fig. 1.1, down to the level of resolution of the experiment (i.e., a few

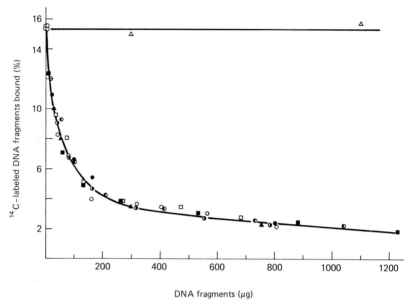

Fig. 1.1. Competition by unlabeled DNA fragments in the reaction of labeled DNA fragments with DNA agar. One microgram of ^{14}C-labeled DNA fragments (2500 cpm/μg) from mouse L cells was incubated with 0.50 gm of agar containing 60 μg of mouse embryo DNA in the presence of varying quantities of unlabeled DNA fragments from various mouse tissues, from mouse L cells, or from *Bacillus subtilis*. The percentage of ^{14}C-labeled DNA fragments bound is plotted against the amount of unlabeled DNA present. ○, mouse L cell; ●, embryo; □, brain; ■, kidney; ◑, thymus; ◐, spleen; ▲, liver; △, *Bacillis subtilis*. From B. J. McCarthy and B. H. Hoyer (1964). *Proc. Natl. Acad. Sci. U.S.A.* **52**, 915.

percent). As far as it goes, this experiment strongly reinforces the conclusion drawn from the evidence we have already considered, namely, that the various differentiated cell nuclei present in an organism are not distinguished by a diverse content of genomic information.

It is now known that animal genomes contain both repetitive and single copy DNA sequences, a fact which was not understood in 1964. In the experiment reproduced in Fig. 1.1, the conditions were such that only the repetitive DNA sequences could have renatured. That is, the effective DNA concentrations and the reaction times employed were insufficient to allow complementary DNA fragments containing sequences present only once per genome to react to form stable duplexes (see discussion of nucleic acid renaturation kinetics in Chapter 6). Since most of the presumptive genetic information is included in the single copy rather than the repetitive class of DNA sequence, the equivalence of single copy sequence among differentiated cell genomes is clearly an important issue. Unfortu-

TABLE 1.1. **Some DNA's in Which Single Copy Sequence Fractions from Various Cell Types of the Same Organism Renature with Similar Kinetics**[a]

Animal	Source of labeled single copy DNA	Source of total unlabeled DNA	Reference
Sea urchin	Cleaving embryos	Sperm	Britten (1972)
Rabbit	Lung cells[b]	Spleen	Schultz *et al.* (1973a); Brown and Church (1972)
Xenopus	Kidney cells[b]	Erythrocytes	Davidson and Hough (1971)
Mouse	L cells (fibroblasts)[b]	Liver	Hahn and Laird (1971)
		Whole embryos	Gelderman *et al.* (1971)
Calf	Kidney cells[b]	Brain, thymus, liver	Kohne and Byers (1973)

[a] As noted in text differences of up to 5–10% in renaturation rate would be undetectable.
[b] Tissue culture cells.

nately the accuracy of available techniques permits such demonstration only to within a range of ±5 to 10%. However, to this level it has now been shown that the same single copy sequences are indeed present in the single copy fraction of the DNA from any tissue of a given organism. Thus labeled single copy sequence isolated from the DNA of tissue culture cells or early embryos reacts at the same rate and to the same extent with total DNA from various adult organs as the single copy sequence in the total DNA reacts with itself. This result has been achieved with the DNA's of several organisms, extracted from a variety of cell types. Some of the clearest cases, together with references, are listed in Table 1.1.

Gene Amplification Cannot Explain Differentiation

Even if all cells contain the DNA sequences originally present in the zygote genome, this does not exclude the possibility that some cells contain extra copies of certain genes, particularly those required for intense, specialized activities. Thus, ribosomal RNA gene amplification in the oocyte constitutes a bona fide case in which selective replication of the ribosomal cistrons represents one way of regulating ribosomal RNA synthesis in response to the requirement for higher synthesis rates during oogenesis. However, this mechanism, if not unique, is at least unusual even for ribosomal RNA regulation. For example, Ritossa *et al.* (1966) showed that DNA extracted from various tissues of the chick always contains the same 0.03% of the genome homologous to ribosomal RNA, even though the rates of ribosomal RNA synthesis vary sharply in these tissues. Following the discovery of ribosomal DNA amplification, the view that structural gene amplification might provide a general explanation for cell differentiation was espoused by several writers. As a result, a detailed search has been made for amplified structural genes in several systems where relatively purified messenger RNA is available. At this writing, the three best known cases concern the structural genes for hemoglobin (Bishop *et al.*, 1972; Packman *et al.*, 1972; Gilmour *et al.*, 1974; Harrison *et al.*, 1974; Ross *et al.*, 1974), ovalbumin (Sullivan *et al.*, 1973), and silk fibroin (Suzuki *et al.*, 1972). These genes are all found to be present in only one or a few copies per haploid genome when the DNA on which the measurement is made is extracted from tissues other than those synthesizing these special proteins. Many experiments have now shown that each gene is present in the same small number of copies in DNA extracted from the active cell types, i.e., erythropoietic cells, oviduct, or silk gland, respectively. Therefore, in these cases at least, gene amplification does not occur. A different kind of experiment was carried out by Kohne and Byers (1973), who isolated single copy DNA sequences (from kidney cell DNA) complementary to the total RNA of three cow tissues: thymus, brain, and liver. They then measured the quantity of these expressed sequences in the DNA's of the different tissues. Their results showed that the expressed DNA sequences of each tissue are present at an equal frequency, whether expressed or not, in the other tissues. Again no evidence for genomic amplification of the expressed DNA sequences was found. In addition, a quantitative argument has been constructed by Kafatos (1972) which indicates that on the basis of known rates of transcription, translation, and messenger RNA turnover, even the enormous accumulation of specific proteins which occurs in some differentiated cells can be adequately explained assuming only a single copy of each structural gene per haploid genome. We may conclude that structural gene amplification is not likely to be a general mechanism for differentiation any more than is irreversible structural gene inactivation or loss.

Given the dependence of functional cell character on the cell genome, the equivalence within any one organism of these genomes leads directly to the proposition that selective variation in gene expression controls cell differentiation. Gene expression could conceivably be regulated post-transcriptionally, that is, by controlling which transcripts appear in the polysomes as messenger RNA's. However, evidence presented below indicates that the fundamental control of gene expression occurs at the transcriptional level. Certain regions of the genome are transcribed in each cell type. The apparently small fraction of the total genomic capabilities which actually materialize in any one cell type thus indicates that structural gene transcription is in general restricted to just that fraction of genes needed to direct that cell's special behavior. The rest of the structural genes in the cell are to be regarded as repressed, or inhibited from synthesizing messenger RNA. These two propositions, that in any differentiated cell only a minor fraction of the structural genes are transcribed and that most of the genome is repressed (reversibly), form the basis of the variable gene activity theory of cell differentiation. Though this concept was briefly discussed along with some other ideas by Morgan in 1934, the serious proposal that variable gene activity could underlie differentiation can be considered to date from the early 1950's, and the writings of Stedman and Stedman (1950), Mirsky (1951, 1953), and others (e.g., Sonneborn, 1950).

Direct Evidence for the Variable Gene Activity Theory of Cell Differentiation

Various forms of evidence now directly support the variable gene activity interpretation of cell differentiation. Our purpose here is to summarize briefly the major forms of this evidence, some of which is reviewed in detail in Chapter 6.

ONLY MINOR FRACTIONS OF DNA SEQUENCE ARE TRANSCRIBED IN ANY GIVEN DIFFERENTIATED CELL TYPE

RNA–DNA hybridization experiments show that in general RNA's hybridizing with less than 10% of single copy sequences are present in many differentiated cells and tissues. Mouse liver RNA, for instance, hybridizes with 2–5% of mouse single copy DNA (Hahn and Laird, 1971; Brown and Church, 1972; Grouse *et al.*, 1972). Examples of some representative values for the fraction of total single copy DNA which can be hybridized by the RNA's of single differentiated cell types (as opposed to organs or whole embryos which may contain many cell types) are 4% for rat ascites cells (Holmes and Bonner, 1974a) and 0.6–0.9% for *Xenopus* oocytes (Davidson and Hough, 1971).

DIVERSE SETS OF SEQUENCES ARE TRANSCRIBED IN EACH DIFFERENTIATED CELL TYPE

The evidence for this statement derives from RNA–DNA hybridization experiments in which RNA sequence populations from different cell types are compared. This approach was first utilized in a quantitative manner by McCarthy and Hoyer (1964), and has since been applied by many other workers. The data now extend to the transcripts of both repetitive and single copy sequence. In Table 1.2 are listed several examples of sequence

TABLE 1.2. Evidence for Transcription Level Regulation from RNA–DNA Hybridization Experiments with Total RNA or Nuclear RNA[a]

Cell types compared	Method	Relative differences in RNA populations compared	Reference
1. Repetitive sequence transcripts			
Mouse L cell, liver, spleen, kidney	Competition experiment[b]; agar-bound DNA	15–40%[c]	McCarthy and Hoyer (1964)
Sea urchin eggs, blastula, gastrula, prism[g]	Competition experiments[b]; filter-bound DNA	~35%[d]	Glišin et al. (1966)
Various stages of embryonic mouse liver	Competition experiments[b]; filter-bound DNA	Up to ~50%[c]	Church and McCarthy (1967)
Xenopus oocytes and blastulae[g]	Competition experiments[b,e]; Nygaard–Hall liquid systems and filter-bound DNA	Total lack of homology[a] (>90%)	Davidson et al. (1968)
Mouse liver and uterus ± estrogen stimulation	Competition experiments[b]; filter-bound DNA	Up to ~25%[c]	Church and McCarthy (1970)
2. Nonrepetitive sequence transcripts			
Mouse liver, brain and spleen	RNA-driven addition reactions with labeled nonrepetitive DNA[f]	Up to ~30%	Brown and Church (1972)
Mouse blastocyst and later stages[g]	RNA-driven complexity measurements with labeled nonrepetitive DNA	Total complexity increases more than 10 times	Church and Brown (1972)
Mouse liver, spleen and kidney	RNA-driven addition reaction with labeled nonrepetitive DNA[f]	>70%	Grouse et al. (1972)
Chick oviduct nuclear RNA, estrogen treated	RNA-driven complexity measurements with labeled nonrepetitive DNA	20%	Liarakos et al. (1973)
Dictyostelium at various stages	RNA-driven addition reaction with labeled nonrepetitive DNA[f]	10–40%	Firtel (1972)

[a] After E. H. Davidson and R. J. Britten (1973). *Quart. Rev. Biol.* **48,** 565.

[b] Competition experiment: labeled "reference" RNA is hybridized to DNA in the presence of unlabeled RNA from another source. If the unlabeled RNA contains similar sequences to the hydridizing labeled sequences, a stoichiometric quantity of the latter will be replaced by unlabeled molecules when the reference RNA is present in excess with respect to the DNA.

[c] This value cannot necessarily be taken as a direct measurement of the amount of qualitative difference in the sequence population, since the labeled RNA was present at far less than saturating levels with respect to the DNA. Thus, new reacting species could have been introduced as the competitor content (and the RNA concentrations) are increased.

[d] Experiment carried out with saturating labeled RNA amounts. Control curves approximate theoretical dilution curve.

[e] Nygaard–Hall system: the hybridization is carried out in liquid medium and the nucleic acids are subsequently trapped on membrane filters.

[f] RNA-driven addition reaction: an experiment in which the amount of labeled nonrepetitive DNA hybridized with the mixed RNA's of two or more tissues is compared to that hybridized with RNA's of each tissue separately. If the RNA sequences present in the tissues are the same, the amount of DNA hybridized will not rise when the tissue RNA's are mixed, while if they are partially different, an increse will be observed.

[g] Experiment discussed in detail in Chapter 6.

homology measurements which show that qualitatively diverse sets of RNA molecules exist in different cell types of the same organism. The RNA's studied in the experiments of Table 1.2 are mainly nuclear RNA's, i.e., the primary transcript population of the cell. Measurements carried out on whole cell RNA in effect concern nuclear RNA, since this is the RNA in the cell which contains by far the largest number of diverse species and consequently is responsible for most of the hybridization. The experiments referenced in Table 1.2 demonstrate that overlapping but clearly distinct sets of RNA sequences are synthesized in the various cell types tested. Only a small fraction of the nuclear RNA transcript is messenger RNA (see Chapter 6). However, since the patterns of nuclear RNA transcription are cell type specific, these observations provide evidence for transcription level regulation of the genome.

DIFFERENTIATED CELL NUCLEI CONTAIN INACTIVE (NONTRANSCRIBED) DNA

It has been demonstrated extensively that transcription in isolated chromatin is restricted to a small fraction of the DNA. Therefore, most of the chromatin DNA is inactive. This has been the result of many measurements of "template activity," that is, of the extent to which the chromatin supports RNA synthesis in an *in vitro* system containing exogenous RNA polymerase (reviewed by Bonner *et al.*, 1968). When chromatin is deproteinized, template activity increases greatly. Progress has recently been made toward separating out fractions of chromatin which are enriched for active and for inactive sequences (see, e.g., Gottesfeld *et al.*, 1974). This suggests that in some of its physical properties active chromatin differs from inactive chromatin.

CYTOLOGICAL MANIFESTATIONS OF VARIABLE GENOMIC ACTIVITY

In many cell types easily stained and densely packed heterochromatic regions of the interphase chromatin can be discerned, and high resolution radioautograph experiments have shown that these regions tend to be inactive in RNA synthesis (see, e.g., Littau *et al.*, 1964). Genetic evidence also indicates that a lower concentration of genes with known function exists in heterochromatic elements of the chromosomes which are visible at metaphase or in the polytene condition. An example is the *Drosophila* Y chromosome. However, in differentiating *Drosophila* spermatocytes the Y chromosome extrudes several lateral loops which prove to be sites of RNA synthesis in these particular cells (Hess, 1966, 1970). Transcription on these loops is absolutely required for spermatogenesis to occur successfully (see Chapter 8 for a more detailed discussion of this case). Other well-known examples include dipteran polytene chromosome puffs, which are the sites of intense RNA synthesis. These structures are also localized to specific chromosomal regions, varying according to the state of differentiation. Their appearance is correlated with the cell type, and the developmental and hormonal state of the cells. Furthermore, in one case, that of the Balbiani rings of *Chironomus* salivary gland cell chromosomes, specific chromosomal puffs have now been associated with the synthesis of specific RNA and protein products (see, for example, Grossbach, 1974; Daneholt and Hosick, 1974). In Fig. 1.2 is reproduced a photograph of one of these puffs, Balbiani ring 2, on chromosome IV. The three large puffs on this chromosome account for about 80% of its RNA synthesis, though they apparently represent only a few percent of the genome in the chromosome. Both the secretory polypeptides whose synthesis depends on the presence of the puffs and the puffs themselves are characteristic only of certain salivary gland cells. Thus the activity of a restricted set of genomic elements is responsible for the differentiated function of these cells.

MEASUREMENTS OF SPECIFIC MESSENGER RNA ACCUMULATION

Intensive studies in a few well-defined systems have now provided some of the most complete and convincing evidence for the variable gene activity theory. Among these are the hemoglobin and ovalbumin synthesizing systems already referred to. The messenger RNA's for these specialized proteins have been isolated. For both cases, developmental sequences are known in which the rate of synthesis of the specialized protein can be followed from a low or undetectable level to the high level characteristic of the mature, differentiated cell type. Measurements have shown that during these processes of differentiation, the increase in specific protein synthesis is wholly due to an increase in the concentrations of the respective messenger RNA's (for hemoglobin, see, e.g., Hunt, 1974; Gilmour *et al.*, 1974; for ovalbumin, see, e.g., Rhoads *et al.*, 1973; Palmiter, 1973; Chan *et al.*, 1973; Harris *et al.*, 1975). An example is illustrated in Fig. 1.3, reproduced from the work of Palmiter. Here it can be seen that in oviduct which has been secondarily stimulated to synthesize ovalbumin by administration of estrogen, the increase in rate of ovalbumin synthesis is quantitatively correlated with increase in the ovalbumin messenger RNA con-

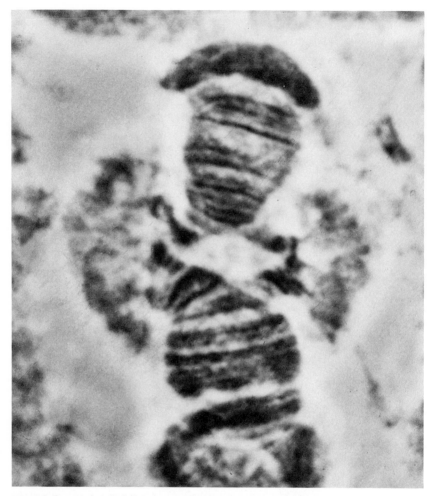

Fig. 1.2. Proximal end of chromosome IV from the salivary gland of *Chironomus pallidivit-tatus* with the large Balbiani ring (giant puff) 2, stained with acetoorcein/acetocarmine. Phase contrast, diameter of the Balbiani ring 25 μm. From U. Grossbach (1974). *Cold Spring Harbor Symp. Quant. Biol.* **38**, 619.

tent during the hours following estrogen treatment. This has been shown both by assay of ovalbumin messenger RNA in a cell-free protein synthe-sizing system as in Fig. 1.3 (Palmiter, 1973), and by titration of the mes-senger RNA sequence in hybridization experiments (Harris *et al.*, 1975). For these systems the conclusion is that structural genes for the special-

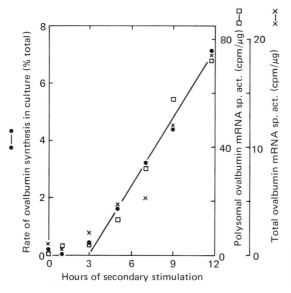

Fig. 1.3. Correlation between the induction of ovalbumin synthesis and ovalbumin messenger RNA (mRNA) accumulation during secondary stimulation with estrogen. Female chicks 2-weeks-old were stimulated with 1 mg/day 17β-estradiol benzoate for 10–12 days and then withdrawn from stimulation for several weeks prior to the secondary estrogen treatment. At the end of this period, the oviduct cells contain little or no ovalbumin mRNA. A dose of 2 mg estrogen was given for secondary stimulation. At the indicated times after secondary stimulation, groups of 10 chicks were killed and the magnum portion of the oviducts removed. The relative rate of ovalbumin synthesis was determined by incubating magnum explants with ^{3}H-labeled amino acids and then measuring the incorporation into ovalbumin and total protein. Other samples of tissue were used to extract either polysomal or total RNA. Aliquots of RNA were tested for ovalbumin mRNA activity using a cell-free, protein-synthesizing system. From R. D. Palmiter (1973). *J. Biol. Chem.* **248**, 8260.

ized proteins are activated when the cells differentiate to the stage where synthesis of these proteins becomes their major activity. Prior to this time the specific messenger RNA's are not present or exist at extremely low levels. The ovalbumin messenger RNA sequence is found at very low concentrations in the total RNA of noninduced oviduct cells and other cell types such as liver (Harris *et al.*, 1975; Axel *et al.*, 1976), and is absent from the RNA of chicken fibroblasts (Groudine and Weintraub, 1975). Thus when the ovalbumin messenger RNA is not present in the polysomes, transcription of the ovalbumin structural gene is repressed. Groudine and Weintraub (1975) showed that hemoglobin messenger RNA sequences are absent from the total RNA of normal chick fibroblasts. However, transformation or long-term growth in tissue culture may result in derepression of the hemoglobin genes (Groudine and Weintraub, 1975; Humphries et al., 1976). It might be supposed that in normal cells not expressing a given structural gene, its messenger RNA precursor is in fact synthesized at the same rate as in cells where it is being expressed, but is very rapidly degraded within the nucleus rather than accumulated in functioning cytoplasmic polysomes. However, at least for normal erythropoietic cells this possibility has been essentially ruled out by experiments on transcriptional specificity in chromatin. Chromatin extracted from reticulocyte cell nuclei transcribes RNA's in vitro which include the hemoglobin messenger sequence while in contrast the in vitro transcripts of chromatin from other cell types lack the hemoglobin message sequence (Gilmour and Paul, 1973; Axel et al., 1973; Crouse et al., 1976). Such a result is illustrated in Fig. 1.4, taken from the work of Axel

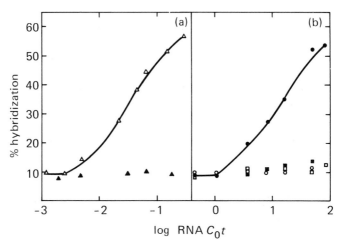

Fig. 1.4. Annealing of hemoglobin cDNA probe to polysomal messenger RNA's and to RNA's synthesized in vitro from chromatin preparations. (a) Annealing to polysomal RNA's of reticulocyte and thymus. cDNA (10,000 cpm) was annealed to 0.3 μg of duck reticulocyte polysomal RNA (△); or to 0.3 μg of calf-thymus polysomal RNA (▲). (b) Annealing to RNA's transcribed in vitro from chromatin with E. coli RNA polymerase. cDNA (10,000 cpm) was annealed to 10 μg of RNA made with reticulocyte chromatin template (●); to 10 μg of RNA made with reticulocyte DNA template (□); to 10 μg of RNA made with liver chromatin template (■); or to 10 μg of RNA made with Escherichia coli DNA template added to control preparations of reticulocyte chromatin (○). Hybrid formation was assayed by resistance to staphylococcal nuclease under conditions where this enzyme specifically degrades only single strand DNA. No background values have been substracted. RNA C_0t = moles of ribonucleotides × second/liter. The reaction of the reticulocyte chromatin transcript with the hemoglobin cDNA is about three orders of magnitude slower than with the polysomal hemoglobin messenger RNA. This is because many sequences other than the hemoglobin sequence are being transcribed in the reticulocyte chromatin. From R. Axel, H. Cedar, and G. Felsenfeld (1973). Proc. Natl. Acad. Sci. U.S.A. 70, 2029.

et al. Here a specific probe for the presence of hemoglobin messenger RNA sequence, a highly labeled DNA complementary to the messenger RNA ("cDNA"), is used to search for these sequences in the chromatin RNA transcripts. The specificity of the probe is demonstrated in Fig. 1.4a where it is shown to react only with polysomal messenger RNA from reticulocytes. Figure 1.4b demonstrates that the cDNA hybridizes with RNA made by reticulocyte chromatin but not with RNA made by liver chromatin. It is clear, therefore, that with respect to the presence or absence of hemoglobin message, the difference between the reticulocyte and the liver cell types is *transcriptional* in nature.

The main purpose of this discussion has been to summarize in very brief form the experimental support for the theory of variable gene activity, its premises, and its consequences. This is one of the most significant concepts of contemporary animal cell biology, and over the last twenty years a vast amount of serious research effort has been directed toward its proof or disproof. The case is now persuasive, and it seems almost an inescapable conclusion that it is in terms of variation in the patterns of genomic function that we must understand cell differentiation, at least to a first approximation. Direct comparisons of messenger RNA populations at various stages of development are in accord with this view, and some of these are reviewed in detail in Chapter 6.

Transcription Level Gene Regulation in Animal Cells

LEVELS OF CONTROL

Regulatory mechanisms control macromolecular interactions at every level between transcription and protein synthesis. Among the stages at which significant control may function are selection of correct DNA sequences for transcription; aggregation of active transcription complexes, probably including cofactors and chromosomal proteins as well as polymerases; post-transcriptional intranuclear "processing" of primary transcripts, including selection and excision of sequences destined for cytoplasmic function; transport from the nucleus; turnover of particular transcripts; and assembly of messenger RNA into polyribosomes. In addition, a complex array of specific translation level mechanisms affects the rates of initiation, translocation, and peptide release. A fairly large amount of knowledge has now accumulated regarding the molecular details of some of these mechanisms, in particular some of those occurring at the translational level. In general, however, we rely for understanding mainly on experimental models, i.e., possibly analogous although distinct biological systems which show how a process might work. For example, it is known that post-transcriptional processing occurs in the preparation of cytoplasmic ribosomal RNA's, and for some years valiant attempts have been made to demonstrate similar processing in the derivation of cytoplasmic messenger RNA from much larger heterogeneous nuclear RNA molecules. Whether messenger RNA's arise by processing of giant nuclear RNA's remains an unresolved question, however, and it is generally realized that in this field arguments by analogy are very unsatisfactory. In some areas we are almost completely naive. Thus we know almost nothing about transport of RNA's from the nucleus, and very little is understood regarding the nature of normal eukaryotic transcription complexes. Though the following discussion is focused on the primary level of control, all of these other levels of regulation must be important as well.

The view taken here is that the mechanism by which transcription is initiated at the correct DNA sequences in the genome is the fundamental process in gene control. It is clear that specific recognition of particular DNA sequences must be involved. Here again experimental models from certain prokaryotic systems are available and have clearly been of great value. However, most workers in the field of animal cell gene regulation are now aware that prokaryote genomes are in several ways very different from those of creatures such as ourselves. Aside from their size, which is one to four orders of magnitude smaller than the genomes of animal cells, prokaryote genomes lack the highly ordered arrangement of repetitive and nonrepetitive sequences characteristic of animal genomes (see below). They also lack the type of histone DNA complexes denoted by the term "chromatin" which are present in all animal cell nuclei, nor are they required to bear complex programs for development and differentiation as do all multicellular animal genomes.

DNA SEQUENCE ORGANIZATION AND TRANSCRIPTION LEVEL REGULATION

Our purpose here does not include a detailed consideration of any of the surviving models for transcription level gene regulation. This subject has recently been reviewed elsewhere (see, e.g., Davidson and Britten, 1973). The state of this field has changed as the result of new information concerning the organization of repetitive and nonrepetitive sequences in animal genomes. All extant models for gene regulation require or predict certain forms of DNA sequence organization (see, e.g., Callan, 1967; Britten and Davidson, 1969, 1971; Crick, 1971; Georgiev, 1972; Darnell *et al.*, 1973). It was discovered in the early 1970's that DNA sequence organization is nonrandom, and enough is known so that it now appears likely that some relation exists between the mechanisms by which structural gene activity is controlled and DNA sequence organization. A brief summary of the relevant evidence follows.

In most animal genomes the major fraction of the repetitive sequences occurs in relatively short elements about 300 ± 150 nucleotide pairs in length. These are interspersed among single copy DNA sequences 800 to several thousand nucleotide pairs long. This form of sequence organization was first discovered in the DNA of *Xenopus laevis* (Davidson *et al.*, 1973; Chamberlin *et al.*, 1975). For ease of communication we refer to the alternating interspersion of short repetitive and nonrepetitive sequences as the "*Xenopus* pattern" of genomic organization. The quantitative features of sequence interspersion in *Xenopus* DNA were first revealed in experiments in which trace quantities of labeled DNA fragments of known average lengths were reassociated with short, unlabeled fragments present in excess. After incubation permitting the renaturation of repetitive sequences only (see Chapter 6 for a discussion of the kinetics of DNA renaturation), the mixtures were passed over hydroxyapatite columns. Under appropriate conditions, hydroxyapatite binds all nucleic acid fragments which include duplex regions. Most labeled DNA fragments 3000 to 4000 nucleotides long (80%) were found to bind to the column even though the fraction of the DNA nucleotides actually present in repetitive sequence regions is only about 25%. Data from the paper of Davidson *et al.* (1973) on *Xenopus* DNA are reproduced in Fig. 1.5. As the diagrams in Fig. 1.5 indicate all the long fragments which bind to the hydroxyapatite contain repetitive sequence elements, now paired with an unlabeled DNA fragment. Most DNA fragments include nonrepetitive sequences as well. Therefore, the repetitive and nonrepetitive sequences must be closely interspersed in the genome. This experiment permits a quantitative conclusion as to the lengths of the interspersed nonrepetitive sequences (see legend to Fig. 1.5). The change in slope of the binding curve at tracer fragment lengths of about 800–1200 nucleotides shows that many of the interspersed nonrepetitive sequences are of this length. Furthermore, it can be concluded that the repetitive sequence elements must be only a few hundred nucleotides long.

The structures shown diagrammatically in Fig. 1.5 can be seen in the electron micrographs reproduced in Fig. 1.6 (Chamberlin *et al.*, 1975). Here typical reaction products obtained by reassociating the repetitive sequences in 2500 nucleotide fragments of *Xenopus* DNA are visualized. The four arms of these structures represent the single-stranded nonrepetitive "tails" emerging from the short duplex regions by which the molecules are joined. About 70% of molecules ⩾ 1500 nucleotides long contain four-ended structures with a central repetitive sequence duplex. Also shown in Fig. 1.6 are several molecules in which two interspersed repetitive sequence elements are visible, separated by longer single copy sequences. Each repetitive sequence element is marked by four single-stranded regions. Chamberlin *et al.* (1975) showed that the length of the

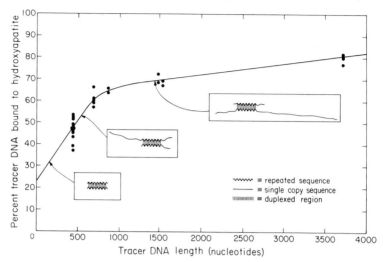

Fig. 1.5. Demonstration of interspersed repetitive and nonrepetitive sequences in *Xenopus* DNA. The ordinate shows the fraction of labeled DNA binding to hydroxyapatite as a result of the renaturation of repetitive sequence elements in the labeled DNA strands with 450 nucleotide long excess unlabeled DNA fragments. The unlabeled DNA was present in 5×10^3- to 10×10^3-fold excess. The DNA samples were annealed to C_0t 50. Renaturation conditions were 0.18 M Na$^+$, 60°C. The ordinate intercept represents the fraction of the DNA which is repetitive sequence, and the increase in binding as fragment length increases from 0 is due to nonrepetitive DNA sequence covalently linked to the repetitive sequence elements. The curve displays a change in slope occurring at 800 to 1000 nucleotides. This change indicates the presence in the DNA of a large class of single copy sequences about a thousand nucleotides long and is terminated by repetitive sequence elements. Other interspersed nonrepetitive sequences are of greater length. The mathematical basis of the quantitative interpretation of this kind of interspersion experiment is given by Graham *et al*. (1974). Enclosed in boxes are schematic diagrams of the structures formed by reassociation of carrier DNA with increasingly longer tracer DNA fragments. From E. H. Davidson and R. J. Britten (1974). *Cancer Res.* **34**, 2034; data from E. H. Davidson, B. R. Hough, C. S. Amenson, and R. J. Britten (1973). *J. Mol. Biol.* **77**, 1.

interspersed repetitive sequences in *Xenopus* DNA is about 300 ± 150 nucleotides.

Several independent lines of evidence exist which confirm 300 ± 150 nucleotides as the length of typical interspersed repetitive sequence elements. One type of evidence derives from experiments in which the repetitive sequences in DNA fragments are reassociated and the duplex-containing structures are separated and thermally melted. It is found that the increase in UV absorbance as the duplex unwinds at high temperatures (hyperchromicity) is directly proportional to the amount of the DNA present in duplex regions. Only a minor fraction of the total length of

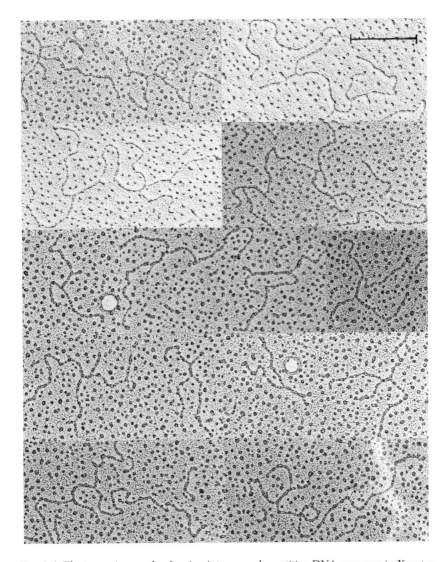

Fig. 1.6. Electron micrographs showing interspersed repetitive DNA sequence in *Xenopus* DNA. The DNA was sheared to a mean length of about 2500 nucleotides and renatured to C_0t 20. The renatured fragments were recovered from hydroxyapatite columns and then spread by a modified Kleinschmidt technique for electron microscopy from 57% formamide (Davis *et al.*, 1971). In this medium both single-stranded and double-stranded regions are extended and can be visualized. The DNA molecules shown contain repetitive duplex regions marked by four single-stranded tails which consist of nonrepetitive sequence. About 60% of the DNA was present in such structures. The molecules shown include one or more interspersed repetitive sequence elements. The mean length of 575 such terminated duplexes was 345 nucleotides. The bar represents 1000 nucleotide pairs. After M. E. Chamberlin, R. J. Britten, and E. H. Davidson (1975). *J. Mol. Biol.* **96**, 317.

DNA in the structures of Fig. 1.6 is evidently present in duplex regions. A proportionately small fraction of the amount of hyperchromicity obtained from melting native DNA results when such structures are melted. Thus analyses of the hyperchromicity of renatured DNA fragments of various lengths also show that the typical repetitive sequences are only a few hundred nucleotides long (Graham et al., 1974; Davidson et al., 1974). Other direct evidence comes from experiments in which a single strand-specific nuclease is used to treat renaturation products such as those shown in Fig. 1.6 (Davidson et al., 1974; Britten et al., 1976). About 75% of the repetitive duplex regions resistant to this nuclease (S1 nuclease from Aspergillus) average 300 nucleotides in length. An additional observation is that the remaining 25% of the repetitive DNA is organized in longer sequence elements. Little is so far known about the internal organization of these sequences.

It is obvious that the Xenopus pattern of sequence arrangement is a highly ordered one. This follows from the fact of repetitive and nonrepetitive sequence interspersion. Furthermore, neither the interspersed repetitive nor the interspersed nonrepetitive sequence lengths are distributed randomly. Extensive studies have been carried out on several animal DNA's in which the Xenopus type of sequence organization has also been demonstrated. Among these are sea urchin DNA (Graham et al., 1974), the DNA of a gastropod mollusc Aplysia californica (Angerer et al., 1975), and human DNA (Schmid and Deininger, 1975). Other observations have shown that the Xenopus pattern of sequence organization is found in the genomes of animals belonging to all major branches of the phylogenetic tree (reviewed by Davidson et al., 1975a). The inference is that this form of sequence organization stems from a remote evolutionary stage antedating the divergence of the metazoa. For example, the DNA of a coelenterate (Aurelia) is organized in approximately the same way as are the DNA's of higher animals (Goldberg et al., 1975). Similarly the genomes of both protostomial and deuterostomial animals, including an extremely primitive acoelomate protostome (Cerebratulus), are organized along the lines of the Xenopus pattern. The protostomial and deuterostomial evolutionary lines diverged before the beginning of the Cambrian fossil record. A possible interpretation is that interspersed sequence organization of the Xenopus type provided part of the basis for the evolution of multicellular forms. This in turn suggests that it has some basic functional significance.

A different pattern of sequence organization has been found in Drosophila DNA (Manning et al., 1975; Crain et al., 1976a) and also in honeybee DNA (Crain et al., 1976b). Here the repetitive sequences have an average length of 6000 nucleotides. Furthermore, the single copy sequence elements extend for at least 10,000 nucleotides on the average without interruption by repetitive sequences (Manning *et al.*, 1975). These conclusions are based primarily on electron microscope data but are supported as well by hydroxyapatite binding studies and by S1 nuclease and hyperchromicity observations. Mysteriously enough, another dipteran, the housefly *Musca domestica*, has a DNA which conforms to the *Xenopus* pattern of sequence organization (Crain *et al.*, 1976b), as do the DNA's of other insects (see, e.g., Efstratiadis *et al.*, 1976). The *Drosophila* form of organization may be related in some way to the extremely small genomes characteristic of this genus, since it is also observed in the small genome of the dipteran *Chironomus* (Wells *et al.*, 1976).

While the general arguments cited above suggest that the ordered sequence interspersion observed in most animal DNA's plays some functional role in the operation of the genome, they provide no clues as to the nature of this role. A number of possible functions have been envisioned, ranging from gene regulation to chromosome folding. To terminate this discussion it is useful to cite two items of evidence which relate sequence interspersion with structural gene function. The first of these concerns the observation that single copy structural genes are located next to interspersed repetitive sequence elements. Fragments of sea urchin DNA were prepared which contained repetitive sequences and flanking single copy regions. About one-third of the total single copy sequence length was located on these fragments. Davidson *et al.* (1975b) showed, however, that 80–100% of the polysomal messenger RNA's in sea urchin gastrulae hybridize with this fraction of the single copy sequence. Most structural genes represented in this system are therefore located nonrandomly in the genome, that is, contiguous (within about 200 nucleotides) to interspersed repetitive sequence elements. Similarly Bishop and Freeman (1974) found that the hemoglobin genes are contiguous to repetitive sequences in duck DNA. In their experiments duck DNA fragments containing repetitive sequences were shown to hybridize preferentially with hemoglobin messenger RNA.

A second observation relevant here is that the repetitive sequences contiguous to those structural genes functional at a given stage of development are a special subset. Davidson *et al.* (1976b) isolated this fraction of the repetitive sequence by hybridizing messenger RNA from sea urchin gastrulae with DNA fragments which were long enough to contain both single copy and repetitive sequences (about 1200 nucleotides). Those fragments including the hybridized regions were separated by isopycnic centrifugation in CsCl. The repetitive sequences thus selected from fragments also bearing expressed structural gene sequence were then recovered. Their diversity was compared to that of the total set of repetitive sequences in the genome. Only 10–20% of the different repetitive sequence families were found to be represented in the selected fraction. Calculations showed, however, that if there were no relation between the location of a specific repetitive sequence and the function of the contiguous single copy region as a structural gene, almost all of the repetitive sequence families in the genome should have been represented in the selected fraction. The result of the experiment shows instead that the set of genes complementary to gastrular messenger RNA tend to share a particular group of repetitive sequences. It follows that the disposition of interspersed repetitive sequences in the genome is functionally significant with respect to structural gene activity. The repetitive sequences located next to the active genes could be a general class consisting of several hundred distinct promoter sequences which will be found next to all structural genes. Or they could be transcribed into processing signals for the excision of messenger RNA's from giant nuclear RNA precursors. Alternatively, they could serve as sequence-specific transcription level regulatory sites. The latter possibility was suggested by Britten and Davidson (1969, 1971) and Davidson and Britten (1973) to account for the coordinate regulation of noncontiguous structural gene sequences. They suggested that "batteries" of genes activated together share homologous repetitive sequences, or "receptors." These would serve as binding sites for diffusible sequence-specific "activators." Thus sequence repetition in the receptor sequences defines the functionally associated genes of a "battery," since all homologous receptor sites possess the capacity to recognize the same activator molecules. This theory requires the existence of interspersed repetitive sequence elements and predicts that the specific locations of particular repetitive sequences are of crucial functional significance. However, it is important to stress that the evidence so far available in no way specifies the actual functional role of the interspersed repetitive sequences, except that some of them are involved in the mechanism of structural gene expression.

To summarize, the repetitive and nonrepetitive sequences in the genomes of metazoa are arranged in a highly ordered fashion. Sequence organization probably provides part of the physical basis for the pattern of gene control encoded in the genome. However, the mechanisms that determine which sets of genes are active in which cells remain unresolved. Nor does knowledge exist regarding the detailed molecular events occurring when previously silent genes are activated. It is certain only that large-scale regulation of structural genes occurs, and that this phenomenon underlies the processes of differentiation and development.

2

The Onset of Genome Control
in Embryogenesis

The first direct experimental studies on the relation between gene function and embryogenesis were carried out on echinoderm species hybrids in the late nineteenth century. In these experiments species with distinct morphogenetic processes evident early in development were crossed, and the point at which paternal traits first appear was regarded as an index of the stage when the embryo genome becomes active. This approach, particularly when applied at the biochemical level, remains powerful and is still in use. Morphological studies on echinoderm, amphibian, teleost, and other species hybrids show that, in general, only maternal characters are evident until gastrular or postgastrular organogenesis. This conclusion is supported by many studies in which enzymes and other proteins of paternal type are first observed at postgastrular stages. However, at least in some echinoderm species, the paternal genome is active during early development since it contributes to histone messenger RNA synthesis and to substances affecting cell surface properties. A general interpretation of the species hybrid experiments is afforded by the concept that much early morphogenesis is controlled by maternal messenger RNA and proteins. The development of enucleated eggs shows that cleavage and in some cases blastula formation, processes which entail complex cytodifferentiations, require only cytoplasmic (maternal) constituents. Significant information on the onset of embryo genome control derives from experiments in which actinomycin is used to block embryo genome transcription. Though these experiments are often difficult to interpret due to actinomycin side effects and other problems, examples exist in which protein synthesis, DNA synthesis, and cell division are shown to be largely unaffected, while RNA synthesis is effectively blocked. Nonetheless, actinomycin-treated echinoderm embryos are able to cleave, and certain embryos can progress through blastulation or gastrulation without significant RNA synthesis. Mammalian embryos are arrested in cleavage by actinomycin, but this is shown to be due to toxic side effects since these embryos successfully complete cleavage when RNA synthesis is instead blocked with α-amanitin. The actinomycin experiments clearly imply the existence of maternal message and suggest that it carries the programs for most pregastrular morphogenesis. An interesting insight derived from actinomycin experiments is that transcription required for gastrulation and organogenesis occurs many hours prior to these morphogenetic events. One possible interpretation is that actinomycin interferes with early cellular interactions. In some cases it is established that these are required for the determination of embryonic cells and subsequent morphogenesis.

It is clearly established for many animal groups that the initial, visible events of embryogenesis are not under the direct control of the embryonic cell genomes. These early events require active cell division, with all the complex biochemical processes entailed, including protein synthesis, membrane formation, mitotic spindle assembly, and chromosomal protein and DNA synthesis. The earliest stages of embryonic life also involve a certain amount of morphogenesis, in particular the construction of characteristic pregastrular structures, such as the hollow blastula of the echinoderm, or the structures demarcating the germinal layers from the nutrient syncytium in meroblastic eggs. Detailed examples of pregastrular differentiation are discussed in Chapter 3. Though specialized cellular structures exist even at these very early periods, it is only following gastrulation that organogenesis occurs, requiring a variety of new, clearly specialized cell types and tissues. *Differentiation* in this discussion is defined operationally as the active manifestation of a specialized function particular to each cell type. Differentiation requires the translation of a particular set of polysomal messenger RNA's. This definition attempts to exclude cells which are different from their neighbors merely by virtue of having passively inherited a different cytoplasm but which are carrying out no detectable special patterns of protein synthesis, even if subtle preparations for a future specialized function might be taking place. The experiments we will now review show that in many species the developmental events occurring during cleavage, blastulation, and even gastrulation are at least partly independent of immediate control by the embryo cell genomes. Only after the onset of functional tissue level differentiation in the postgastrular period is development clearly dependent on the embryonic cell genomes.

The First Species Hybrid Experiments and Their Conceptual Background

NINETEENTH CENTURY SPECIES HYBRID EXPERIMENTS

Effective investigation into the role of embryo genome control in morphogenesis can be said to have begun in 1889, with the first successful interspecific sea urchin hybrid experiments of Theodor Boveri (1893). Boveri and his followers realized that appropriate investigations carried out on hybrid embryos might yield information on genomic control over the observable events of early development. Boveri fertilized normal eggs and enucleated egg fragments of *Spherechinus granulatus* with sperm of a species belonging to a different genus, *Echinus* (= *Parechinus* = *Psammechinus*) *microtuberculatus*. The experiment was undertaken to determine if the nuclear substance alone is the bearer of hereditary qualities. Boveri reported that while true (diploid) hybrids between these species developed skeletal structures of a phenotypically hybrid character, the (haploid) hybrid merogones formed by fertilizing enucleated eggs developed strictly in accordance with paternal type. These results, he believed, demonstrated the nuclear nature of the hereditary determinants active in embryogenesis, since the sperm contributes the only nuclear components in the hybrid androgenetic merogone. The experiment explicitly indicated embryo genome control over later development, morphogenesis, and differentiation. Boveri repeated the experiment in later years, and in his last paper, which was published posthumously in 1918, he partially qualified his earlier results, pointing out several sources of error unknown in the 1890's. Later workers, using far better methods, have learned much about hybrid sea urchin merogones that was not known in Boveri's time. Some of the most important of these investigations have been carried out by Boveri's former students such as Baltzer [see reviews by Hörstadius (1936) and von Ubisch (1954)]. Even taking into account the various artifacts and interpretative difficulties pointed out by Boveri

(1918) and later writers, the early conclusions are in general correct, though real androgenetic haploid hybrids between the species used by Boveri do not display the range of developmental capacities he originally reported. In any case the Boveri experiments opened the way to an extensive investigation of the role of the embryo genome by means of morphological studies on species hybrids. In these studies hybrids are formed between species whose normal development differs sufficiently so that it is possible to determine whether the course of development follows a maternal, a hybrid, or a paternal pattern.

ORIGINS OF THE CHROMOSOME THEORY OF
CELLULAR INHERITANCE

Both the technical and the conceptual developments which made the first species hybrid experiments possible had taken place only a very short time previously. Technically, the species hybrid experiments rested on the work of Hertwig and Hertwig (1887). Boveri carried out his first hybrid merogone studies during the period in which he was associated with the laboratory of R. Hertwig. The Hertwigs had developed methods for the formation of normal and merogonal sea urchin hybrids only a few years previously. Conceptually the species hybrid experiments depended on the view that both male and female parents contribute equally to the hereditary characters of the offspring. Kölreuter had shown this as early as 1761, but his demonstration apparently did not significantly influence nineteenth century workers in cellular embryology. The writings of Nägeli in the 1880's drew attention to Kölreuter's early experimental study, and by this time his conclusions were already assumed by many investigators. In large part the modern embryological concept of equal parental contribution to inheritance grew out of cytological observations on pronuclear fusion and fertilization. Pronuclear fusion was apparently reported first by Warneck, who observed it in a snail egg in 1850, and by Bütschli (1875) who described fusion in both nematode and snail eggs. Auerbach (1874) independently described pronuclear fusion in Ascaris, as did Hertwig (1876) and Fol (1877) in the sea urchin [see Fol (1878) for an extensive consideration of earlier and contemporary references]. Shortly thereafter Strasburger (1877) described pronuclear fusion in plants. These observations were of very great significance in the intellectual development of the field, since they produced the conviction that the nuclei of the male and female gametes carry the parental hereditary determinants.

Figure 2.1 shows the pronuclei of a human egg as viewed in the electron microscope and also illustrates the apparent equality of the egg and sperm pronuclei. This was the feature which was so suggestive to the early observers. The true significance of the pronuclear fusion phenomenon did

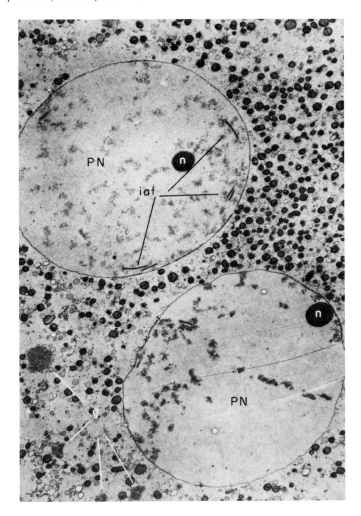

Fig. 2.1. Region of the penetrated human ovum with male and female pronuclei (PN). Nucleoli (n) and intrapronuclear annulate lamellae (ial) are in evidence. Note the numerous organelles which populate the cytoplasm adjacent to the pronuclei. (g) Golgi complex. × 5400. From L. Zamboni, D. R. Mishell, Jr., J. H. Bell, and M. Baca (1966). *J. Cell Biol.* **30,** 579.

not become completely clear until 1883, with the publication of Van Beneden's careful observations of chromosomal movements before, during, and after fertilization in *Ascaris.* Several of the essential plates from Van Beneden's classic 1883 paper are reproduced in Fig. 2.2. The use of *Ascaris megalocephala* contributed enormously to the correct interpreta-

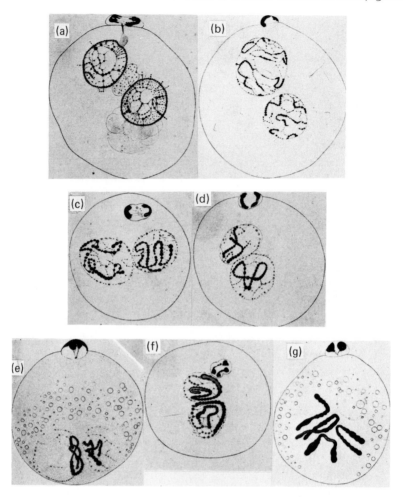

Fig. 2.2. Successive stages of pronuclear fusion and first cleavage mitosis in *Ascaris* as given by Van Beneden. The chromosomes become visible while the pronuclei are still separate (a)–(e). As fusion occurs the four chromosomes remain clearly identifiable (f) and (g), and can still be observed as the first cleavage metaphase plate forms and mitosis is carried out (h)–(m). From E. Van Beneden (1883). *Arch. Biol.* **4**, 265.

tion of fertilization, for reasons which are clear from these figures. In contrast to the case in the fertilized human (Fig. 2.1) or the sea urchin egg, the individual chromosomes can be seen clearly before, during, and after the actual fusion of the gamete pronuclei. In *Ascaris megalocephala*, furthermore, there are only two chromosomes per haploid set at this

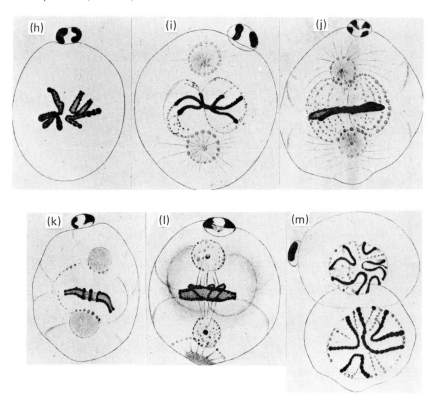

Fig. 2.2 *(continued)*

stage. Accurate observations were possible even with the relatively primitive cytological procedures in use in the 1880's. With Van Beneden's study it became evident that the chromosomes are the particular nuclear components which are contributed equally by both parents to the zygote, and that perfect replicates of these chromosomes are distributed in equal fashion to the two blastomeres as the first cleavage occurs (Fig. 2.2). From this point onward the development of the chromosome theory of heredity occurred with great rapidity. The year after the publication of Van Beneden's paper Nägeli (1884) proposed that every cell contains an "idioplasm" which includes all the hereditary characters of the species. At about the same time O. Hertwig (1885), who studied the fusion of egg and sperm pronuclei in sea urchin fertilization, Strasburger (1884), who had already seen meiosis in plant material, and Roux (1883) and Weismann (1885) all arrived at the conclusion that the "idioplasm" (i.e., the genome) must be located in the chromosomes themselves. In this manner the cellular theory of chromosomal heredity came into being, and the stage was set for the study of how the genomic determinants might operate in differentiation and development.

Evidence for Delayed Onset of Embryo Genome Control from Echinoderm Species Hybrid Studies

MORPHOGENESIS IN ECHINODERM SPECIES HYBRIDS

Out of the hundreds of echinoderm hybrid experiments reported in the literature we can consider briefly only a few. Many of the hybrid crosses studied by Boveri and his followers resulted in the early death of the hybrids. It has been shown that in many cases the early arrest and death of hybrids are associated with mitotic failures and the elimination of a significant portion of the chromosomes. This phenomenon was first established by Baltzer (1910) in a study of hybrids formed by fertilizing *Spherechinus* eggs with *Strongylocentrotus* sperm. These hybrids rarely survive up to the pluteus stage, and when they do they display maternal, rather than hybrid, skeletal characteristics. Baltzer showed that these results are due to the elimination of most of the paternal chromosomes in the course of the cleavage mitoses. Early arrest, lethality, and failure of true diploid hybrid formation stemming from mitotic irregularities occur in many other hybrid crosses as well. Even in cases where obvious chromosome loss cannot be observed, partial elimination of the paternal genome may occur. An example is the hybrid between *Paracentrotus lividus* (female) and *Arbacia lixula* (male) which arrests at gastrulation, though no chromatin elimination has been reported. Denis and Brachet (1969) have shown that *Arbacia* DNA sequences are significantly underrepresented in the cellular genomes of this lethal hybrid. Other sea urchin hybrids develop into the pluteus stage or beyond, and in these both sets of chromosomes are preserved and replicated. Many morphological studies have shown that in such cases development eventually displays the expected hybrid phenotype, with the characteristics of both parental species being evident. However, it is striking that early morphogenesis occurs strictly in accordance with the maternal patterns. In echinoid hybrids the influence of the paternal genome is not generally evident until the primary mesenchyme cells have been given off and gastrulation is under way. It was frequently pointed out by early writers that since pregastrular development in such species hybrids is maternal in form it is probably not controlled by the embryo cell genomes, as these are composed equally of maternal and paternal components. The conclusions drawn from diploid sea urchin hybrid experiments are also derived from studies of androgenetic hybrid merogones. Thus the particular influence of the foreign nucleus is evident only during or after gastrulation in both classes of hybrid. The reader is referred to reviews by Fankhauser (1956) and Chen (1967) for summaries of echinoderm hybrid studies relevant to this point. Reviews of the earlier classical experiments on echinoderm hybrids are to be found in Tennent (1922) and Morgan (1927).

An experiment designed explicitly to indicate the time of onset of embryo genome control over morphogenesis, i.e., the point at which morphogenesis ceases to follow a strictly maternal form, was carried out by Tennent in 1914. Tennent fertilized eggs of *Cidaris tribuloides* with sperm of *Lytechinus variegatus* and compared the time required by the hybrids to form an archenteron, and the site of primary mesenchyme cell formation, with the corresponding developmental parameters in the parental species (Table 2.1). The experiment apparently shows that up to the moment at which the primary mesenchyme cells are produced, development is of the *Cidaris* pattern, which is to say that the course of morphogenesis fails to reflect the hybrid composition of the embryonic genome. At this point, however, the effect of the hybrid genome begins to be seen in the mode of primary mesenchyme formation. This follows the paternal rather than the maternal pattern.

Table 2.2 presents data summarized from one of the most elegant of the early sea urchin species hybrid experiments, published by Driesch in 1898. Among the many hybrids considered in this work are the relatively viable combinations which can be formed among the genera *Echinus*, *Spherechinus*, and *Strongylocentrotus*. These hybrids develop at least to the pluteus stage [many of these results were verified subsequently by Hörstadius (1936)]. Driesch reported that the skeletal rods in the hybrid plutei were of intermediate form, thus showing the influence of both parental species. However, as Table 2.2 shows, counts of the primary mesenchyme cells in the parental and hybrid embryos revealed that *only*

TABLE 2.1. Development of *Cidaris* (♀) × *Lytechinus* (♂) **Hybrids**[a]

	Archenteron invagination (hours)	Mesenchyme formation (hours)	Site of origin of primary mesenchyme cells
Cidaris (♀)	20–33	23–26 (follows invagination)	Archenteron tip
Lytechinus (♂)	9	8 (precedes invagination)	Archenteron base and sides
Hybrid	20	24 (follows invagination)	Archenteron base and sides

[a] Collated from Tennent, 1914.

TABLE 2.2. Primary Mesenchyme Cells in Sea Urchin Hybrids[a]

Egg		Sperm	Average No. primary mesenchyme cells[b]
Echinus	×	Echinus	55 ± 4
Spherechinus	×	Spherechinus	33 ± 4
Spherechinus	×	Echinus	35 ± 5
Strongylocentrotus	×	Strongylocentrotus	49 ± 3
Spherechinus	×	Strongylocentrotus	33 ± 3

[a] Collated from Driesch, 1898.
[b] The mesenchyme cells of 15, 25, 47, 15, and 22 embryos were counted in the five samples, respectively. The average and *range* of the counts are given.

the number of primary mesenchyme cells characteristic of the maternal species appears in these hybrids. The experiments of Table 2.2 and many other subsequent experiments suggest that the onset of embryo genome control over morphogenesis in the sea urchin occurs after the partitioning off of the future mesenchyme cells. This is an important point, since it indicates the complicated and precise nature of the maternal program which controls early development. The initial details of primary mesenchyme cell formation are evidently directed by this maternal program (see Chapter 3 for a description of primary mesenchyme differentiation).

BIOSYNTHETIC PROCESSES IN ECHINODERM SPECIES HYBRIDS

Modern studies of interspecific sea urchin hybrids have shown that many biochemical parameters appear to follow the same patterns as the morphological parameters so far mentioned (Chen, 1967; a general review of earlier biochemical studies on sea urchin species hybrids is to be found in this reference). An interesting example of maternal control of a pregastrular synthetic process is the appearance of hatching enzyme, which has been studied in an intrageneric species hybrid by Barrett and Angelo (1969). This enzyme, as its name implies, frees the blastula by digesting the enveloping membrane. It is one of the earlier specific enzyme activities known to appear during sea urchin embryogenesis. Barrett and Angelo showed that the hatching enzymes of *Strongylocentrotus purpuratus* and *S. franciscanus* can be distinguished by their different sensitivity to inhibition by Mn^{2+}. The hatching enzyme which appears in the hybrid blastulae of these two species is always of the maternal type, for both reciprocal crosses. In these same hybrids it has also been shown that

the time at which the synthesis of echinochrome begins is that characteristic of the maternal parent for both reciprocal crosses (Chaffee and
Mazia, 1963). This activity occurs during gastrulation in S. *purpuratus*.
Hybrids between S. *drobachiensis* and S. *purpuratus* synthesize both maternal and paternal forms of echinochrome in the pluteus stage (Griffiths,
1965; Ozaki, 1975). By this stage they also display hybrid isozyme patterns.
Thus, just as seen in morphological studies with viable sea urchin species
hybrids, the influence of the embryo genome on these synthetic activities
becomes detectable only at postgastrular stages. While this is indeed the
predominant result achieved in the echinoderm species hybrid experiments, several examples of pregastrular embryo genome action have also
been discovered. One example is the synthesis of aryl sulfatase, a cell
surface enzyme which increases sharply in sea urchin embryos from the
blastula stage on. Fedecka-Bruner *et al.* (1971) showed that in viable
hybrids of S. *purpuratus* and *Allocentrotus fragilis* the increase in enzyme
is intermediate between that occurring in the two parental species, and
concluded that the enzyme increase probably results from new messenger
RNA transcription in the embryo genomes. An early gene action appears
to determine cell surface properties which affect embryonic cell adhesion. McClay and Hausman (1975) showed that cells from hybrids between *Lytechinus variegatus* and *Tripneustes eschulentis* begin to display
affinities for cells of the paternal species in an *in vitro* aggregation test
system as early as the mesenchyme blastula stage. This change clearly
requires expression of the paternal genome.

Another interesting case concerns interordinal hybrids, made between
the sand dollar *Dendraster excentricus* and a sea urchin S. *purpuratus*.
Not surprisingly, these hybrids block during gastrulation or at the prism
stage, depending on which species provides the egg. In this cross the DNA
content per cell remains relatively constant in the hybrid embryos
(Brookbank and Cummins, 1972). Furthermore, sequence homology experiments show that the hybrid genomes apparently retain equal complements of both parental DNA's (Whiteley and Whiteley, 1972). Yet only
maternal forms of several enzymes can be detected, including hatching
enzyme and cytoplasmic malate dehydrogenase (Ozaki and Whiteley,
1970; Whiteley and Whiteley, 1972; Ozaki, 1975). Immunological studies
show that protein antigens of paternal type cannot be detected in these
hybrids (Badman and Brookbank, 1970). Both of these approaches are of
course relatively insensitive, since they require accumulation of the
specific enzymes or antigenic proteins before detection is possible. On the
other hand, paternal as well as maternal forms of histone are synthesized
in the hybrid blastulae (the f1 histone of these species is detectably different). Thus the paternal genome is not only present but is capable of

being properly transcribed (Easton et al., 1974). Lack of paternal genome effect on other characteristics measured therefore cannot be easily attributed to deficiencies in the state of the paternal genome itself. These observations are best understood in terms of the general reliance of most, if not all, pregastrular developmental events on the store of maternal gene products already present in the egg at fertilization. Thus, for example, respiratory rate and DNA synthesis rate, which have been studied by Whiteley and Baltzer (1958), both follow the maternal pattern.

Embryo Genome Control in the Development of Species Hybrids in Chordates

AMPHIBIAN HYBRIDS

Interspecific amphibian hybrids of both diploid and androgenetic haploid type have been widely studied. An extensive series of hybrid crosses among various species of the anuran genus *Rana* was described by Moore (1941) who observed various parameters of early development in the hybrids, in particular the rate at which they attain given stages of development at various temperatures. Moore's hybrids fall into two major classes: those which arrest at the onset of gastrulation and those which proceed beyond. Developmental arrest and death in the former group can probably be attributed to mitotic abnormalities (see, e.g., Schönmann, 1938; Hennen, 1963), as in analogous cases with sea urchin species hybrids. Moore showed that morphogenesis always conforms to the maternal rate until gastrulation in hybrids which arrest at this point. In hybrids progressing further deviations from the maternal rate of development are not observed until neurulation. According to the Boveri–Driesch interpretation of the species hybrid experiments, this would suggest that at the gross level of observable morphogenesis, embryo genome control is not established until neural plate formation.

It is noteworthy that even in lethal urodele and anuran crosses the embryos may proceed through cleavage and blastulation and do not arrest until gastrulation [see the summary tables presented in Fankhauser (1956) and Chen (1967)]. For example androgenetic haploids formed between urodeles of separate genera, viz., *Triton palmatus* and *Salamandra maculosa*, actually manage to develop as far as the late blastula. Similarly, diploid hybrids between these species suffer massive mitotic disorders but nonetheless manage to proceed to the beginning of gastrulation (Schönmann, 1938). Throughout cleavage and blastulation the respiration rate of lethally crossed hybrid frogs remains maternal, with deviations occurring only as the stage of arrest approaches. This is so, for instance, in the *Triton* × *Salamandra* cross (Chen, 1960). DNA synthesis follows the same pattern as respiratory activity. Thus, according to Gregg and Løvtrup (1960), DNA accumulation in the lethal combination *Rana pipiens* × *Rana sylvatica* continues to occur at the normal maternal rate up to the time when the controls have begun to neurulate, despite the fact that morphological development in these hybrids arrests hours earlier at gastrulation. The hybrid embryos survive for 4 or 5 days following developmental arrest, and there is evidence for the continuation of many biosynthetic activities during this period. Johnson (1971) has reported the presence of many apparently differentiated structures in these embryos, including multinucleate muscle cells, banded myofibrils, and collagenous basement membranes. Furthermore, a paternal form of lactate dehydrogenase appears in the blocked gastrulae at the normal number of hours postfertilization, when controls are in the heartbeat stage. Other new enzymes which would have appeared in normal embryos by this time are not detectable in the lethal hybrids. Johnson (1969) reported as well that cell contact interactions normally characterizing gastrular cells are lacking in the cells of these hybrid embryos. As in the lethal echinoderm hybrids considered above, the embryo genomes of the *Rana pipiens* × *Rana sylvatica* hybrid remain partially functional, though a complete or correct postgastrular transcription program evidently fails to operate.

   In hybrids between *Xenopus laevis* and *Xenopus mulleri*,* a cross which gives rise to viable but generally infertile offspring, the ribosomal RNA genes of *X. mulleri* origin are always repressed relative to those of *X. laevis* origin (Honjo and Reeder, 1973). This is observed irrespective of whether the *X. mulleri* genome was contributed by the maternal or paternal parent. Honjo and Reeder (1973) concluded that both *X. laevis* and *X. mulleri* egg cytoplasm repress the *X. mulleri* ribosomal genes. Thus a species difference at the DNA level may result in sharp differences in the extent to which a gene is transcribed. The inference is that there is an incompatibility between the DNA and some molecular constituents of the egg cytoplasm required for ribosomal gene transcription.

   Numerous studies have been carried out in which the activity of specific enzymes is measured in hybrid frog embryos. The general purpose of these experiments is to determine when embryo structural gene function begins to affect the enzyme complement of the embryo. For such studies it is desirable to use closely related congeners or even subspecific variants in order to obtain viable hybrids, and yet exploit the advantages of elec-

---

* The organism used in these studies has since been identified as *X. borealis*, not *X. mulleri*.

trophoretically distinct forms of enzyme. The enzymes which have been investigated include malate dehydrogenase (MDH), lactate dehydrogenase (LDH), 6-phosphogluconate dehydrogenase (6-PDH), glucose-phosphate isomerase (GPI), isocitrate dehydrogenase (IDH), and glutamic-oxaloacetic transaminase (GOT) (see, e.g., Wright and Subtelny, 1971; Johnson and Chapman, 1972; Gallien et al., 1973). No paternal forms of these enzymes have been detected prior to the heartbeat stage. Paternal enzyme variants begin to appear at this point. Since sufficient enzyme must be accumulated to permit detection, the actual onset of transcription of the embryo structural genes must occur prior to the time at which the product first can be measured. However, if the rates of synthesis and quantities of these enzymes are typical, this should not require more than several hours. Therefore, it is unlikely that active synthesis of the paternal enzyme forms begins more than a day before the heartbeat stage. At least for MDH, LDH, 6-PDH, and IDH it is also clear that the maternal forms detected in the hybrid embryos represent proteins stored from earlier periods rather than proteins newly synthesized from maternal messenger RNA. This has been shown nicely by Wright and Subtelny (1971) in experiments on haploid androgenetic hybrid merogones in which sperm nucleus provides the only genome present. It is found that paternal forms of the four enzymes appear at the same stages in the androgenetic hybrids as in their diploid counterparts. However in diploid embryos *new hybrid forms of the enzymes* themselves appear at the same time as the paternal forms, while in the androgenetic hybrids only the new paternal forms are seen. Such an experiment is shown in Fig. 2.3. The hybrid enzyme (6-PDH), which migrates in the intermediate position between the two parental variants in Fig. 2.3, contains subunits of both parental origins. The significance of the experiment is that such hybrid proteins can form only when the enzyme is being synthesized. The absence of the hybrid enzyme in the androgenetic embryos shows that the maternal forms are not being synthesized even after stage 19. That is, no maternal messenger RNA for this enzyme is being translated. Therefore, rather than being synthesized *de novo* throughout early development, the maternal enzyme is already present in the egg at fertilization.

TELEOST AND ASCIDIAN HYBRIDS

Extensive species hybrid studies have also been carried out with teleost embryos [earlier work is reviewed by Morgan (1927)], and the results, overall, bear close resemblance to the amphibian species hybrid results. The rate of early development of teleost hybrids is generally maternal.

6-PGD

(+)

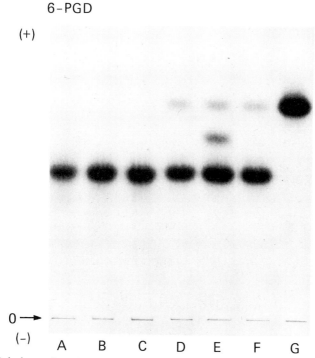

0 →

(-) A B C D E F G

Fig. 2.3. Gel electrophoresis patterns of 6-phosphogluconate dehydrogenase (6-PGD) in embryos resulting from the diploid cross *Rana pipiens* (pip) ♀ × *Rana berlandieri* (ber) ♂ and the androgenetic haploid cross pip ♀ × ber ♂. Column A is 6-PGD from the androgenetic haploid cross pip ♀ × ber ♂ at the time when the diploid hybrid cross (column C) is at Shumway stage 11 (midgastrula). Columns D and F are 6-PGD isozymes from androgenetic haploid crosses at Shumway stages 15 and 16 (late neurula) which are the same chronological age as the diploid cross shown in column E, Shumway stage 19 (heartbeat). Column G is 6-PGD from *R. berlandieri* liver. From D. A. Wright, and S. Subtelny (1971). *Dev. Biol.* **24,** 119.

Newman (1914) showed that for teleost hybrids this criterion may be unreliable, however, and in certain cases at least the cleavage rate may be altered in either direction by the foreign sperm. Developmental arrest in either intrageneric or intergeneric teleost species hybrids occurs only during gastrulation, or later, according to Newman (1915), and the initial signs of paternal genome influence are detectable only with gastrulation. Chromatophore development, the patterns of optic cup, circulatory system formation, and other aspects of embryonic organogenesis in teleosts

suggest control by the hybrid embryo genomes, as expected. Summarizing his own many studies, Newman observed (1914):

> It is doubtless during the process of gastrulation that the first steps in
> differentiation take place, and it is very interesting to note that in so many
> heterogeneric crosses the developmental stoppages occur at the onset of
> or during the process of gastrulation. The conclusion would seem to be
> obvious that any teleost spermatozoon may play a role in cleavage equiv-
> alent to that of agents that are successful in artificial parthenogenesis . . .
> (but the paternal genome) fails to exercise any really hereditary function
> until the embryo begins to differentiate tissues and organs.

With some exceptions this summary remains equally pertinent to the
interspecific echinoderm and amphibian hybrids we have already discussed, except that clear paternal effects can usually be perceived only
after gastrulation in most of these hybrids. Another point of similarity
between teleost hybrids and both echinoderm and amphibian hybrids is
the fact that teleost crosses resulting in lethal developmental arrests frequently display massive chromosome elimination and other mitotic abnormalities (Morgan, 1927).

According to the studies of Minganti (1959b) interspecific ascidian hybrids show a similar behavior, with abrupt gastrular arrest following the
onset of severe mitotic abnormalities and the elimination of what is probably the paternal set of chromosomes. Minganti (1959a) found that androgenetic haploid ascidian merogones may arrest at gastrulation even
without gross loss of chromosomes. Other ascidian hybrids, both haploid
and diploid, can gastrulate successfully, suffer no chromosome elimination, and encounter difficulty only at the stage of larval differentiation (i.e.,
organogenesis). For example Minganti studied androgenetic hybrid merogones formed by fertilizing enucleated eggs of *Ascidia malaca* with sperm
of *Phallusia mamillata*. He found that the adhesive papillae of the swimming tadpole were of the maternal morphology, even though the only
genes in the embryo were paternal in origin. In the species contribut-
ing the egg cytoplasm in this case the swimming tadpole appears as early
as 9 hours after fertilization, and it is perhaps a consequence of this rapid
rate of development that detailed maternal influence over the course of
morphogenesis extends to such a late stage.

Interpretation of the Species Hybrid Experiments

The molecular approaches to which most of this book is devoted have
greatly deepened our comprehension of maternal programming in early
development and of genomic control in later differentiation. In this context the species hybrid experiments remain important in several ways.

These experiments have played an interesting and significant role in presenting the whole problem of gene action in embryological development and thus foreshadowed some of the most essential of our present concepts. Among these are the idea of maternal templates, and the direct relationship between embryo genome function and the onset of extensive cellular differentiation. The species hybrid experiments also provide some of the best data yet available on the lack of embryo genome control early in development. With respect to this point the basic conclusions of the hybrid experiments have been reinforced by other kinds of data, such as the time of appearance of parental isozymes where *intraspecific* allelic variants are available. Investigations of this nature have been carried out with both chordate and invertebrate material. An example is the study of Wright and Shaw (1970) on *Drosophila*, where it is found that paternal enzyme variants appear only at hatching of the embryos.

As a general summary the species hybrid and related experiments appear to show that:

(a) Early morphogenesis is programmed mainly or exclusively by maternal components already present in the egg at fertilization

(b) The presence of typical catabolic enzymes as well as the appearance of special proteins, such as hatching enzyme, similarly depend on maternal components rather than on new embryo transcripts

(c) The last two statements cannot be interpreted simply as the result of total repression of the paternal genome early in development, since early embryo structural gene transcripts such as histone messenger RNA can in fact be detected

In themselves the species hybrid experiments cannot provide complete proof for these statements, however, since the species hybrid approach is vulnerable to several kinds of objection. Before leaving the subject it is worth considering some of the caveats surrounding the interpretation of these experiments. We know, for example, that intolerance of the paternal genome by the recipient egg cytoplasm causes the complete destruction and elimination of the paternal chromosomes in certain crosses. It is possible that failure to show paternal characters could result from other discriminatory though less obvious effects on the paternal chromosomes in interspecies hybrid embryos. These effects might preferentially inhibit certain paternal chromosome functions until the responsible cytoplasmic factors disappear or are diluted out. In this case the absence of hybrid phenotype in early development would scarcely constitute a reliable index of maternal cytoplasmic control.

Another argument concerns the effect of the taxonomic distance separating the crossed species. Hybrids between closely related species may be the most likely to succeed (though this is not always true), but they are also the least likely to display early hybrid genome control over morphogenesis, since except for temporal adaptations early development will tend to be more similar the more closely related are the species. An example may be the experiments of Driesch and of Tennent cited above: In Driesch's experiment the two genera involved, *Spherechinus* and *Echinus* (= *Parechinus* = *Psammechinus*), are members of different families belonging to the same order, Camarodonta, while in Tennent's experiment the parental genera, *Cidaris* and *Lytechinus*, belong to completely different orders with diverse patterns of development, viz., Cideroidea and Camarodonta (reviewed by Hyman, 1955). It thus seems predictable that paternal genomic effects would become manifest earlier in the *Lytechinus* × *Cidaris* hybrids than in the *Echinus* × *Spherechinus* hybrids. This is what is observed in the altered mode of primary mesenchyme formation in the *Lytechinus* × *Cidaris* cross. It will be recalled that the *Spherechinus* × *Echinus* hybrid displayed a mode of primary mesenchyme elaboration identical with that of the maternal parent and that paternal genome effect is detected only much later in this cross, in the hybrid form of the skeletal spicules formed by the mesenchyme cells.

Despite these and other objections which could be raised, the broad conclusions drawn from the species hybrid literature appear valid. Though current knowledge is based mainly on other evidence, the species hybrid experiments remain a source of valuable information and provide a logical framework for the subjects taken up later in this book. The more penetrating and sophisticated molecular methods of our era are now being applied to the very same questions raised so long ago in the species hybrid experiments of Boveri, Driesch, and their contemporaries—the nature of maternal programming in early development, and the functional role of the embryo genome in morphogenesis.

Development in Physically Enucleated Embryos

In order to investigate directly the dependence of early embryogenesis on new gene activity, there have been attempts made to study the "development" of embryos lacking any nuclear genome whatsoever. Only those morphogenetic events which continue to occur are meaningful in such drastic experiments, but in fact certain complex early processes do continue to be carried out in the total absence of nuclei. Harvey's 1936 experiments on the fate of parthenogenically activated enucleate sea urchin egg fragments mark an important point in the history of this class of experiment, though there were many direct forerunners in lethal hybrid and prior enucleation experiments. Harvey reported that the absence of the nucleus does not prevent a certain amount of *cleavage* from taking

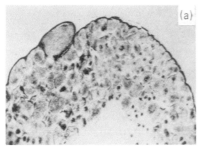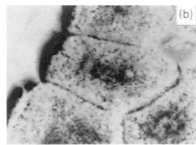

Fig. 2.4. Achromosomal partial blastula, 21 hours (enucleated *pipiens* egg × irradiated *pipiens* sperm). (a) Section through cleaved animal hemisphere. Cells are intact and show well-defined boundaries throughout most of the cleaved area. ×65. (b) Enlarged view of amphiastral figure in same blastula as that shown in (a). The figure contains no Feulgen-positive material. Note alignment of pigment granules between the centers. ×500. From R. Briggs, E. U. Green, and T. H. King (1951). *J. Exp. Zool.* **116**, 455.

place. The complete lack of any nuclear material in the experimental eggs was verified by the absence of a Feulgen reaction (Harvey, 1940).

In amphibians essentially similar observations had been reported by Fankhauser (1934), who observed cleavage in abnormal embryonic cells of the salamander *Triton* which are completely without chromatin. In 1951 Briggs *et al.* reported a classic series of experiments in which eggs of *Rana pipiens* were enucleated after fertilization with lethally irradiated sperm. The lethality of the irradiation was checked by fertilizing *R. pipiens* eggs with irradiated sperm of *Rana catesbeiana*. These eggs develop as typical gynogenetic haploids, showing that the lethal effect of the foreign chromatin normally to be expected in this particular interspecific cross is abolished by the irradiation. Enucleated eggs prepared in this fashion carry out extensive cleavage and even succeed in forming partial blastulae, one of which is shown in Fig. 2.4. As can be seen in this figure, cleavage amphiasters are formed in the blastomere cytoplasm despite the absence of chromosomes. The enucleation experiments (many more of which could be listed) have demonstrated that the division of the egg mass into cells separated by membranes, the most obvious function of the cleavage process, is dependent on maternal cytoplasmic factors rather than on the embryo genome.

"Chemical Enucleation": Development of Actinomycin-Treated Embryos

It is scarcely surprising that eggs subjected to so severe an operation as physical enucleation do not cleave normally. With the discovery of actinomycin D, which binds to the DNA and prevents RNA synthesis, it appeared possible to effect a more subtle "chemical enucleation," as it were. There followed in the middle and late 1960's a spate of experiments in which embryos of various species were treated with actinomycin D and the effects on morphogenesis and other parameters monitored. It is now known that actinomycin has many undesirable effects aside from blocking RNA synthesis. This complicates the interpretations of actinomycin experiments to the point where they often cannot be considered to provide solid evidence. For example, Singer and Penman (1972) and Goldstein and Penman (1973) have shown that actinomycin interferes with initiation of protein synthesis, thus resulting in polysome disaggregation. Obviously the effect of actinomycin on the early embryo would be compound in this case, including the inhibition of maternal messenger RNA translation. Another problem is that actinomycin D does not easily penetrate some kinds of eggs and thus often fails to block RNA synthesis rapidly. This is mainly because of the impermeable outer membranes characteristically insulating fertilized eggs and embryos from the external environment. Other factors are probably involved as well, e.g., the relatively enormous quantity of cytoplasm and the presence in eggs of cytoplasmic elements which also may trap actinomycin D [cf. Steinert and Van Gansen (1971) who report actinomycin binding by vitelline platelets in amphibian eggs]. In what follows we mainly consider experiments in which protein synthesis appears to continue at a normal rate, though RNA synthesis is shown to be seriously affected, and where DNA synthesis and cell division are not also blocked. Despite this safeguard it is today impossible to treat most actinomycin experiments as other than suggestive or corroborative of data obtained by other means.

ACTINOMYCIN EFFECTS ON ECHINODERM EMBRYOS

The first important utilization of actinomycin to investigate the role of gene action in early development was in the experiments of Gross and Cousineau (1963a, 1964). It was reported that sea urchin embryos could be treated with sufficient actinomycin to block 94% of the RNA synthesis during the first 5 hours of development without preventing cleavage. Cleavage in these heavily treated embryos is irregular and delayed. However, even at the high doses of actinomycin (24–100 μg/ml) used to effect this near complete repression of RNA synthesis, DNA synthesis continues, though at a reduced rate. In these experiments the small fraction of the early RNA synthesis which is actinomycin resistant cannot be regarded as responsible for what morphogenesis does occur, since it has been shown that the actinomycin-resistant incorporation is merely end-group turnover in transfer RNA (Gross *et al.*, 1964). At lower doses of actinomycin development advances beyond cleavage in the sea urchin and irregular blastulae form. Significant cellular differentiation never occurs in the actinomycin-treated embryos, however, and these are always unable to gastrulate.

Greenhouse *et al.* (1971) have shown by autoradiography and extraction of labeled actinomycin from treated embryos that the drug is able to enter prehatching embryos at a sufficient rate and in sufficient concentration. Therefore it cannot be argued that the apparent insensitivity of early sea urchin embryos to actinomycin is due only to their impermeability relative to later stages. Furthermore, demembranated sea urchin embryos, which lack the major permeability barrier, respond to actinomycin just as do normal embryos (Summers, 1970). That is, they carry out cleavage, and then arrest at the blastula stage. De Vincentiis and Lancieri (1970) have also studied the effects of a close analog of actinomycin D (desaminoactinomycin C3) which is unable to bind to DNA and does not block RNA synthesis but which might be expected to display similar general toxicity. Unlike actinomycin D this agent fails to affect development of *Paracentrotus* embryos. Figure 2.5 from the work of De Vincentiis and Lancieri, illustrates the effect of actinomycin D and C3 on these embryos.

The most striking aspect of the results obtained with actinomycin by Gross and his associates is that the total protein synthesis rate seems not to be greatly affected by the drug, even at doses where virtually all messenger RNA synthesis is cut off. This remains true for cleavage stage sea urchin embryos for some hours (Gross and Cousineau, 1964; Gross, 1967; Stavy and Gross, 1969). The cleavage divisions themselves require protein synthesis, as has been shown for echinoid embryos by the use of various protein synthesis inhibitors (e.g., Hultin, 1961a; Karnofsky and Simmel, 1963). Since these divisions occur in the presence of actinomycin, sufficient protein synthesis to support cleavage is evidently carried out despite the blockage of new messenger RNA synthesis. From these experiments the important conclusion was drawn that *protein synthesis in early embryos occurs on preformed templates, i.e., maternal messenger RNA.* It has now been found that in actinomycin-treated embryos about 50% of histone synthesis is repressed, but the remaining synthesis includes all species of histone (Kedes *et al.*, 1969; Ruderman and Gross, 1974). The synthesis of tubulin (Raff *et al.*, 1971) and of hatching enzyme (Barrett and Angelo, 1969) also continue in actinomycin-blocked embryos. All these proteins, and by implication many others required for pregastrular morphogenesis, are apparently coded on stored maternal messenger RNA. Whatever the possible and real shortcomings of the actinomycin approach, this conclusion has proved correct as reviewed in Chapter 4. Like the species hybrid

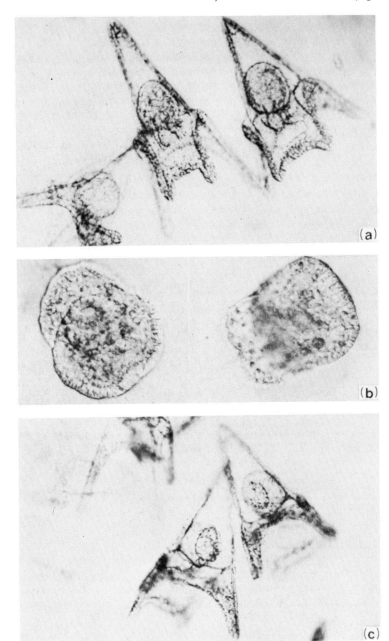

experiments, the actinomycin experiments argue strongly for maternal programming of early protein synthesis, though the uncertainties implicit in the use of this toxic inhibitor require for definitive proof other forms of data which are discussed below. At least for the sea urchin embryo the side effects of actinomycin on protein synthesis are limited. Sargent and Raff (1976) used enucleated sea urchin half-eggs to show that actinomycin had no effects on the level of protein synthesis for up to 12 hours. Nor was the spectrum of proteins synthesized in enucleated eggs treated with actinomycin detectably different from that synthesized in untreated merogones.

Actinomycin has been used to treat a variety of other embryos besides sea urchins, as noted above. In most cases, no secure evidence on the effect of the drug with respect to RNA and protein synthesis has been obtained, and the results are hence difficult or impossible to interpret. Qualitatively it is generally observed that development proceeds past gastrulation and until organogenesis but not beyond. Examples include *Nassaria* (*Ilyanassa*) *obsoleta*, the marine mudsnail (Feigenbaum and Goldberg, 1965; Collier, 1966) and ascidian embryos (Reverberi, 1971c). Furthermore, Newrock and Raff (1975) showed that actinomycin treatment of cleavage stage *Ilyanassa* embryos sufficient to block all RNA synthesis nonetheless does not interfere with gastrular epiboly. In addition, the treated eggs maintain a normal rate of protein synthesis. In order to bypass one major source of uncertainty actinomycin has been microinjected into the eggs of several species. For example Lockshin (1966) studied eggs of coleopteran insects in this way and reported that development blocks after blastema formation.

ACTINOMYCIN EFFECTS ON CHORDATE EMBRYOS

The microinjection method has been used most extensively on amphibian eggs. Development of *Pleurodeles* and *Xenopus* after actinomycin injection was studied by Brachet and Denis (1963) and Brachet *et al.* (1964), and a similar experiment with similar results was described by Wallace and Elsdale (1963). Brachet and Denis reported that cleavage is "completely unaffected" by actinomycin, though gastrulation and neuru-

---

Fig. 2.5. Effect of actinomycin on sea urchin embryos. Observations made 53 hours after fertilization. (a) Embryos developed in seawater (controls). (b) Embryos developed in the presence of actinomycin D. The embryos were treated with actinomycin D (12 μg/ml) for 14 hours (from 9 hours to 23 hours after fertilization); (c) Embryos developed in the presence of desaminoactinomycin C3. The embryos were exposed to desaminoactinomycin C3 for 14 hours and then transferred into seawater and kept there for 30 hours. From M. De Vincentiis and M. Lancieri (1970). *Exp. Cell Res.* **59**, 479.

lation are blocked. In the teleost *Fundulus* it is found that exposure of embryos to actinomycin D during the first hour after fertilization does not interfere with normal cleavage and blastulation, though subsequent gastrular axiation is blocked. Experiments demonstrating this have been reported by Wilde and Crawford (1966) and Crawford and Wilde (1966) using actinomycin at levels which inhibit only 50% of total precursor incorporation into embryo RNA, and similar results are obtained if the embryos are poisoned with cyanide. These observations, like Driesch's demonstration that the exact number of primary mesoderm cells in the sea urchin embryo may be cytoplasmically programmed, indicate the detailed complexity of the maternal developmental program carried in the egg cytoplasm.

A number of workers have reported that mammalian embryos respond differently to actinomycin D than do other embryos. It is observed that at least for doses in the range of 0.01 to 0.1 μg/ml, treatment with actinomycin arrests the development of mouse embryos during cleavage (see, e.g., Mintz, 1964; Skalko and Morse, 1969; Monesi *et al.*, 1970; Golbus *et al.*, 1973). The implication is that in mammalian embryos protein synthesis and perhaps ribosomal RNA content as well become dependent on embryo gene transcription at earlier developmental stages than in other embryos. However, more recent studies indicate that this inference is likely to be in error, and that some actinomycin side effect, rather than inhibition of RNA synthesis, is responsible for early developmental arrest. Thus Tasca and Hillman (1970) showed that protein synthesis rates are not affected, even by high doses of actinomycin, within a 3-hour period. The same result was reported by Manes (1973) for 1-day rabbit embryos exposed to the drug for periods of up to 24 hours. During this time RNA synthesis was severely inhibited. In this study even more stringent inhibition of RNA synthesis was achieved by the use of the RNA polymerase inhibitor α-amanitin. Nonetheless, no effect on protein synthesis could be observed. While actinomycin at the dose used caused immediate arrest of development, the α-amanitin treated embryos were able to both initiate and continue cleavage. All but about 0.5% of the RNA synthesis was blocked by α-amanitin, which affects only nonribosomal RNA synthesis. Golbus *et al.* (1973) also showed that α-amanitin does not interfere with cleavage in mouse embryos, nor does it affect protein synthesis. Therefore, the ability of actinomycin to block ribosomal RNA synthesis cannot be blamed for the difference in response to these two inhibitors. It follows that an actinomycin toxicity other than inhibition of transcription is responsible for mammalian embryo arrest. We conclude tentatively that mammalian embryos are not significantly different from other embryos in the independence of their early protein synthesis from new transcription.

Consistent with this view is a report of Chapman *et al.* (1971) who also found that until day 5 of preimplantation mouse embryogenesis (late blastocyst), only maternal forms of glucose-6-phosphate isomerase are found. These experiments were carried out using hybrids between inbred strains which differ in their GPI isozyme forms.

Effects of Actinomycin and Other Treatments on Early Morphogenesis

DELAYED MORPHOGENETIC RESPONSE TO EXPERIMENTAL TREATMENTS

By administering actinomycin at progressively later periods of development a series of interesting results have been obtained which suggest that transcription may occur hours earlier than the morphogenesis which it controls. Though these experiments are all subject to alternative interpretations, they are reviewed here because they are in accord with several other forms of evidence, also treated below, and because they suggest an interesting conclusion. Developmental effects of actinomycin added at various times after fertilization were described by Barros *et al.* (1966) in a study carried out with the starfish *Asterias forbesii*, and experiments along similar lines with the sea urchin *Paracentrotus lividus* have been reported by Giudice *et al.* (1968). In the starfish it is found that by interfering with RNA synthesis during the period from 5 hours after fertilization to 11 hours after fertilization gastrulation is blocked. Gastrulation does not normally begin in this organism until 15 hours, and midgastrula normally occurs at 18–19 hours, with the first primary mesenchyme cells being released after 20 hours. Yet if actinomycin treatment is delayed until after 11 hours, gastrulation is able to take place up to the stage of primary mesenchyme formation. A period during which there occurs some synthetic activity needed for gastrulation which is blocked by actinomycin appears to be delineated from about 6 to 11 hours postfertilization. In *Paracentrotus* the normal sequence of events is the reverse of that in the starfish, and primary mesenchyme cells appear well in advance of gastrular invagination. Here it is found that actinomycin treatment as early as 6–11 hours after fertilization, i.e., during the hatching blastula–early mesenchyme blastula period, blocks gastrulation. Gastrulation does not normally take place until after 18 hours. An interesting additional finding reported by Giudice *et al.* (1968) is that appearance of mesenchyme cells seems impervious to actinomycin treatment sufficient to block 73–77% of all RNA synthesis. This result is of course consistent with the classical species hybrid experiments indicating maternal rather than embryo genome control of primary mesenchyme determination (Table 2.2). In normal sea urchin embryos several of the enzymes controlling mobilization of deoxyribonucleotides for DNA synthesis normally decrease in activity during early development, and actinomycin prevents this decrease (De Petrocellis and Monroy, 1974). Paradoxically, the drug leads to a sharp increase in the activity of dCMP aminohydrolase. However, as in the above cases, this actinomycin effect is noticed only when the drug is added some hours prior to the time when the change in dCMP aminohydrolase activity occurs. Along the same lines are reports from Czihak (1965) and Czihak and Hörstadius (1970) regarding the effect of treatment of sea urchin embryos with 8-azaguanine at the 16-cell stage. The treated embryos differentiate normally for many hours, forming a blastula with primary mesenchyme cells and spicules, but much later, gastrulation is blocked as a result of failure to develop an archenteron.

A similar pattern of events seems to occur in *Ilyanassa* embryos, where it is also found that some actinomycin-sensitive activity required for the early events of embryonic differentiation is carried out long before the point when these differentiations become manifest (Collier, 1966). Actinomycin treatment between the fourth and fifth day of development in *Ilyanassa* prevents the differentiation of eyes, which normally appear at 6.5–7 days, for example, and if the embryos are exposed to actinomycin only after 5 days, eye formation is unaffected (though morphogenesis of other structures is now affected). The same pattern of events holds for shell gland, esophagus, intestine, and other organ primordia, with the sensitive period occurring 1–2 days before the respective morphogenesis.

There are in the literature several earlier studies which also may be susceptible to the interpretation that a biosynthetic activity required for a particular type of differentiation occurs long in advance of the appearance of the differential cells. In 1933 Gilchrist published an unusual experiment in which *Rana* eggs were exposed to lateral temperature gradients at various stages in early development. This was done by orienting the eggs in a water bath containing a constant hot-to-cold gradient. After given periods of exposure the gradient was reversed in order to compensate for faster cell division at the higher temperatures on one side of the egg, a procedure which is useful mainly at the earlier stages when the egg is symmetrical. As a result of the high temperature treatments, various abnormalities in specific areas of the embryo occur, but only much later. Figure 2.6 summarizes the temporal pattern linking time of treatment with time and location of effect. The general resemblance between this pattern and that observed in the actinomycin studies is evident. For instance, in the midblastula stage treatment which blocks gastrulation, the "determina-

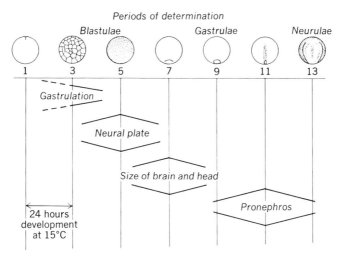

Fig. 2.6. Periods of determination as shown by reversed thermal gradients. From F. G. Gilchrist (1933). *J. Exp. Zool.* **66**, 15.

tion" to which Gilchrist refers (Fig. 2.6) could be the result of early transcriptional processes required for gastrulation in amphibians. Of course many alternative interpretations are also possible, such as direct effect of heat on the regional translational apparatus. Another interesting experiment which may be relevant here is that of Neyfakh (1964), who used relatively low dosages of X-irradiation to "inactivate" the nuclei of teleost embryos, supposedly without interfering with cytoplasmic functions. This interpretation of the irradiation effect is supported in Neyfakh's account by the observation that androgenetic haploids formed by irradiating whole eggs (including both cytoplasm and nucleus), and then fertilizing, develop exactly the same as do gynogenetic haploids formed by irradiating the sperm only. That is to say, the irradiation of the egg does not interfere with the ability of the cytoplasm to direct cleavage and blastulation, though in either case it does destroy one parental genome and result in a haploid individual. Neyfakh irradiated *Misgurnus fossilis* embryos at successive intervals and reported a 2.5-hour sensitive period ending 8.5 hours after fertilization. Irradiation during this period blocks gastrulation, although that process occurs only at 9–18 hours of development.

INTERPRETATION OF MORPHOGENETIC EFFECTS

The actual effects of 8-azaguanine, heat shock, irradiation, actinomycin, etc., in these various experiments can only be a matter of conjecture.

The actinomycin experiments last discussed are the more difficult to disregard. In each case the drug is shown to have some effect so that it is clear it gets into the embryos, sometimes rapidly, but not a general toxicity effect, since development continues for many hours before blocking at some later stage. The cellular divisions and interactions and the particular protein synthesis patterns needed for the intervening morphogenesis evidently continue in the presence of the drug. If a specific transcriptional (and/or translational) inhibition is responsible for the delayed actinomycin effect, any of several mechanisms could be involved. These include failure to synthesize a set of specific regulatory or structural gene products used later, or perhaps interference with cell interactions which have to occur during the time of treatment in order for subsequent differentiations to take place. In this connection the classic experiments of Hörstadius (1939) should be recalled. These experiments showed that the normal fate of sea urchin blastomeres can be altered by juxtaposing them in artificial combinations. The general conclusion is that the ultimate behavior of given blastomeres and their lineal descendants depends in part on what other blastomeres are next to them. Another way to state this is that cellular interaction plays a determinative role in cleavage stage embryos for processes of development manifest only much later. Some of the effects of blastomere removal and recombination are mimicked by various agents, for example, lithium chloride, as has long been known (Herbst, 1892). Lithium-treated embryos behave somewhat like vegetal half-embryos, producing a number of abnormalities including archenterons which are enlarged or sometimes supernumerary or sometimes evaginated, and diminished oral regions. Runnström and Markman (1966) and de Angelis and Runnström (1970) reported that cleavage stage exposure to actinomycin suppresses the lithium effect, and also alters the development of isolated animal half-embryos. Thus the treatment with lithium and other such agents, and isolation of embryo parts modify processes which normally depend on certain cleavage stage intercellular interactions. Since actinomycin treatment during cleavage affects the results of such treatments, we can derive the argument that the actinomycin interferes with essential cellular interactions (or responses to such interactions) beginning in cleavage. This is also suggested by the impairment of normal cytological structures connecting contiguous cells in lethal hybrid amphibian embryos (Johnson, 1969). Cellular interaction in the sea urchin continues to be of crucial importance to developmental processes during the blastula stage. De Petrocellis and Vittorelli (1975) have shown, for example, that disaggregated sea urchin blastula cells which remain normal in their rates of respiration, RNA synthesis, and amino acid incorporation undergo striking changes in the activity of various enzymes. Thus the activities of

DNase, thymidine kinase, and thymidylate kinase decrease in these cells, while DNA polymerase activity doubles and normal changes in dCMP aminohydrolase are blocked after disaggregation.

The data considered to this point indicate that the complex processes of early morphogenesis are at least in part independent of embryo gene activity. However, as documented in Chapter 5 the embryo genome is actively transcribed during these early stages. Some indication of this has already been noted in the foregoing discussion of the species hybrid experiments, and has also been inferred from the actinomycin studies here reviewed. While the early morphogenetic events programmed in the egg cytoplasm are taking place new transcripts required for later differentiation are evidently being synthesized. Characteristically it is not until organogenesis, when a complex, organized multicellular structure has formed that the switchover from egg cytoplasmic to embryo nuclear direction becomes easily demonstrable, e.g., in species hybrid experiments. Not until this point can the embryo be considered to have assumed complete genomic control of its own development.

3

First Indices of Differential Embryo Cell Function

Gel electrophoresis has been used to compare the spectrum of proteins synthesized at various stages during the embryogenesis of many animal species. Data are reviewed for mammalian, echinoderm, amphibian, molluscan, and other embryos. In all cases important differences are observed early in development, during cleavage and blastulation. In mammals most of the changes reported to take place during preimplantation development occur during early cleavage. The electrophoresis methods used are sensitive only to changes in about ≤400 relatively prevalent protein species, however, and these probably represent only a few percent of the total set of proteins being synthesized. In sea urchins and amphibian eggs some of the pregastrular change in protein synthesis appears to result from post-transcriptional modulations in the use of maternal messenger RNA. This conclusion is derived from gel electrophoresis experiments on proteins extracted from actinomycin treated (echinoderm) and enucleated (amphibian) eggs. Actinomycin experiments, however, do not provide reliable evidence on when in sea urchin development changes in structural gene transcription determine the changes in protein synthesis patterns. It is concluded that in sea urchin embryos changes in the sets of structural genes transcribed which affect the observed protein synthesis patterns do not occur much before gastrulation. In contrast, new transcriptional activity may affect protein synthesis patterns during cleavage in mammalian embryos. At least one enzyme whose activity changes during cleavage, hypoxanthine-guanine phosphoribosyltransferase (HGPRT) is shown to be synthesized as a result of early embryo structural gene activity in the mouse. However, mammalian eggs require more time to traverse cleavage and morula stages than do many lower organisms to complete gastrulation and embark on organogenesis. Post-transcriptional control of maternal messenger RNA utilization may be significant for a certain period of time in all species. The duration of this period should probably be measured in terms of the amount of biosynthesis which has occurred. Thus the developmental stage when changes in embryo transcription dominate the synthesis pattern may vary. Pregastrular morphogenesis involves many highly complex cytodifferentiations. Junctional specializations, changes in cytoarchitecture of the blastomeres, and cell motility are among these. Cytodifferentiation implies that complex and specific patterns of biosynthesis are functionally important in early embryos. However, few histospecific proteins are known in embryos during the preorganogenesis period, though many are known for later stages. Some examples are discussed and a review is provided of the detailed events occurring during the differentiation of primary mesenchyme cells in echinoderm embryos. These cells display a variety of differentiated characteristics from the early blastula stage on. Their differentiation is at least partially dependent on embryo genome action.

Though we are aware that cells of the early embryo differentiate, it has been a difficult task to associate the synthesis of particular proteins with the appearance of cell specificity. Examples of the few cases which exist are considered in the last section of this chapter. A number of particular proteins are of course known to be synthesized in early embryos, but almost none of these can be regarded as cell type specific. Rather they are generally occurring proteins which belong to a class loosely described as "housekeeping" proteins. By this term is meant proteins required by most or all cells in the organism. Though housekeeping proteins may be necessary in order for differentiated cells to carry out their specialized functions, they are not themselves the constituents which distinguish one functional cell type from another, and they are not confined to one or a few cell types. Current examples include DNA polymerase, ribosomal proteins, many metabolic enzymes, tubulin, and probably the RNA polymerases. All of these proteins are present in significant quantities in newly fertilized eggs, remain present throughout early development, and either sooner or later are further synthesized on embryo polyribosomes (see Chapter 4 for quantitative details and references). It should be noted that in principle the class of *maternal proteins*, defined as those stored in the mature egg and inherited by the embryo, may be larger than the class of housekeeping proteins. This will be true to the extent that the class of maternal proteins may include specific proteins required only by certain embryonic cell types.

We begin this chapter with an examination of evidence for change in the state of early embryonic cells, both biosynthetic and cytological. It is apparent from this review that at least in some systems cell differentiation may be initiated very early in embryogenesis, perhaps even during cleavage. However, it cannot be concluded that embryo genome activity is responsible for precocious differentiation. Cell type- or stage-specific proteins and messenger RNA's, as well as mechanisms controlling their later use, are apparently all inherited preformed in the egg. Further evidence on this point is considered below and in the next chapter, in conjunction with our earlier conclusions on the onset of embryo genome control.

Qualitative Changes in Patterns of Protein Synthesis Correlated with Developmental Stage

CHANGES IN PROTEIN SYNTHESIS PATTERN STUDIED BY HIGH RESOLUTION GEL ELECTROPHORESIS

The first effective attempts to measure changes in the sets of proteins synthesized during early embryogenesis were those of Terman and Gross in 1965. Soluble proteins were extracted from sea urchin embryos and partially resolved by gel electrophoresis. Though no quantitation was available it was clear that the radioactively labeled proteins being synthesized by the embryos in general migrated differently from the bulk constituents of the egg, i.e., the major maternal proteins. A main conclusion of Terman and Gross was that a different spectrum of proteins is synthesized during gastrulation than during cleavage. This conclusion was in accord with other contemporary reports relying on methods of lower resolving power. Among these were the investigations of Spiegel *et al.* (1965) who also examined labeled sea urchin embryo proteins by gel electrophoresis. Westin *et al.* (1967) applied immunological methods to this problem and demonstrated changes in several newly synthesized antigens as a function of developmental stage. In recent years higher resolution gel electrophoresis procedures have been developed and are now being used to study alterations in embryonic protein synthesis patterns. We now discuss some current examples of the application of this approach to embryos of several species.

In Fig. 3.1 is reproduced a high resolution exponential gel separation of newly synthesized soluble embryo proteins (Van Blerkom and Manes, 1974). These proteins were obtained from labeled preimplantation rabbit embryos at different stages of development. After electrophoresis the gel was overlain with X-ray film, resulting in the pattern shown. The authors of this study note that most of the visible bands are themselves heterogeneous, each representing many proteins of similar electrophoretic mobility in the one-dimensional system used. This is a most important point in interpreting experiments such as those illustrated here. Comparison of the number of bands seen even in such relatively high resolution procedures to the number of diverse messenger RNA species being translated in embryo polysomes (see Chapter 6) also shows that the electrophoretic separations resolve only a few percent of the probable number of different proteins being synthesized. Since translation rates for diverse messenger RNA's are generally similar (e.g., see Kafatos, 1972), any intensely labeled relatively homogeneous bands represent those proteins being synthesized on relatively large numbers of messenger RNA's. Therefore the gel electrophoresis experiments are sensitive only to changes in the synthesis of that few percent of protein species translated on the most prevalent messenger RNA species (these, of course, would likely include a large fraction of the mass of the newly synthesized protein). The result may be a large underestimate of the actual amount of qualitative change in the set of proteins synthesized.

Only a few of the bands in Fig. 3.1 can be tentatively identified, including myosin (band f), tubulin (band k), and actin (band l). Certain other bands which are present at one stage but not at later or earlier stages are also marked (see legend to Fig. 3.1). Inspection of the figure shows that while many of the proteins continue to be synthesized throughout the preimplantation stages, distinct synthetic changes accompany the morphological development. During the period of the experiment the embryos progress through cleavage (0.5–2.5 days past fertilization) and into the blastocyst stage. Implantation occurs at day 6. Most of the observable change in synthesis pattern seems to occur during the cleavage stages. Van Blerkom and Brockway (1975) carried out similar experiments on mouse embryos. About 110 protein bands could be resolved, and as in the rabbit embryo all major changes in the synthesis pattern were observed to occur between fertilization and the 4- to 8-cell stage (day 3).

THE DOUBLE LABEL METHOD

In order to quantitate differences in electrophoretic patterns between two preparations it is now customary to label one of the samples with a

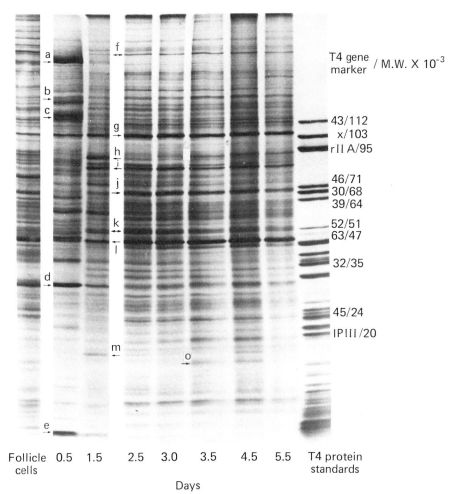

Follicle cells 0.5 1.5 2.5 3.0 3.5 4.5 5.5 T4 protein standards

Days

Fig. 3.1. Qualitative changes in the pattern of protein synthesis during cleavage (day 0.5 to day 2.5) and blastocyst development (day 3 and onward) in preimplantation rabbit embryos growing *in vivo* are demonstrated in this autoradiograph. The protein synthetic pattern of follicle cells is presented in the first column and should be compared with the protein patterns of 0.5- and 1.5-day-old embryos. Molecular weight values may be estimated by comparing embryonic bands with prereplicative bacteriophage T4 proteins shown in the last column. The proteins were labeled by incubating the embryos with ^{35}S-methionine. Labeling was for 0.5–3 hours depending on stage. Bands f, k, and l probably represent or include myosin, tubulin, and actin, respectively. The other marked bands are examples of proteins or groups of proteins whose presence depends on developmental stage. The proteins are displayed by exponential SDS-acrylamide gel electrophoresis. From J. Van Blerkom and C. Manes (1974). *Dev. Biol.* **40,** 40.

^{14}C-amino acid and the other with the same ^{3}H-amino acid. The two samples are then coelectrophoresed and the ratio of counts plotted as a function of mobility in the gel. This method provides a very sensitive assay for change in synthesis pattern, one which is as quantitative as the resolution of the gels and the gel slicing procedures permit. An example is shown in Fig. 3.2a, which also concerns preimplantation mammalian embryos (Epstein and Smith, 1974). This figure displays electrophoretic comparisons of proteins synthesized in day 2 versus day 1, day 3 versus day 2, and day 4 versus day 3 mouse embryos. In the lower portions of Fig. 3.2a are the actual count profiles (normalized to the same number of total ^{3}H and ^{14}C counts), and in the upper portion the isotopic ratio is plotted. For identical samples the normalized ratio is 1.0, as shown by the horizontal line. It can be seen that the greatest amount of change occurs during cleavage between the first and second day after fertilization, consistent with the impression gained from Fig. 3.1. During this time the embryo does not increase in mass but divides into 8–16 cells. A contrasting pattern is seen in Fig. 3.2b, which concerns amphibian rather than mammalian material. Here comparisons between soluble proteins in axolotl embryos are presented (Lützeler and Malacinski, 1974). Between 2-cell and blastula stage almost no change in the pattern of protein synthesis can be discerned save in one rapidly migrating set of proteins. Other evidence suggests that the latter are histones (see Chapter 4 for a discussion of embryonic histone synthesis). However, as panel 2 of Fig. 3.2b shows, the pattern of protein synthesis in gastrulae differs greatly from that found in blastulae. Panel 3 of Fig. 3.2b, where ^{14}C- and ^{3}H-labeled proteins from gastrulae are compared, provides a control on the other experiments shown.

Electrophoretic analyses of newly synthesized proteins have demonstrated changes in protein synthesis pattern early in the development of other animals as well. Stage-specific protein synthesis patterns are known from the earliest periods of frog embryogenesis. Thus Ecker and Smith (1971) reported that the patterns of protein synthesis in 2-cell *Rana* embryos already display distinct differences from those of ovulated body cavity eggs. Further changes occur between cleavage and gastrulation. Alteration in protein synthesis patterns occurs at least as early as the period between blastulation and gastrulation in killifish (*Fundulus heteroclitus*) embryos (Schwartz and Wilde, 1973). In the marine mud snail *Nassaria* (*Ilyanassa*) *obsoleta* a sharp change in the spectrum of proteins synthesized has been observed in the postgastrular period of development (Teitelman, 1973). Visible differentiation of various organs is already occurring by this time. Other observations on protein synthesis pattern in *Ilyanassa* embryos discussed in Chapter 7 (Donohoo and Kafatos, 1973;

Newrock and Raff, 1975) show that regional and stage-specific protein synthesis patterns first occur early in *Ilyanassa* embryogenesis, during cleavage and blastulation. Another report of very early changes in protein synthesis pattern concerns eggs of the clam *Spisula solidissima* (Nadel *et al.*, 1976). Comparison by the double label method of the proteins made before and after fertilization in these eggs shows that dramatic alteration in the spectrum of proteins synthesized occurs even before cleavage begins.

A high resolution qualitative study of protein synthesis in sea urchin embryos of various stages has been reported by Brandhorst (1976). The method of analysis used in this experiment was two-dimensional gel electrophoresis, and about 400 individual newly synthesized proteins were resolved. Very few changes in the protein synthesis pattern thus defined are evident in comparing unfertilized eggs, fertilized eggs, or blastulae. However, the set of proteins synthesized by these embryos changes markedly by the gastrula stage. Further data regarding protein synthesis changes in sea urchin embryos are reviewed in the next part of this chapter.

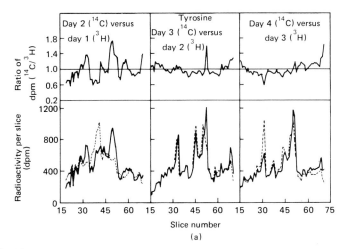

Fig. 3.2a. Changes in protein synthesis during development. (a) Comparisons of incorporation of tyrosine into mouse embryos at different stages of development. Left, day 2 (8–16 cells) versus day 1 (2 cells); center, day 3 (early blastocyst) versus day 2; right, day 4 (late blastocyst) versus day 3. Earlier-stage embryos, incubated with ^{3}H-tyrosine, and later-stage embryos, incubated with ^{14}C-tyrosine, were mixed and then solubilized and analyzed electrophoretically on acrylamide gradient gels. The normalized dpm (lower) and ^{14}C/^{3}H ratios (upper) are shown. A ratio of greater than 1.0 indicates a relative increase in the rate of synthesis; less than 1.0 indicates a relative decrease. Solid line, ^{14}C; dashed line, [3]H. From C. J. Epstein and S. A. Smith (1974). *Dev. Biol.* **40**, 233.

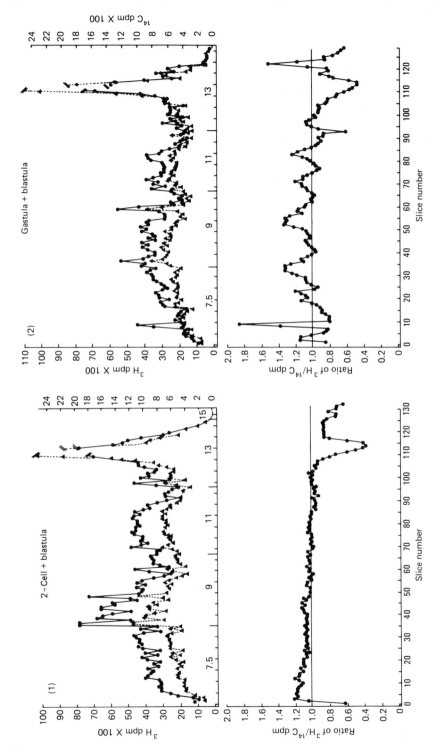

64

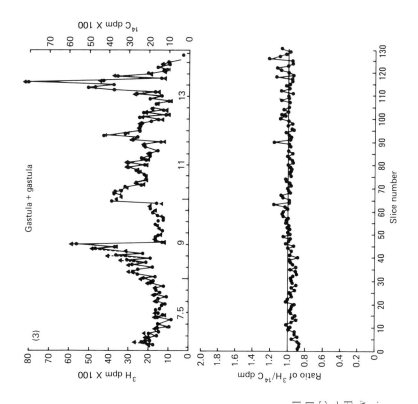

Fig. 3.2b. Comparisons of incorporation of leucine into axolotl embryo proteins at various stages of development. Panel 1, [3]H 2-cell stage and ^{14}C blastula stage proteins; panel 2, ^{3}H gastula and ^{14}C blastula proteins; panel 3, control experiment showing co-electrophoresis of ^{3}H gastrular and ^{14}C gastrular proteins. The normalized dpm profiles are shown above and the ratios of these profiles below in each panel. From I. E. Lützeler and G. M. Malacinski (1974). *Differentiation* **2**, 287.

An important and general conclusion follows from all of these studies. This is that in every organism which has been carefully investigated changes in the sets of specific proteins being synthesized occur by the gastrula stage of development, and in many organisms as early as the cleavage stage.

Are Changes in Protein Synthesis Pattern in Early Embryos Post-transcriptional?

The 1965 study of Terman and Gross suggested that the pattern of protein synthesis observed in cleavage stage sea urchin embryos is impervious to the effects of actinomycin. In contrast, their experiments indicated that the pattern of protein synthesis in gastrula stage embryos is extremely sensitive to actinomycin. It is known that actinomycin penetrates sea urchin embryos and effectively blocks RNA synthesis (Greenhouse et al., 1971). However, the actinomycin-blocked embryos carry out protein synthesis at approximately normal rates (see Chapter 2; other data on this point are to be found in Gross, 1967; Kedes et al., 1969; Stavy and Gross, 1969; Fry and Gross, 1970a). The conclusion was drawn by Terman and Gross (1965) that the spectrum of proteins synthesized at the gastrula stage depends on messenger RNA synthesis in the embryo genomes, while the pattern of protein synthesis at early cleavage stages depends on preformed messenger RNA's. In 1967 an additional report claimed that even in actinomycin-blocked embryos the protein synthesis pattern undergoes specific alterations as early as the interval between fertilization and the blastula stage (Gross, 1967).

In Fig. 3.3 some further experiments on the effect of actinomycin on protein synthesis patterns are reproduced (Terman, 1970). The method of analysis is again that shown in Fig. 3.2, namely, ratio counting of soluble protein preparations labeled with different isotopes and then coelectrophoresed. Figure 3.3a demonstrates that the pattern of protein synthesis changes greatly between fertilization and the swimming blastula stage. From the control experiment (dotted line) it is clear that normal and actinomycin-treated fertilized eggs display identical synthesis patterns. In Fig. 3.3b the effect of continuous exposure to actinomycin from 1 hour before fertilization on the *change* in synthesis pattern between fertilization and blastula stage is demonstrated. The dotted line in Fig. 3.3b shows that the actinomycin-blocked blastulae translate a spectrum of proteins somewhat different from that of control blastulae. This is interpreted to mean that certain types of messenger RNA being translated are the product of new transcription. The dashed line in Fig. 3.3b shows that even blastulae

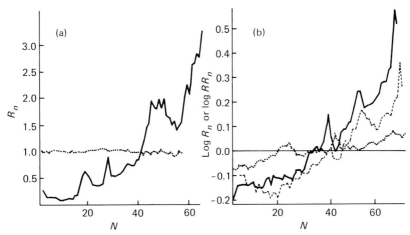

Fig. 3.3. Comparisons of proteins synthesized by early sea urchin embryos in the presence and absence of actinomycin. (a) Dotted line: normal and actinomycin-treated zygotes compared. Solid line: normal hatched blastulae and normal zygotes compared. (The normalized ratio of ^{3}H/^{14}C counts, R_n, is plotted versus fraction number, N.) (b) Solid line: normal hatched blastulae and normal zygotes compared. Dashed line: actinomycin-treated hatched blastulae and actinomycin-treated zygotes compared. Dotted line: normal hatched blastulae and actinomycin-treated hatched blastulae compared. In order to retain consistency when either ^{3}H or ^{14}C is arbitrarily administered, the reciprocal of R_n, RR_n, is plotted on the same graph as R_n. Since $\log R_n$ is $-\log RR_n$ both are equally distant from the line at $\log 0$. Thus the logs are plotted rather than the linear representations. The proteins were electrophoresed in urea acrylamide gels. From S. A. Terman (1970). *Proc. Natl. Acad. Sci. U.S.A.* **65**, 985.

which have been grown continuously in actinomycin have carried out a change in their synthesis patterns which is almost the same as that displayed by equivalent normal embryos (compare solid and dashed lines in Fig. 3.3a and b). Similar results were obtained in comparisons of 4- and 8-hour cleavage stage embryos. These also differ specifically in their patterns of protein synthesis, irrespective of whether the embryos are grown in actinomycin (Terman, 1972). It was reported by Terman (1972) that these effects are all independent of cell division, since they occur in the presence of the mitotic inhibitor colchicine.

These experiments and others similar to them have led to the view that a *post-transcriptional* developmental program qualitatively controls protein synthesis in early embryos. This program is considered to determine the set of protein species in synthesis at each stage. According to this interpretation, as the embryo develops a progressively larger fraction of its translational program depends directly on embryo genome transcription, as shown by the increasing disturbance of the synthesis pattern in actinomycin-treated embryos. By gastrula the observable protein synthesis is considered to be dominated mainly by new species of messenger RNA deriving from the embryo nuclei.

Since these conclusions are of some significance they deserve a cautious appraisal. Obviously it is the case that by some point in the *postgastrular* period embryo genome transcription does determine the spectrum of messenger RNA's present. This is shown by the isozyme and species hybrid experiments reviewed in Chapter 2 as well as by the general logic of the principles of gene control. We noted in Chapter 2, however, that the experiments discussed there provide little evidence for genome control of either protein synthesis or morphogenesis prior to the end of gastrulation, in either sea urchin or frog embryos. The proposition of *qualitative transcriptional regulation* of protein synthesis as early as hatching blastula and beginning gastrula in the sea urchin depends almost wholly on actinomycin experiments. Actinomycin is now known in other systems to affect protein synthesis initiation, and some messenger RNA's may be more sensitive to this interference than others. However, as noted in Chapter 2, Sargent and Raff (1976) have shown that protein synthesis carried out in enucleated sea urchin egg cytoplasm during the first few hours of development is not detectably affected by actinomycin, either quantitatively or qualitatively. At least for this system, the translation-level effects of this drug thus seem to be limited. Nonetheless, transcription of some messenger RNA's may be less affected by actinomycin than is transcription of others; RNA synthesis is not totally eliminated by the treatments used. The longer the embryos remain in actinomycin the more they differ from "normal" in both morphological and molecular terms. For these reasons, it cannot be concluded that qualitative differences in protein synthesis pattern between actinomycin-treated and normal embryos (e.g., the dotted line in Fig. 3.3b) actually demonstrate *transcriptional control* over protein synthesis. That is, these experiments fail to exclude the possibility that *all* of the species of messenger RNA transcribed even during blastula and gastrula stages are also present in the fertilized egg at the beginning of development. This issue is taken up in more detail subsequently, in the context of comparison of messenger RNA sequence content at various stages of embryogenesis (Chapter 6).

It is more difficult to escape the conclusion that post-transcriptional control is responsible for much of the change in early protein synthesis patterns in sea urchin embryos. Here the possible toxicity of actinomycin for processes other than transcription cannot be blamed, since the result obtained is that *similar* changes in protein synthesis pattern occur in the presence of actinomycin as in its absence. As used, the drug does block most messenger RNA transcription, and experiments such as those in Fig. 3.3b (dashed line) indeed appear to demonstrate a qualitative post-transcriptional control process during early sea urchin development. An experiment carried out with frog eggs by Ecker and Smith (1971) directly demonstrates such a process without the use of actinomycin. These authors showed that the electrophoretic patterns of proteins synthesized in oocytes during the maturation process (*in vitro*) change greatly between 24 and 48 hours. Remarkably, the same pattern of changes was discovered in oocytes whose nuclei had been manually removed at the beginning of the experiment as in normal oocytes. This is an important point for the present argument, since it demonstrates the existence of post-transcriptional control processes affecting protein synthesis patterns in a completely independent way.

We conclude tentatively that the changing patterns of protein synthesis characteristic of many early embryos (i.e., prior to postgastrular organogenesis) are indeed to some extent controlled by post-transcriptional mechanisms. All the components required, messenger RNA, ribosomal factors, etc., including regulatory elements, are evidently maternal, since their presence does not require new transcription. It remains possible that the changes observed are largely quantitative, rather than representing the actual appearance or disappearance in the translation apparatus of various messenger RNA species. However, this is still a matter of conjecture, as insufficient direct evidence yet exists. Most of the experiments on this subject have been carried out on sea urchin embryos, and there are as yet little relevant data available for other forms. Different mechanisms may be dominant in mammals, where striking changes in protein synthesis occur early in cleavage, and thereafter the synthesis pattern appears to remain constant until implantation. The amount of time required for mouse and rabbit embryos to traverse cleavage (2–3 days) is greater than that required by many sea urchin and amphibian embryos to complete gastrulation. Perhaps post-transcriptional regulation of maternal message usage remains important only for a certain length of time, after which transcriptional regulation becomes more prominent. Thus, the morphological advancement of the embryo at the stage when this switchover occurs may be less significant than the time which has elapsed and the amount of biosynthetic activity which has occurred since fertilization.

First Morphological Indications of Differential Cell Function

It is impossible to review here the vast descriptive literature on early morphogenesis, and instead we focus attention on a few interesting examples. Unfortunately, the relation of protein synthesis to these early morphological changes is still almost wholly obscure. Protein synthesis is required for cell division and cleavage, including the synthesis of several

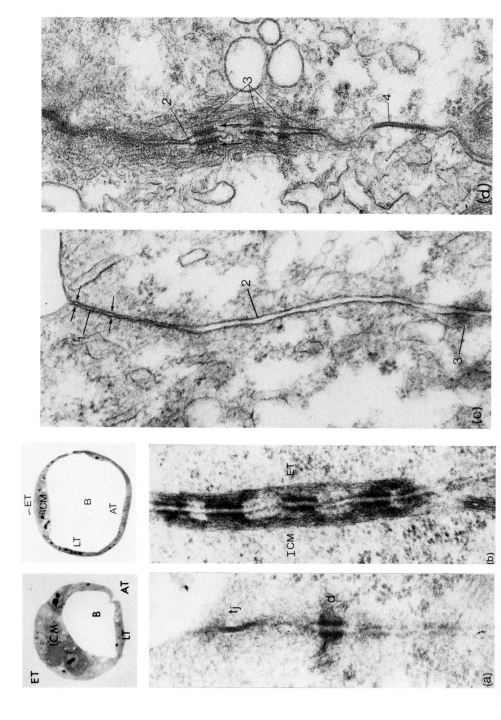

specific "housekeeping" proteins, such as tubulin and histone. However, even if molecular analyses of morphological processes are still beyond reach, early cytodifferentiation can be regarded as an indication of specific macromolecular reorganization, if not *de novo* synthesis. In general it is not yet known whether newly synthesized or preformed maternal protein molecules, or both, are involved in the first differential cell activities which can be observed in the electron microscope.

INTERCELLULAR JUNCTIONS IN EARLY EMBRYOS

An ultrastructural change associated with development in several chordate organisms is the appearance of intercellular junctional specializations. Since these are localized to particular stages and sites in the

Fig. 3.4. Cell contact specializations in early chordate embryos. (a) A junctional complex between adjacent trophoblast cells in an early mouse blastula. The stage and general morphology of this embryo is shown in the light micrograph (top). The blastocoelic cavity (B) is formed but not fully expanded. The inner cell mass cells (ICM) are closely apposed to the overlying embryonic trophoblast cells (ET). Both the lateral (LT) and abembryonic trophoblast (AT) cells are elongated. ×400. In the electron micrograph (bottom) a tight junction (tj) and a desmosome (d) are shown. In a tight junction the outer cell membranes of both cells are directly fused, eliminating the intercellular space. Desmosomes are characterized by dense cytoplasmic plaques and microfilaments, and an intercellular space of about 200 Å. No defined intermediate structure can be seen between the desmosome and the tight junction. ×66,000. (b) Junction between two cells of a later stage mouse blastula. The stage of the embryo is indicated in the light micrograph (top). Here it is seen that the blastocoel (B) is expanded compared to that seen in (a), and all the trophoblast cells (embryonic, ET; lateral, LT; abembryonic, AT) are elongated. Intercellular spaces are infrequently present in the inner cell mass (ICM). ×300. The electron micrograph (bottom) shows a cluster of desmosomes connecting an ICM cell with an embryonic trophoblast (ET) cell. Neither desmosome clusters nor tight junction–desmosome complexes are found between adjacent ICM cells. ×74,600. (a) and (b) from M. Nadijcka and N. Hillman (1974). *J. Embryol. Exp. Morph.* **32**, 675. (c) Junction between surface blastomeres of *Fundulus* (killifish) bastula (stage 8). In the apical portion of the junction (1), the membranes are closely apposed but generally separated by a narrow intercellular space. Below this, the membranes are parallel and separated by a larger space of about 100–200 Å (2). At a deeper level, localized masses of moderately dense cytoplasmic material are symmetrically apposed to the plasma membranes (3). The intercellular space shows an increase in density in this region. These densities appear to represent the first stage in desmosome development. ×67,000. (d) Junctional complex between two cells of a *Fundulus* late gastrula (stage 14). The four elements of the complex are an apical tight junction (sectioned obliquely for most of its length) (1), an intermediate zone (2), desmosomes (3), and a gap junction (4). The gap junction is 180 Å in overall width and is shorter than in midgastrulae. The first desmosome is well developed. The intercellular material is bisected by a dense line. The cytoplasmic plaques of dense material are separated from the membranes by a thin line of low density. Some short filaments (arrows) extend out from the dense plaques into the prominent bundle of filaments running parallel to the lateral borders. ×63,000. (c) and (d) from T. L. Lentz and J. P. Trinkaus (1971). *J. Cell Biol.* **48**, 455.

embryo, they can be regarded as markers of embryonic cell differentiation even though the functions of some of the various types of junctions are unknown or remain questionable. Junctional markers appear very early in the developmental process. An example is the mouse embryo studied by Nadijcka and Hillman (1974). As early as blastocoel formation, the embryonic trophoblast cells are connected to each other by tight junctions and by desmosomes. Tight junctions have also been reported between these cells by others (e.g., Calarco and Brown, 1969). A typical junctional region between trophoblast cells of the early blastula is illustrated in Fig. 3.4a (Nadijcka and Hillman, 1974). The inset shows the morphology of the whole embryo at this point in development. Clusters of desmosomes connect the inner cell mass cells with adjacent trophoblast cells, as shown in Fig. 3.4b. However, neither desmosome clusters nor complexes of desmosomes and tight junctions are observed between inner cell mass cells. Later the inner cell mass cells become separated from each other by large extracellular spaces. The characteristics of the intercellular junctional structures in the mouse embryo are thus reflective of the particular embryonic stage and cell type.

Junctional specializations also occur in the early embryos of teleosts, at the other end of the chordate taxonomic scale. These have been described by Trinkaus and Lentz (1967) and Lentz and Trinkaus (1971) in embryos of *Fundulus heteroclitus*, the killifish. Figure 3.4c shows a region of this embryo where two surface blastula cells come in contact, illustrating characteristic junctional areas of these cells. The highly organized structure shown in Fig. 3.4d is a junctional complex typical of gastrular cells. A group of desmosomes and a tight junction can be seen. In *Fundulus* the embryonic cells are electrically coupled throughout development (Bennett and Trinkaus, 1970). The structures probably responsible for electrical coupling are gap junctions (Bennett, 1973), though other forms of junctions could be involved as well. Data on electrical coupling of early embryonic cells now exist for a large number of chordate as well as other species (see, e.g., Furshpan and Potter, 1968; Bennett, 1973; Dicaprio *et al.*, 1975).

Cell surface specializations such as those displayed in Fig. 3.4 are almost certainly of functional significance. An interesting correlation exists between the ability to gastrulate in interspecific amphibian hybrids and the ability to form normal intercellular contacts (Johnson, 1969, 1970). Hybrid embryos between several pairs of *Rana* species which do not successfully develop beyond the gastrula stage also fail to develop normal junctional structures. By the early gastrula stage, presumptive mesoderm cells are normally applied to each other very closely, with gap junctions in which the cells are separated by only about 20 Å. However, in the hybrid embryos, the same cells show relatively little close junctional contact. For example, in a typical contact region of presumptive mesoderm cells in *Rana pipiens/R. clamitans* hybrid embryos, the gaps separating the cell membranes are in the range 120–450 Å. The correlation between junctional structure and developmental capacity in these hybrids is indirectly supported by studies of the contact behavior of the presumptive mesoderm cells *in vitro*. Johnson (1969) showed that cells from hybrid combinations unable to complete gastrulation also cannot reaggregate, while cells from viable hybrids or normal embryos can.

OTHER EARLY CYTODIFFERENTIATIONS

Many other early morphological specializations are known which imply differential cellular activities related to stage and cell type. Among these are secretory activity in the blastocoel forming cells of late cleavage *Xenopus* embryos (Kalt, 1971). Secreted materials include chains of spherical glycogen particles (Van Gansen and Schram, 1969), accounting for the high blastocoel glycogen content. Secretory activity in cleavage stage cells is known in many other embryos as well, e.g., sea urchin embryos (Motomura, 1966). In cleavage and morula stage mouse embryos scanning electron micrographs reveal a variety of unusual surface specializations, including microvilli and intercellular ridges (Calarco and Epstein, 1973). The density and form of these structures is related to their location and to the stage of the embryo. An important change in this embryo occurring at the 8-cell stage is *compaction*, the flattening of some cells in the early stages of blastocyst formation. Compaction is marked by the formation of new tight and gap junctions and by an increase in cellular adhesivity (Ducibella and Anderson, 1975). Compaction and the cytodifferentiations which it involves can be regarded as specific cellular activities required for morphogenesis at the 8-cell stage. Van Blerkom and Brockway (1975) have stressed the temporal correlation between early cytodifferentiation in the mouse embryo and the changes in the pattern of protein synthesis which also take place during cleavage.

Another class of morphologically detectable cell specializations is the development of cell motility, and in particular the capacity for coordinated migration. Cell migration is the dominant mode of formation of the embryonic axial structures in teleosts, as shown for *Salmo gairdneri* (Ballard, 1973a,b). In this embryo cells of each presumptive embryonic region are recruited from large areas of the deep layers of the early blastodisc. Cellular specializations which permit the all important migratory behavior of these "deep cells" have been described by Trinkaus (1967, 1973), Trinkaus and Lentz (1967), and Lentz and Trinkaus (1971) for *Fundulus* embryos. During blastulation these cells are characterized by

transitory blebs and they also develop lobopodia. Time-lapse micro-
cinematography shows that cell locomotion, which relies on these blebs
and lobopodial structures, begins in the midblastula stage, and continues
into gastrulation. The processes by which locomotion occurs are complex
and not well understood, but it is probable that the mechanism involves
cellular contractility, surface adhesiveness, and appearance and disappearance of specialized cell–cell contacts including tight junctions and
gap junctions.

The rate of locomotion can be quite high in these migratory cells, up to
30 μm/minute in one case and averaging about 10 μm/minute. This
amounts to translocation of a distance roughly equal to about the diameter of the cell per minute. Trinkaus concludes "it is clear from all this that
. . . cellular differentiation may begin well before the onset of gastrulation. Indeed, in *Fundulus* it is the differentiation of locomotory activity
during the blastula stage that makes possible the morphogenetic cell
movements of gastrulation." Lobopodia are known also in particular cell
types in early amphibian embryos (e.g., Nakatsuji, 1974), and migratory
cell specializations have been studied extensively in the chick embryo
(Trelstad et al., 1967).

A much greater range of specific cytodifferentiations will no doubt be
discovered as investigations of early embryogenesis continue. Another
problem altogether is analysis of the molecular basis for the observed
cytological changes. It seems safe to suppose that these are ultimately
complex phenomena, which probably involve a considerable number of
specific macromolecular constituents.

Appearance of Histospecific Proteins

While certain early embryonic cell types can be characterized morphologically, little is known of early histospecific protein synthesis patterns. Histospecific proteins are those associated with particular cell types
or tissues. They have been identified in many embryos at stages when
functional tissues are differentiating during postgastrular organogenesis.
Numerous cases could be cited, such as the appearance of hemoglobin in
blood island differentiation, and appearance of digestive enzymes in differentiating gut. For example, the period when hemoglobin first appears
in the blood island cells of the chick embryo was observed to be at the 7
somite stage (Wilt, 1965). Histogenesis is well advanced in many regions of
the anterior end of the chick embryo by this point. In mouse blastocysts,
in which trophoblast cells and inner cell mass cells are clearly distinct (see
Chapter 7), there are several distinct proteins characteristic of each of

these two embryonic cell types (Van Blerkom *et al*., 1976). The amphibian isozyme studies reviewed in Chapter 2 include related cases in that enzyme forms probably involved in organogenesis appear first in the tailbud and heartbeat stages. Such enzyme change has been demonstrated as well in various nonchordate embryos. Morrill and Norris (1965) showed that there is a class of hydrolytic enzyme activity in embryos of the snail *Nassaria* (*Ilyanassa*) *obsoleta*, the appearance of which coincides with extensive organogenesis (days 4 to 7 of development). During this time the larval kidney, gut, heart, etc., differentiate. The enzyme activities measured include various phosphatases, esterases, sulfatases, β-glucuronidase, and others. In another snail, *Physa fontinalis*, it was shown by Morrill (1973) that there is a histospecific distribution of ten electrophoretically distinguishable enzymes, all displaying some form of phosphatase activity. These appear at diverse stages, some as early as gastrula. The role played by these enzymes in the differentiation or function of the tissues in which they occur remains unknown, and furthermore, the enzyme activities rather than synthesis of the enzyme proteins were measured. Nonetheless, taken together the data on histospecific enzyme appearance seem to support the expected correlations between postgastrular organogenesis and molecular differentiation. What is more difficult is to find evidence for the appearance of localized or cell lineage-specific protein synthesis in earlier periods of development, prior to the widespread appearance of obviously differentiated tissues. Some examples of early distinctions among cell types are now reviewed.

APPEARANCE OF HISTOSPECIFIC PROTEINS BEFORE ORGANOGENESIS

The synthesis of collagen begins in *Xenopus* embryos during the gastrula stage and cannot be demonstrated at all during cleavage (Green *et al*., 1968). After gastrulation collagen synthesis increases at least a hundredfold, presumably associated with connective tissue differentiation. Another interesting case in amphibian embryos concerns the appearance of histospecific enzymes in differentiating neural crest and neural plate cells. In the frog *Rana pipiens*, the neural crest cells give rise to differentiated melanophores by the hatching stage (Smith-Gill *et al*., 1972). A histospecific biochemical activity of these cells is the conversion of tyrosine to melanin, via the enzyme tyrosine-DOPA oxidase (DOPA = dihydroxyphenylalanine). Smith-Gill *et al*. (1972) demonstrated this activity qualitatively by an *in situ* histological procedure. Neuronal cells use the same enzyme, in conjunction with DOPA decarboxylase, to prepare catecholamines. Caston (1962) reported that in *Rana* catecholamines appear in differentiating neural crest cells as early as the neurula stage, immediately following gastrulation. DOPA decarboxylase activity appears at about the same time (stage 15) (Benson and Triplett, 1974a). The synthesis of new tyrosine-DOPA oxidase can be detected in early neurulae (stage 13), according to Benson and Triplett, who utilized radioimmunoprecipitation to study actual enzyme synthesis. Direct evidence for the *de novo* synthesis of this protein was also obtained by identifying polysomes which contain nascent tyrosine-DOPA oxidase (Benson and Triplett, 1974b). By this criterion as well, translation of this protein begins at the neurula stage, and prior to this the polysomes contain no detectable tyrosine-DOPA oxidase. Presumably, therefore, the messenger RNA for this enzyme appears *de novo* at the neurula stage and is not previously present in the polysomal apparatus. In support of this view, Benson and Triplett also found that injection of actinomycin (at midgastrula stage) almost completely prevents the accumulation of tyrosine-DOPA oxidase protein by the neural fold stage (75–100% inhibition) while interfering with only about 15% of the total protein synthesis. The authors point out that a transcriptionally dependent process resulting in the activation of a preformed message cannot be excluded by these results, but this seems a less likely possibility. The observations of Benson and Triplett therefore probably signify the histospecific activation of the tyrosine-DOPA oxidase structural gene. An additional interesting feature of this system is that the enzyme begins to be synthesized at the neurula stage, well in advance of the appearance of enzyme activity in melanophores at the hatching stage. The explanation is that the initial product of synthesis is a proenzyme which can be broken down into the active form *in vitro* by trypsin. Something similar to this apparently occurs in the embryo, accounting for the interval between the onset of translation and the onset of pigment formation.

Many enzyme activities are known to rise (and fall) during early development. This subject is reviewed for sea urchin embryos by Giudice (1973) and for mouse embryos by Biggers and Stern (1973) and Brinster (1973). Enzymes of nucleic acid metabolism have been the subject of several recent studies (in sea urchin, e.g., see De Petrocellis and Vittorelli, 1975). In the absence of direct observations involving the enzyme proteins, however, changes in enzyme activity alone are difficult to interpret. Hence, these data are not further discussed here. An informative case was described by Epstein and Daentl (1972), who used genetic manipulations to show that the appearance of a new enzyme activity in mouse embryos is dependent on *de novo* genomic activity. The activity of this enzyme, hypoxanthine-guanine phosphoribosyltransferase (HGPRT) rises about eightfold between days 2 and 3 of development. The gene for HGPRT is apparently on the X chromosome. Epstein and Daentl (1972) showed that XO females produce eggs which at the 2-cell cleavage stage have only about half the HGPRT activity as do eggs from normal XX mothers. However, in 3-day morulae, embryos from XO and XX mothers have almost the same HGPRT activity. The simplest explanation is that HGPRT is synthesized during cleavage as a result of embryo structural gene transcription. Had the enzyme been made on maternal templates, or by processing a maternal proenzyme, the differences observed at the 2-cell stage would have persisted.

How early are *regionally specific* protein synthesis patterns set up? One item of evidence indicates that biosynthetic differentiation may exist from the earliest time that cell lineages with distinct fates can be defined, even if this occurs at first cleavage. In the gastropod *Ilyanassa*, only one of the two first cleavage blastomeres retains the capacity to give rise to coelomic mesoderm and all of its various derivations (see Chapter 7 for an extensive discussion of cell lineage and cell fate in this organism). Donohoo and Kafatos (1973) isolated preparations of these two cells, the AB and CD blastomeres, and labeled their proteins during a 100-minute period *in vitro*. Coelectrophoresis of the two preparations revealed a very different pattern of protein synthesis in the two blastomeres. Whether this is due to differential inheritance of preformed messenger RNA's or to differential synthesis of new messenger RNA's is as yet unknown. However, this issue does not obscure the demonstration that extensive *regional* distinctions in macromolecular synthesis patterns exist from the very beginning of development, at least in this organism.

DIFFERENTIATION OF PRIMARY MESENCHYME CELLS IN THE SEA URCHIN

To conclude this discussion, we consider one of the few pregastrular differentiation processes which has been studied in some detail. This example is the differentiation of the primary mesenchyme cells of the sea urchin embryo. These cells are derived from the micromeres sectioned off early in cleavage (by the 16-cell stage). During the blastula stage they appear within the blastocoel as individual spherical cells, the approximate number of which is a species characteristic (see Table 2.2). By midgastrulation they have taken up specific positions along the blastocoel wall, where they begin the formation of triradiate spicules. Eventually they give rise to the branched skeleton of the pluteus larva.

The initial signs of differentiation of the primary mesenchyme cells, even before their intrusion into the blastocoel, is the formation of distinct pulsatory lobes projecting into the blastocoel (Gustafson and Wolpert, 1963; Gibbins et al., 1969). In their subsequent migratory phase the cells develop pseudopodia with which they explore the inner blastocoel wall. It has been believed since the nineteenth century that the ectoderm of the blastocoel wall plays a critical role in the localization of these cells along the path where they are to lay down the skeletal elements. Recent observations have reinforced this view. Thus, Gustafson and Wolpert (1963) refer to the ectoderm as a "template for the mesenchyme pattern . . . reflected in the distribution of mesenchyme cells." As the primary mesenchyme cells align themselves on the ectodermal wall, their pseudopodia *fuse* to form oriented syncytial "cables," within which the skeletal matrix is laid down (Wolpert and Gustafson, 1961; Gustafson and Wolpert, 1963; Hagström and Lönning, 1969; Gibbins et al., 1969). According to Hagström and Lönning, the ancestors of these cells, the micromeres, show a tendency to form syncytial combinations as early as the 16-cell stage. Pseudopodial fusion is shown in Fig. 3.5, which is a series of time lapse photographs taken *in situ* by Okazaki (1965). A region including one of the mesenchymal cables with several associated mesenchymal cell bodies is shown in the electron micrograph of Fig. 3.6a. Here the arrow points to a process extended between the mesenchymal cell complex and the underlying ectodermal "template." Gibbins et al. (1969), from whose study Fig. 3.6 is reproduced, believe that the oriented organizations of microtubules within the cables and the stalks connecting the cables to the cell bodies are a key feature of the morphogenetic activity of these cells. Additional evidence for this was obtained in a study in which microtubules were caused to dissasemble by hydrostatic pressure or colchicine treatment (Tilney and Gibbins, 1969). Cable syncytia were no longer found, and the pseudopodia of mesenchyme cells were reduced. Within the syncytial cables of treated embryos the skeletal elements are deposited in membrane-bound vacuoles, as is beautifully illustrated in Fig. 3.6b (Gibbins et al., 1969). Okazaki showed in 1960 that the skeleton is deposited as an organic matrix together with an inorganic calcareous element which is known to be $CaCO_3$. In addition to the biosynthetic activities involved in skeletal secretion, acetylcholinesterase has also been detected as a specific marker of differentiation, localized in at least some of the skeleton forming mesenchyme cells (Ozaki, 1974).

The primary mesenchyme cells thus display a variety of differentiated characteristics from the time of their first invasion of the blastocoel or possibly even earlier. Their cytological, behavioral, and biosynthetic characteristics all mark them as a specific early differentiated cell type, one which is clearly distinct long before gastrulation. These cells provide

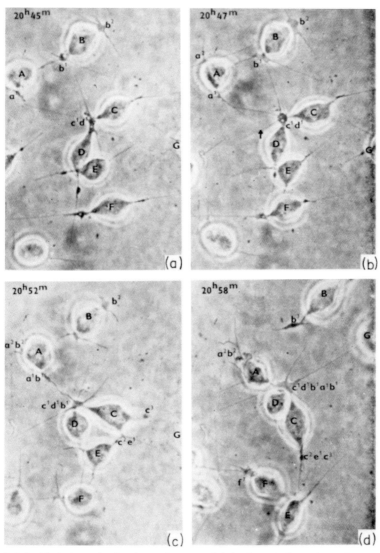

Fig. 3.5. Serial photographs showing early stages of skeletal matrix formation in larvae of the sea urchin *Mespilia globulus*. (a) represents the earliest stage. A matrix is present at c^1d^1. In (b) a fine process coming out of a^1 is passing across the upper left process of c^1d^1 and just touching, at the lower left, another process of c^1d^1. This contact is indicated by the arrow in (b). Though not evident in the photograph, the process of a^1 touched the lower process of c^1d^1 and moved along that process toward c^1d^1 until it touched the top of a short third process. Hereupon the process from a^1 fused with c^1d^1 in (c). In (d) further growth of the matrix is shown. Processes from all the cells are now connected. Numerals in the upper left-hand corner of the figures indicate the times of photographing. ×950. From K. Okazaki (1965). *Exp. Cell Res.* **40**, 585.

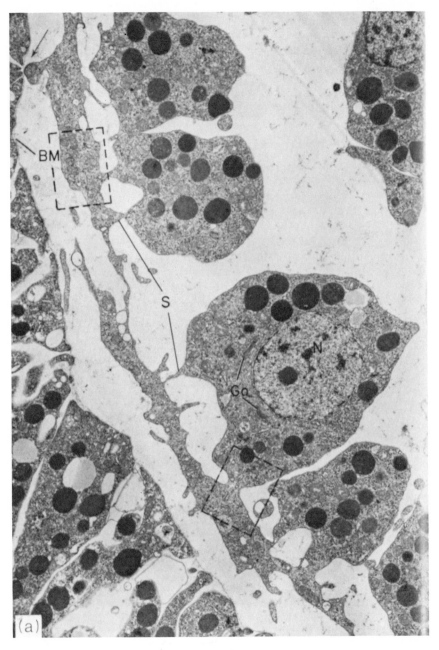

Fig. 3.6 (a)

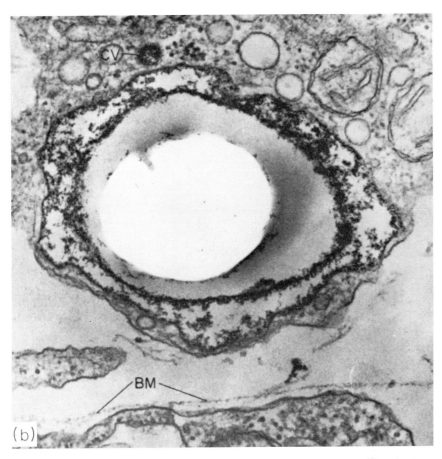

Fig. 3.6. Skeletal formation by primary mesenchyme cells in the sea urchin. (a) Low-magnification electron micrograph illustrating the form and arrangement of the primary mesenchyme of sea urchin embryos. The pseudopodia of several primary mesenchyme cells have fused to form a cable syncytium. The relation of the cell bodies to the cable cytoplasm and the relation in turn of the cable cytoplasm to the ectoderm can be easily observed. Each cell body is connected to the cable by one or more stalks (S), the cell bodies being without exception on the side of the cable opposite the basement membrane (BM) which underlies the ectoderm. Short processes extend from the cable cytoplasm towards the ectoderm (see arrow), but these have not been seen to penetrate the basement membrane. Within the cell body the nucleus (N) is in a central position. The Golgi zone (Go) lies on one side of the nucleus, and the other formed elements occupy the remainder of the cytoplasm of the cell body. Fine extracellular fibrils are present throughout the blastocoel. (b) Transverse section through the cytoplasm of a cable which is in the process of forming the skeleton. The skeleton forms within a membrane-limited vacuole in the cytoplasm of the cable. Within this vacuole is some electron-opaque material. The center of the vacuole, containing the skeleton, disappears from the section, leaving a hole in the Epon. A small coated vesicle (CV) containing the same dense material as that in the skeletal vacuole is present near the cable. The basement membrane (BM) and a portion of an ectodermal cell lie at the lower edge of the micrograph. ×60,000. From J. R. Gibbins, L. G. Tilney and K. R. Porter (1969). *J. Cell Biol.* **41**, 201.

what is probably the best known case of preorganogenesis differentiation. Mesenchyme cell differentiation (i.e., skeleton formation) is not completed in actinomycin-treated embryos. Furthermore, in viable species hybrids it is clear that skeletal form in the pluteus stage is hybrid in character (see Chapter 2 for references). Probably, therefore, at least the later portion of the primary mesenchyme differentiation pathway is under the control of the mesenchyme cell genomes (and/or the genomes of the ectodermal "template" cells).

Knowledge of primary mesenchyme cell differentiation exists in part because of their conspicuous location and their discrete character. Were other cell types as easily distinguished, a greater variety of evidence of early differentiated behavior might exist for sea urchin embryos. Thus it is not unlikely that as additional cell types are studied evidence of preorganogenesis differentiation will accumulate. Such certainly is the implication of the remarkable result obtained by Donohoo and Kafatos (1973) with *Ilyanassa* embryos.

Conclusions

The examples treated in this chapter can be considered to lead to two conclusions. First, it is clear that differential protein synthesis and the appearance of histospecific proteins can be demonstrated as expected in postgastrular stages of many organisms. This is the period when organogenesis is manifest. From the results of species hybrid and actinomycin experiments, as well as the basic principle that variable gene activity underlies differentiation (Chapter 1), it can be presumed that organogenesis sooner or later requires new programs of structural gene transcription. We note, however, that direct evidence for this inference remains rare.

A second and more paradoxical conclusion is also suggested. We have seen that striking changes in protein synthesis pattern seem to occur in early sea urchin and amphibian embryos even in the absence of transcription. This finding indicates the existence of a far-reaching pattern of post-transcriptional control. Furthermore, the species hybrid, enucleation, protein synthesis, and isozyme experiments all show that many of the events of early development are independent of embryo genome action. This is the case at least through gastrulation in the sea urchin and at least into gastrulation in amphibian embryos. It will be recalled that in close congeneric sea urchin hybrids the first histospecific paternal influences, either morphological or molecular, are observed only at pluteus stage. Similarly, in the frog, paternal isozyme forms appear only in the tailbud stage and thereafter. These findings must be considered in light of the data reviewed in this chapter, which demonstrate involved pregastrular specializations at the cellular level. Both temporal and spatial differences in cellular activity occur during this period. It follows that these early processes of cell specialization must be largely programmed, operated, and controlled by developmental systems which were stored in the egg at fertilization. Such maternally loaded systems thus may be of high complexity. Their responsibilities include regional cytodifferentiation, direct or indirect specification of the position and division rates of hundreds or thousands of cells, and determination of a changing pattern of protein synthesis. These statements are not meant to imply that embryo structural gene transcription plays no role in the preorganogenetic period of development. In fact messenger RNA synthesis is known to occur even in cleavage (see Chapter 5). However, it is thus far impossible to show for many species of animal that new transcription exercises a determinant effect on *changes* in protein synthesis and cell function in the early phases of embryonic organization. The slowly cleaving mammalian embryo may represent an exception, in that here it is not unlikely that embryo structural gene function exercises a greater degree of control over biosynthetic processes during the cleavage and morula stages.

4

Quantitative Aspects of Protein Synthesis in Early Embryos: The Role of Maternal Components

Absolute rates of protein synthesis in early sea urchin embryos are calculated from several kinds of data, and are collated in Table 4.1. Measurements based on free amino acid pool specific activity agree well with those obtained by other means, including determinations of the aminoacyl-tRNA specific activity. In the sea urchin most amino acid pools are not significantly compartmentalized. Absolute protein synthesis rates have also been determined for early amphibian embryos, which contain about 10^3 times more ribosomes than do sea urchin embryos. The absolute rates of protein synthesis per embryo ribosome are similar in these two systems. The ribosomes of early embryos are largely maternal in origin, as are the tRNA's. This is probably also true of initiation and translocation factors and the enzymes needed for protein synthesis. Cell-free embryo ribosome systems can support protein synthesis from both exogenous and endogenous messenger RNA's. The competence of the oocyte protein synthesizing system is best demonstrated in experiments in which exogenous messenger RNA is injected into *Xenopus* oocytes and shown to be translated efficiently and faithfully. These experiments show that the oocyte contains all necessary components of the protein synthesis apparatus. Injected messenger RNA's are very stable in the oocyte, suggesting that the egg cytoplasm is an environment in which endogenous polysomal RNA's could survive for long periods.

Maternal messenger RNA has been extracted from unfertilized eggs and translated in cell-free protein synthesis systems. Quantitative estimates of the content of maternal messenger RNA are derived for sea urchin and amphibian eggs (Table 4.2). The maternal message is probably stored in ribonucleoprotein particles smaller than ribosomes. Maternal messenger RNA's for the histones have been extracted from such particles, as well as for many other unidentified protein species. In the sea urchin embryo, as in some others, the rate of protein synthesis increases manyfold after fertilization. This is shown to be due to release of maternal message and its mobilization in polysomes. In other organisms, such as amphibians, increase in polysome content due to mobilization of maternal message begins during the maturation period, well before fertilization, and continues into early development. Little change in protein synthesis rate occurs at fertilization in such cases. Provision during embryogenesis of four classes of protein is considered. These are the histones, the DNA polymerases, the RNA polymerases, and microtubule proteins. The rate of histone synthesis in sea urchin embryos is closely related to the rate of DNA synthesis. It is maximum during early blastula when the rate of DNA synthesis is maximum. Histone messenger RNA is synthesized by the embryo and is also inherited as maternal message. Some of the histones appearing after gastrulation differ from those present and synthesized in cleavage. The early form of histone I is synthesized from maternal messenger RNA and the postgastrular form from newly transcribed messenger RNA. The rates of histone synthesis, the quantities of histone messenger RNA, and other parameters relating to histone synthesis in sea urchin and amphibian eggs are compared (Table 4.3). In amphibian eggs histone synthesis rate is not dependent on DNA synthesis rate, but remains high from before fertilization. Here newly synthesized histones are translated mainly from maternal messages. There is in addition a large stockpile of presynthesized histones stored in amphibian eggs, in contrast to sea urchin eggs. DNA and RNA polymerases are required by embryos in large quantities, but synthesis of these proteins during early development plays no significant role. Instead they are synthesized and stored during oogenesis for use during development. Microtubule proteins are similarly inherited from oogenesis, and a large pool of these proteins exists in amphibian and sea urchin eggs. Microtubule proteins are synthesized actively on maternal messenger RNA's in sea urchin eggs as well. Thus, histones and microtubule proteins both have a dual source during embryogenesis. There are maternal plus newly synthesized messenger RNA's for histones in the sea urchin and maternal messenger RNA plus stored proteins for microtubule proteins in sea urchin embryos and for histones in amphibian embryos.

The existence of maternal messenger RNA follows logically from much of the data presented in the previous two chapters. Maternal message is strongly implied by the continuation of complex patterns of protein synthesis in actinomycin-treated and enucleated eggs, though other explanations such as persistence of a small trickle of new message synthesis can in some cases be entertained. Maternal messenger RNA also provides the simplest general explanation for the species hybrid experiments, since protein synthesis is required for early cytodifferentiation and cell division. The purpose of the present chapter is to examine the direct, molecular evidence for maternal messenger RNA and to consider the quantitative contribution of maternal messenger RNA to the synthesis of embryonic proteins. We begin with a review of data from which rates of total protein synthesis during early embryogenesis can be calculated.

Absolute Rates of Protein Synthesis

FREE AMINO ACID POOLS IN SEA URCHIN EMBRYOS

The basic problem in measuring absolute protein synthesis rates is to know the specific activity of the true precursor pool. Taken literally, this would require measurement of aminoacyl-tRNA or peptidyl-tRNA specific activities, but until recently the technical means to accomplish such measurements conveniently have not been available. In practice it has generally been assumed that the soluble or free amino acid pool is the true precursor pool, i.e., that the pools are not compartmentalized. Thus most measurements of protein synthesis rate have been based on direct or indirect estimates of the free amino acid pool specific activity. Often this requires that the behavior of the pool during the labeling period is known. One method used successfully with sea urchin eggs relies on preloading the egg or embryo with large amounts of labeled amino acid (Berg, 1970; Berg and Mertes, 1970). For amino acids with relatively small internal pool sizes, such as valine, proline, histidine, threonine, and phenylalanine, significant pool expansion results from uptake of the exogenous amino acid. Eventually the pools approach the external precursor specific activity. In the study of Berg and Mertes (1970) the embryos were exposed for 30–40 minutes to relatively high concentrations of one of the labeled amino acids whose pool is expandable. After 10 minutes the rate of labeled amino acid incorporation into protein becomes constant. Pool expansion and constant incorporation rate also can be obtained using alanine, which is present in a relatively huge internal pool but has a very high rate of inward transport. Extensive data on pool expansion in sea urchin eggs were reported by Fry and Gross (1970b). In one experiment the endogenous leucine pool, about 6×10^{-8} μmoles leucine per embryo, was expanded 400% to 25×10^{-8} μmoles per embryo by exposure for 1 hour to 3 μg/ml exogenous leucine. Fry and Gross (1970b) showed that, in general, the normal endogenous amino acid pools remain constant throughout embryogenesis. For most amino acids the pools are in the range 2×10^{-8} to 40×10^{-8} μmoles per embryo for both Arbacia and Strongylocentrotus, though the size of the individual pools differs markedly between these species. Pool sizes for certain amino acids tend to be comparatively large, e.g., glutamic acid, glycine, lysine, and arginine. Together with taurine these comprise over 80% of the mass of the free amino acid pool (Fry and Gross, 1970b).

ABSOLUTE RATES OF PROTEIN SYNTHESIS IN SEA URCHIN EMBRYOS

Using the preloading technique, Berg and Mertes (1970) found that the absolute rate of protein synthesis changes little between mesenchyme blastula and gastrula stages. Protein synthesis rates measured for Lytechinus anamensus gastrulae average 0.84 ng hr^{-1} per embryo. The measurements of Fry and Gross (1970b), also made under conditions of moderate pool expansion, yield average values for cleavage stage Strongylocentrotus purpuratus embryos of about 0.4 ng hr^{-1} per embryo, and similar though slightly higher values were obtained by these authors for embryos of Arbacia punctulata. An additional measurement based on exogenous amino acid incorporation is that of Seale and Aronson (1973a). These authors also relied on relatively large external leucine concentrations to control intracellular leucine pool specific activities. Their data include measurements of both pool specific activity and incorporation into protein. The results agree well with those just discussed, as a calculated protein synthesis rate of about 0.45 ng hr^{-1} per embryo was obtained for both late cleavage and blastula stage Strongylocentrotus. These and other data on absolute protein synthesis rates in sea urchin embryos are assembled in Table 4.1.

TABLE 4.1. Absolute Rates of Protein Synthesis in Sea Urchin Embryos

Species	Stage	Protein synthesis rate (ng hr^{-1} per embryo)	Source of data
Strongylocentrotus purpuratus	Cleavage	0.4[a]	Fry and Gross (1970b)
		0.45	Seale and Aronson (1973a)
	Blastula	0.44	Seale and Aronson (1973a)
	Gastrula	0.64[b]	Galau *et al.* (1974, 1976a)
	Gastrula	0.52[c]	Regier and Kafatos (1976)
	Gastrula	0.89[d]	Regier and Kafatos (1976)
	Gastrula	0.64[c]	Regier and Kafatos (1976)
	Gastrula	0.92[d]	Regier and Kafatos (1976)
Lytechinus pictus	Very early cleavage	0.13[e]	Humphreys (1971)
Lytechinus anamensus[f]	Blastula	0.77	Berg and Mertes (1970)
	Gastrula	0.84[g]	Berg and Mertes (1970)
Arbacia punctulata	Cleavage	0.72	Fry and Gross (1970b)

[a] This value is calculated directly from raw incorporation data presented by Fry and Gross (1970b) who gave the rate of amino acid incorporation as about 6×10^{-9} μmole min^{-1} per embryo (averaging the leucine and valine results). The simple assumption is made that in the moderate pool expansion conditions of the experiment, the specific activity of the internal pool is given by that of the exogenous amino acid. The authors reduced their data differently by application of an expression relating uptake and incorporation rate and pool size, but arrive at very similar values in the range 0.28–0.52 ng hr^{-1} per embryo.

[b] According to Galau *et al.* (1974) about 60% of the ribosomes of the gastrula of this species may be found in polysomes. Assuming 4% of the polysomal RNA is messenger RNA (Galau *et al.*, 1976a), 3.3 ng of total RNA per embryo and 80% of the RNA in ribosomes, the total length of message is 1.4×10^{11} nucleotides. We assume a translational velocity of about 1 codon sec^{-1} per ribosome at 15°C, and spacing of 140 nucleotides per ribosome [for these parameters, see data listed by Kafatos (1972) and Kafatos and Gelinas (1974)]. From the amount of message and the translational velocity, about 3.8×10^{14} daltons protein is polymerized per hour, or about 0.64 ng hr^{-1} per embryo.

[c] This value is obtained from measurements of leucine pool specific activity and leucine incorporation.

[d] This value is based on measurements of the specific activity of the embryo leucyl-tRNA, using the same material as in footnote c.

[e] This value is calculated similarly to that in footnote b, i.e., from translational rate parameters and the content of polysomal RNA reported by Humphreys (1971) for cleavage stage of *L. pictus*. These embryos contain about 12 pg of messenger RNA per embryo, assuming that 4% of the polysomal RNA is messenger RNA (Galau *et al.*, 1976a).

[f] *Lytechinus pictus* and *L. anamensus* may be the same species (Giudice, 1973).

[g] This value is the average of many individual determinations using six different amino acids (valine, proline, alanine, histidine, phenylalanine, and threonine). The determinations agree within ±30%.

Two measurements of protein synthesis rate which depend on totally different forms of data can be compared to the values deriving from uptake studies (Table 4.1). Regier and Kafatos (1976) isolated leucyl-tRNA from labeled sea urchin embryos and measured its specific activity directly. The rate of protein synthesis for *Lytechinus* gastrula stage embryos was estimated by this method at about 0.92 ng hr^{-1}, in close agreement with the other values listed in Table 4.1. Regier and Kafatos (1976) also calculated the protein synthesis rate in the same embryos from measurements of leucine pool specific activity. This measurement yielded a value of 0.64 ng hr^{-1}. Another calculation can be made from the amounts of polysomes present in sea urchin embryos, assuming average rates of translation. For example, in *Strongylocentrotus* embryos at the gastrula stage, 45–60% of the ribosomes are in polysomes (Infante and Nemer, 1967; Galau *et al.*, 1974). The 60% value leads to calculation of a rate of 0.64 ng hr^{-1} per embryo. Details of this calculation are given in the notes to Table 4.1, and the result is again in excellent agreement with the other rates shown in Table 4.1. The same calculation can be applied to the very early cleavage stage embryos of *Lytechinus*. From the measurements of Humphreys (1971) these embryos contain about 12 pg of polysomal messenger RNA [assuming that 4% of the polysomal RNA is message, as is found for *Strongylocentrotus* embryos (see Chapter 5)]. This is approximately one-fifth the mass of polysomal messenger RNA present in *Strongylocentrotus* gastrulae. The expected absolute rates of protein synthesis are correspondingly lower (Table 4.1). Since *Lytechinus* eggs have less RNA than *Strongylocentrotus* eggs, the calculated value, 0.13 ng hr^{-1} per embryo, is acceptable. The agreement between the absolute rates calculated from free amino acid pool specific activities and those obtained by other means proves that in sea urchin embryos the amino acid pools are not significantly compartmentalized, contrary to earlier suggestions (e.g., Berg, 1968), and this was also shown by direct studies of precursor uptake kinetics by Berg (1970). The most direct evidence on this point is that of Regier and Kafatos (1976), since the protein synthesis rate calculated from the direct leucyl-tRNA precursor specific activity is less than a factor of 1.5 different from that based on total intracellular leucine specific activity. As Table 4.1 shows, this comparison is available for both *Strongylocentrotus* and *Lytechinus* gastrula stage embryos.

Table 4.1 shows that the absolute protein synthesis rate has been satisfactorily measured for sea urchin embryos, at least within a factor of two. The amount of protein synthesized per hour in these organisms is about 1% of the total protein content. By following the loss of label from newly synthesized proteins with time, Berg and Mertes (1970) also showed that approximately the same fraction of newly synthesized protein, 0.8% per hour, is lost through turnover. The total protein content of the embryo thus remains almost constant throughout early embryogenesis.

ABSOLUTE RATES OF PROTEIN SYNTHESIS IN AMPHIBIAN EMBRYOS

Data on absolute protein synthesis rates also exist for amphibian embryos. Ecker (1972) studied protein synthesis rates in *Rana pipiens* eggs which had been induced to mature *in vitro* with progesterone and microinjected with labeled amino acids. Similar experiments were carried out earlier by Ecker and Smith (1966, 1968), and data on oocytes and early embryos were obtained more recently by Shih (1975), also using the microinjection method. By using the microinjection technique the problems associated with limited permeability to exogenous amino acids are circumvented. Synthesis rates are calculated by estimating the free amino acid pool specific activity from the amounts of radioactive amino acid injected, from the rate of disappearance of label from the free amino acid pool, and from the rate of its appearance in labeled proteins. Amino acid pools in these comparatively large eggs evidently are fed by the progressive hydrolysis of stored yolk, and the pool can be considered a steady state system with the flow out equal to incorporation into protein. All the amino acid pools are somewhat compartmentalized, but to a quantitatively insignificant extent. For example, about 10–20% of the free leucine pool is apparently compartmentalized (Shih, 1975). The absolute rates of protein synthesis calculated are about 20–40 ng hr^{-1} for meiotic *Rana* oocytes (Ecker, 1972; Shih, 1975) and 50 ng hr^{-1} for the two-cell stage of cleavage (Shih, 1975).

Another estimate of absolute protein synthesis rates in an amphibian egg was obtained by Woodland (1974). In this study, which is discussed at more length later in this chapter, the fraction of ribosomes included in the polysomal structures of *Xenopus* embryos was measured. Absolute protein synthesis rates were also measured in oocytes from the incorporation into protein of microinjected ^{3}H-histidine, assuming a histidine content in the proteins of about 3%. The rate obtained was about 4 pmoles of histidine hr^{-1}, or 19 ng of protein hr^{-1}. About 1.5% of the oocyte ribosomes are in polysomes at this stage (Woodland, 1974) and applying appropriate translation rate parameters (see footnotes to Table 4.1), a protein synthesis rate of about 25 ng hr^{-1} is obtained. The agreement suggests that the histidine pool is not significantly compartmentalized. Between blastula and neurula stage (a period of about 25 hours) the polysome content remains essentially constant, at about 15% (Woodland, 1974). The absolute synthesis rate during this period is calculated to be on the order of 200 ng hr^{-1} per embryo. This is about 450 times the rate of synthesis in early embryos of S. purpuratus. However, the Xenopus egg contains about a thousand times more ribosomes than does the S. purpuratus egg. Protein synthesis rates per polysomal ribosome in these two species are thus comparable.

Maternal Ribosomal and Transfer RNA's

It has long been evident that maternal ribosomal RNA represents the bulk of the RNA of early embryos. This follows immediately from measurements of total RNA content, since 80–90% of the total RNA is ribosomal RNA. Total RNA content is essentially invariant in Xenopus embryos from fertilization at least into gastrulation (Bristow and Deuchar, 1964; Brown and Littna, 1964). At the latter stage this embryo contains over 5×10^4 cells. Similarly, little change in total RNA content occurs in sea urchin embryos during early development (e.g., Tocco et al., 1963). New ribosomal RNA synthesis does not contribute significantly to the ribosomal RNA complement of either early amphibian or sea urchin embryos (see Chapter 5 for details). Since by gastrulation an important fraction of the total embryo ribosome pool, e.g., 15% in Xenopus (Woodland, 1974) and 40–60% in Strongylocentrotus (Infante and Nemer, 1967; Galau et al., 1974) is in polysomes, it follows that embryonic translation must occur on maternal ribosomes.

The role of maternal ribosomes in early embryogenesis was demonstrated most unequivocally by Brown and Gurdon (1964, 1966). Their investigations utilized the anucleolate (o nu) mutant of Xenopus, homozygous bearers of which lack the capability to synthesize ribosomal RNA. The o nu mutation is a deletion of the nucleolar organizer region of the chromosomes, and was first described by Elsdale et al. (1958). In the homozygous form it is lethal, but heterozygous individuals are able to develop and function normally. The mutation can be detected cytologically in heterozygotes by the presence in the cell nuclei of only one nucleolus rather than the usual two or more. Wallace and Birnstiel (1966) found that the genomes of homozygous o nu embryos contain no detectable DNA hybridizable with ribosomal RNA; i.e., o nu homozygotes lack the ribosomal RNA cistrons. Homozygous o nu embryos fail to synthesize any new ribosomal RNA, though they do synthesize other RNA's (Brown and Gurdon, 1964, 1966). Several partial anucleolate mutants are also known, and in combination with the o nu genotype, these have reduced numbers of ribosomal genes and reduced ribosomal RNA synthesis (Knowland and Miller, 1970; Miller and Knowland, 1970). The heterozygous mothers of the o nu homozygotes are normal with respect to their capacity to synthesize ribosomal RNA, and they shed eggs which contain a normal complement of ribosomes. Homozygous *o nu* embryos, therefore, provide an opportunity to study the role of maternal ribosomal RNA as embryogenesis progresses. Brown and Gurdon (1964) found that *o nu* homozygotes differentiate and develop all the way to the swimming tadpole stage. This provides a direct demonstration of the extended conservation and use of maternal ribosomes. It is evident that there is a long period during which new ribosomes are not actually necessary, for the development of *o nu* homozygotes becomes retarded only after hatching, when the embryo contains more than 5×10^5 cells. It has also been shown in an independent study that the maternal ribosomes of homozygous *o nu* tadpoles are able to translate newly synthesized messenger RNA (Gurdon and Ford, 1967).

In the sea urchin embryo the maternal ribosomal RNA content actually declines slightly during early development. This decline was noted by Comb and Brown (1964), Comb *et al.* (1965), and Nemer and Infante (1967a), who made use of a trait carried by certain individual females of *Strongylocentrotus purpuratus* to study directly the fate of the inherited ribosomal RNA. Eggs from these females contain ribosomal RNA in which the 18 S component can be split into two 13 S fragments by heating briefly to 60°C. This thermal fragility characteristic may be used as a marker for the maternal ribosomal RNA. The concentration of the aberrant ribosomal RNA species in total ribosomal RNA was found not to alter from cleavage through the mesenchyme blastula stage. Therefore, there is no dilution of the maternal ribosomal RNA by newly synthesized embryo ribosomal RNA during this period. As expected, the ribosomes of eggs and blastulae of sea urchins prove to be identical in various physical properties, and the subunits hybridize freely with one another (Kedes and Stavy, 1969).

Transfer and 5 S RNA's are also known to be inherited from oogenesis. The late vitellogenic (i.e., nearly mature) oocyte of *Xenopus* contains over five tRNA molecules per 28 S ribosomal RNA molecule (Ford, 1971) and over seven 5 S ribosomal RNA molecules. Detailed studies on the synthesis of these classes of low molecular weight RNA during oogenesis have been carried out and are briefly reviewed in Chapter 8. For now our purpose is merely to note that at the time of fertilization the egg is supplied with both tRNA and 5 S RNA and that these are present in functional form. Arguments similar to those used above with respect to ribosomal RNA apply here as well. That is, the first few hours of development, during which sea urchin embryos mobilize a significant fraction of their preexistent ribosomes for protein synthesis, represent far too short a time to synthesize sufficient *new* 5 S RNA to service the active ribosomes. In

Xenopus oogenesis 5 S RNA begins to be synthesized and accumulated in immature oocytes even before ribosomal RNA synthesis becomes active, in order to provide the amount of RNA needed in the mature egg ribosomes (Ford, 1971). Even at maximum rates, synthesis of the final complement of 5 S RNA requires a month or more in this species. Yet within a few hours after fertilization, many of the 5 S RNA's are in use in the embryo polysomes. It follows that the active ribosomes include mainly or exclusively maternal 5 S RNA's, in both *Xenopus* and sea urchin embryos.

If data on their synthesis kinetics were available, similar arguments could probably be applied to all the various macromolecular components needed for protein synthesis: including initiation, elongation, and termination factors; charging enzymes; and ribosomal proteins. What is clear in any case is that all the necessary factors are already present in mature eggs. For instance a number of aminoacyl-tRNA synthetases have been shown to be present in sea urchin eggs at fertilization (e.g., Maggio and Catalano, 1963; Nemer and Bard, 1963; Ceccarini and Maggio, 1969; Molinaro and Farace, 1972), as of course are the tRNA's themselves (O'Melia and Villee, 1972). However, the most important demonstrations of the functional completeness of the maternally prepared protein synthesis system have come from experiments in which exogenous messenger RNA is introduced. These fall into two classes, those carried out with cell-free systems of egg or oocyte origins, and those in which messenger RNA is injected into oocytes or eggs and shown to be translated *in situ*. We now review briefly both forms of evidence.

Translational Capacities of Mature Oocytes and Early Cleavage Stage Embryos

IN VITRO PROTEIN SYNTHESIS ON EMBRYO RIBOSOMES

The first significant experiments employing cell-free protein synthesis systems derived from eggs were those of Hultin (1961b). Within the next few years several groups showed that homogenates of sea urchin eggs could translate "synthetic messenger RNA's," such as polyuridylic acid. As is now understood, under the high Mg^{2+} conditions used in many of these early experiments, the normal translational initiation mechanisms are bypassed. Polynucleotide stimulated translation was used widely to demonstrate that unfertilized sea urchin egg ribosomes could carry out translation. It was concluded that the ribosomes themselves are functional and must contain or have available many of the various factors required for protein synthesis. Among the workers who contributed to this finding in the early 1960's were Nemer (1962), Wilt and Hultin (1962), Brachet *et al.* (1963a), Nemer and Bard (1963), Tyler (1963), Maggio *et al.* (1964), and Monroy *et al.* (1965). Stavy and Gross (1969) showed that the capacity of sea urchin embryo ribosomes to accept polyuridylic acid as a template remains more or less constant throughout early development. A further improvement along these lines from the standpoint of demonstrating faithful translation mechanisms was reported by Clegg and Denny (1974). These workers showed that both fertilized and unfertilized sea urchin egg ribosomes can correctly translate rabbit α- and β-hemoglobin messages. However, this was intended as a test of the activity of sea urchin egg ribosomes, not as a test for the presence of specific initiation factors, and in this experiment soluble factors of the Krebs II ascites cell-free synthesis system were added to the sea urchin ribosomes. The proteins synthesized in the hybrid system in response to added globin message were verified as α- and β-hemoglobin by carboxymethyl cellulose chromatography. It is also to be noted that in many of the earlier studies cited above, cell-free sea urchin embryo systems were shown to translate endogenous messenger RNA's, though no comparisons to normally synthesized proteins were possible. Such activity may have represented mainly completion of already initiated chains. Taken together, the cell-free translation experiments show that sea urchin eggs are supplied with many of the essential elements of the protein synthesis system, that is, other than messenger RNA which is not assayed in these studies. One possible caveat is that the experiments so far reviewed would not have assayed message-specific initiation factors, if these exist and play a role in early sea urchin embryos. We may conclude that the mature egg must contain functional ribosomes, tRNA's, aminoacyl synthetases, some initiation factors, and all translocation factors and enzymes needed for protein synthesis. Since these components already exist in the egg at fertilization, they are the result of biosynthetic processes occurring during oogenesis.

INJECTION OF EXOGENOUS MESSENGER RNA INTO OOCYTES

The capacity of mature ovarian oocytes to translate exogenous messenger RNA's has been measured directly by Gurdon and his associates. In 1971 Gurdon *et al.* reported that rabbit hemoglobin messenger RNA injected into *Xenopus* eggs or oocytes is translated on the oocyte ribosomes to yield complete globin polypeptide chains. Calculated on the basis of the total messenger RNA injected, the rates of translation observed are as high as 30 β-globin molecules synthesized per messenger RNA hr^{-1} at 19°C, only a factor of about five lower than the translation rates observed in reticulocytes at 37°C (Gurdon *et al.*, 1973). Considering that these calculations are based on the premise that all the injected messenger RNA is functional, the translation of this message can be considered very efficient. The translational product has been clearly verified as hemoglobin by carboxymethyl cellulose chromatography and by other methods as well (Lane *et al.*, 1971; Marbaix and Lane, 1972; Gurdon *et al.*, 1973). These include tryptic peptide mapping. In Fig. 4.1a α- and β-hemoglobins (dotted lines) are cochromatographed together with proteins newly synthesized in *Xenopus* oocytes injected with mouse hemoglobin messenger RNA and ³H-histidine (Gurdon *et al.*, 1973). The control in Fig. 4.1b shows that no hemoglobin is synthesized without the exogenous messenger RNA. Hemoglobin is not the only protein which can be translated in *Xenopus* oocytes, as numerous additional experiments have shown. Others include, for example, the messenger RNA's for immunoglobulin light chains (Smith *et al.*, 1973) and thyroglobulin (Vassart *et al.*, 1975).

Figure 4.1a illustrates another feature of the oocyte translational system, namely, that it displays some quantitative translational selectivity. Thus α-hemoglobin message is translated only 20% as efficiently as the β-hemoglobin message. Figure 4.1c shows, on the other hand, that in an appropriate cell-free system the two messages are both present in the reticulocyte RNA preparation and are translated with equal efficiency. Thus the relative number of ³H-histidines incorporated into each is close to their relative histidine contents. It is not understood why the oocyte translational apparatus prefers the β-hemoglobin message over the α-hemoglobin message, and the explanation may not lie in the initiation mechanism. Thus the polysomes bearing the α-globin message contain about the same number of ribosomes as do those bearing the β-globin message (Lingrel and Woodland, 1974).

Once injected, exogenous messenger RNA's survive and continue to be translated in *Xenopus* oocytes for long periods. This is illustrated in Fig. 4.1d, where hemoglobin synthesis is shown to occur in oocytes injected with messenger RNA 12–13 days earlier. In these oocytes hemoglobin translation still continues at 70% of the initial rate. Nor do fertilization and development interfere with the long-term survival of the injected messages. Thus Gurdon *et al.* (1974) have shown that injected hemoglobin messenger RNA continues to be translated at close to a constant rate throughout early development, at least to the swimming tadpole stage 8 days later. A conclusion of general interest which can be drawn from these experiments on the stability of exogenous messenger RNA's is that the internal milieu of the oocyte and embryo is conducive to long-term message stability. Presumably this may refer as well to some of the endogenous polysomal messages.

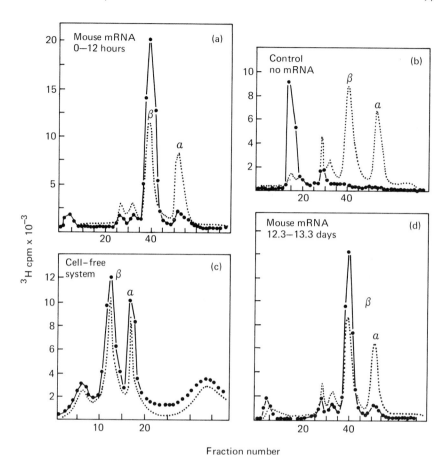

Fraction number

Fig. 4.1. Translation of injected hemoglobin messenger RNA in *Xenopus* oocytes. (a) Car-boxymethyl cellulose chromatography of globin synthesized by 10 oocytes injected with 7 ng of hemoglobin messenger RNA, and labeled for 12 hours. Dotted lines represent authentic hemoglobin marker present on the same column and identified by absorbance. Oocytes were labeled by incubation in medium containing 5 μM histidine including ^{3}H-histidine at 250 μCi/ml. (b) Eleven oocytes were injected with saline solution, but no RNA, and were labeled and analyzed as in (a). Labeling was for 1–18 hours. The radioactive proteins were again cochromatographed with hemoglobin markers (dotted lines). (c) Mouse hemoglobin messenger RNA translated in a cell-free reticulocyte system for 1 hour. Newly synthesized proteins were extracted and analyzed together with marker hemoglobins (dotted lines). (d) Ten oocytes were injected with 7 ng of mouse hemoglobin messenger RNA and labeled from 12.5–13.3 days later. Analysis was again as in (a). From J. B. Gurdon, J. B. Lingrel, and G. Marbaix (1973). *J. Mol. Biol.* **80**, 539.

For our present purposes the most interesting feature of this system is the insight it provides into the capacities for protein synthesis possessed by the mature oocyte. It is clear that all the necessary components of the protein synthesis apparatus, including initiation factors for at least some messenger RNA's, are present in *Xenopus* oocytes. Furthermore, the fact that exogenous messenger RNA's can be accepted by the oocyte translational system tells us that newly appearing messages of endogenous origin could also be accepted, whether their advent is due to new transcription or to release from a previously sequestered state. The quantity of exogenous messenger RNA which can be translated is quite large. Thus Moar *et al.* (1971) showed that in mature oocytes the rate of hemoglobin synthesis continues to increase proportionately to the amount of injected messenger RNA up to an intracellular concentration of about 20 ng of hemoglobin message per oocyte. As discussed in detail below, the mature *Xenopus* oocyte contains 50–70 ng of putative messenger RNA [i.e., poly(A)-containing RNA]. Assuming that most of the injected messenger RNA is translated, these experiments suggest that the oocyte translational apparatus is capable of accepting an appreciable fraction of the total quantity of maternal message. After fertilization the amount of injected hemoglobin message which can be translated drops about twofold to 8–9 ng per egg (Moar *et al.*, 1971).

Direct Demonstration of Maternal Messenger RNA

To prove the existence of maternal messenger RNA in an early embryo, it is necessary to demonstrate messages which do not derive from transcriptional activity in the embryo genome(s) but are already present in the egg at fertilization. Evidence of this kind is now available for several systems. Thus enucleated eggs and egg fragments have been shown to synthesize protein, as have actinomycin- and α-amanitin-treated eggs. Some of this evidence was discussed earlier, and only a brief review of several of the enucleate cytoplasmic protein synthesis experiments is given here. However, it is important to stress that with appropriate controls these experiments provide incontrovertible evidence for the presence of maternal message.

PROTEIN SYNTHESIS IN ENUCLEATED EGG CYTOPLASM

The first enucleated egg protein synthesis experiments were carried out by Brachet *et al.* (1963b) and Denny and Tyler (1964). Both laboratories reported that enucleated halves of unfertilized sea urchin eggs prepared by centrifugation can synthesize protein at control rates when parthenogenically activated. Tyler (1965) later showed that the amino acid composition of the total proteins synthesized in enucleated sea urchin egg fragments is the same as that of normal eggs. Further studies were carried out with embryos of the snail *Nassaria* (*Ilyanassa*) *obsoleta*, where a natural opportunity to obtain enucleated egg cytoplasm occurs during polar lobe extrusion at first cleavage. The lobe contains about a third of the total egg volume and is attached to one of the blastomeres by only a thin strand of cytoplasm (see Chapter 7 for a detailed discussion). Thus it can easily be removed. Clement and Tyler (1967) showed that isolated polar lobes synthesize protein for at least 24 hours. The responsible messenger RNA's must have been stored in that portion of the egg cytoplasm which is partitioned into the polar lobe. This experiment does not prove that the polar lobe messages are maternal, since they could have been synthesized after fertilization, but it shows that messages can survive for long periods in egg cytoplasm.

Mitochondrial templates cannot be the source of any significant fraction of the messenger RNA being translated in either sea urchin or *Ilyanassa* cytoplasmic fragments. Electron microscope autoradiographs (Geuskens, 1969) indicate that the newly synthesized proteins in the polar lobe are mainly in the vicinity of the cytoplasmic polyribosomes. A secure biochemical demonstration of the nonmitochondrial nature of most protein synthesis in enucleated sea urchin egg cytoplasm was accomplished by Craig and Piatigorsky (1971). Mitochondrial biosynthesis in enucleated egg halves was severely inhibited with ethidium bromide, without the slightest effect on the electrophoretic spectrum of the proteins being synthesized. The complexity of this set of proteins is in any case far too high for a significant part of it to be coded by mitochondrial genomes.

Maternal messenger RNA has also been demonstrated in enucleated amphibian eggs (Smith and Ecker, 1965, 1969a; Ecker *et al.*, 1968; Ecker, 1972). In this series of investigations ripe *Rana pipiens* oocytes at various stages of maturation and shortly after fertilization were manually enucleated, and the absolute rates of protein synthesis compared to controls. Throughout the maturation period, which lasts for many hours, the measured rates of incorporation of injected leucine remain the same in enucleated as in normal eggs. That is, the same scheduled changes in absolute rate occur whether or not the nucleus is present. It will be recalled, furthermore, that the important qualitative changes in the pattern of protein synthesis which occur during maturation also take place in enucleated eggs (Chapter 3). Smith and Ecker have thus shown that all or almost all of the protein synthesis occurring during the maturation and fertilization period in frog eggs is coded by maternal messages already stored in the cytoplasm of the ripe ovarian oocyte.

ISOLATION AND QUANTITATION OF MATERNAL
MESSENGER RNA

The most direct molecular proof of maternal message has come by its extraction from mature oocytes and just-fertilized eggs, and its partial characterization in cell-free translation systems. The first successful attempt to extract a template active RNA from unfertilized sea urchin eggs was that of Maggio *et al.* (1964) who used an unfractionated rat liver translation system. Slater and Spiegelman (1966) also isolated the RNA of unfertilized sea urchin eggs and assayed its template activity in an *Escherichia coli* system relative to that of a phage RNA. They concluded that some 4% of the total RNA of the unfertilized egg is template active. Similar measurements were carried out with RNA extracted from mature ovarian oocytes of *Xenopus* (Davidson *et al.*, 1966), again using an *E. coli* cell-free system calibrated with a phage messenger RNA. About 2–3% of the total RNA in the *Xenopus* oocyte was scored as template active by this experimental criterion, suggesting the presence of 50–100 ng of putative messenger RNA per egg. The same value was obtained for *Xenopus* eggs by Cape and Decroly (1969) in very similar experiments. The latter authors also showed that the total quantity of RNA which is translationally active in the *E. coli* cell-free system remains constant throughout early development all the way up to hatching. However, it is now clear that prokaryotic translation systems cannot necessarily be relied upon to provide quantitative estimates for eukaryotic messenger RNA. Modern approaches have relied on use of 3'-poly(A) tracts as a diagnostic for messenger RNA, and on eukaryotic cell-free systems which translate exogenous animal messages with good fidelity.

Careful measurements of the amount of poly(A) in unfertilized sea urchin egg RNA have been made by several groups, and we now consider the amounts of maternal message consistent with these measurements if the poly(A)RNA of the oocyte is indeed message. Unfortunately no direct measurement of the amount of poly(A)RNA in sea urchin eggs is available. Measurements of the quantity of poly(A) agree closely on values of 0.032–0.042% of total egg RNA as poly(A) tracts (Slater *et al.*, 1972, 1973; Wilt, 1973; Mescher and Humphreys, 1974). The stored poly(A)RNA's display a length distribution typical for animal messenger RNA's in general, and for sea urchin embryo messages in particular, with a mean (i.e., number average) size of about 2000 nucleotides (Wu and Wilt, 1973; Nemer *et al.*, 1974; Slater and Slater, 1974; Fromson and Duchastel, 1975). If the poly(A) tracts are an average length of 100 nucleotides or so (Wu and Wilt, 1974), the quantity of poly(A)-containing stored message is then on the order of 0.6–0.85% of the total egg RNA, a smaller number than the 3–5% measured by the *E. coli* template activity studies. However, it is possible that the mean length of poly(A) tracts on maternal message is smaller than 100 nucleotides. Furthermore, nonpolyadenylated messenger RNA is known to be synthesized during early sea urchin development (references are listed in Chapter 5) and at most stages to amount to around half of the newly synthesized messenger RNA. It is also present in unfertilized eggs. Ruderman and Pardue (1976) demonstrated roughly similar quantities of poly(A)+ and poly(A)− messenger RNA in *Lytechinus* eggs in cell-free translation experiments. Slightly more poly(A)+ RNA seems to be present in *Arbacia* eggs, according to these authors. If about half the maternal messenger RNA is polyadenylated, the total maternal message content based on the mass of poly(A) maternal message would be about 1.5% of total egg RNA. An additional factor which may have to be taken into account is that the mature egg contains a relatively large amount of RNA which is polyadenylated only following fertilization. Thus the total quantity of poly(A) rises about twofold within several hours of fertilization (Slater *et al.*, 1972, 1973; Wilt, 1973; Mescher and Humphreys, 1974). Increase in poly(A) content occurs quantitatively in enucleated, activated merogones; i.e., it is a purely cytoplasmic phenomenon (Wilt, 1973). Newly synthesized RNA can be distinguished from maternal RNA by measurement of ^{3}H-uridine incorporation after fertilization, and by this method several groups have shown that much or all of the newly appearing poly(A) is added to preformed maternal RNA's (Wilt, 1973; Slater *et al.*, 1973; Slater and Slater, 1974). However, it is not known what fraction of the recipient maternal RNA's are totally nonadenylated. If all are, and if the poly(A)RNA is all messenger RNA, the result of the twofold increase in poly(A) content would be to establish the situation which occurs later in development, when about half the total message is polyadenylated. In this case the poly(A)RNA of the unfertilized egg would include less than half the total maternal message by mass, providing, as a maximum estimate, about 3% of the oocyte RNA as maternal message.

After fertilization much of the poly(A)RNA is found associated with the embryo ribosomes (Wilt, 1973). In addition, Jenkins *et al.* (1973) showed that the poly(A)RNA of sea urchin eggs is active in a rat sarcoma cell-free system, and that it gives rise to completed and released polypeptide chains with an apparent efficiency per unit mass of RNA equal to that of animal virus messages. By this measurement at least 3% of the total RNA is active as message, and additional activity is present in the nonpolyadenylated RNA fraction. The results of these measurements are thus quite similar to the earlier estimates of Slater and Spiegelman (1966), and given the above considerations, they are not inconsistent with the measurements of egg poly(A) content just reviewed. If we accept 1.5–3% of total egg RNA as an estimate, the amount of stored maternal message present in the sea urchin egg is about 0.05–0.1 ng. These values appear in Table 4.2, where they can be compared to those derived for *Xenopus* eggs.

Rosbash and Ford (1974) measured the content of mature *Xenopus* oocyte RNA which can be reacted with poly(U) and reported that about 1% of the total RNA, or 40 ng, is poly(A)RNA. Other measurements relying on different procedures were reported by G. J. Dolecki and L. D. Smith (personal communication) and these yield a slightly larger value, 70 ng per oocyte. Figure 4.2, from the study of Rosbash and Ford (1974), shows that the poly(A)RNA of *Xenopus* oocytes is also of typical messenger RNA size with a mean length of about 2000 nucleotides. The poly(A)RNA has other messenger-like characteristics. Thus it is primarily transcribed from nonrepetitive DNA sequences. Its complexity, i.e., the amount of nonrepetitive sequence in the RNA, is approximately that of embryo messenger RNA populations, as discussed in detail in Chapter 6. The quantity of poly(A)RNA revealed by recent measurements lies within a factor of two of the amount of template active RNA estimated in the earlier cell-free protein synthesis studies of Davidson *et al.* (1966). Most convincing is the fact that the oocyte poly(A)RNA can be translated in the wheat germ cell-free system to yield many discrete protein species (Darnbrough and Ford, 1976; Ruderman and Pardue, 1976). According to these authors the *Xenopus* egg RNA active in the cell-free protein synthesis systems is mainly polyadenylated. It therefore seems reasonable to consider the maternal messenger RNA content of *Xenopus* eggs to be about 40–70 ng. Table 4.2, which includes these values, also lists the content of ribosomal RNA. There it can be seen that the ratio of maternal messenger RNA to ribosomal RNA is the same for sea urchin and *Xenopus* eggs, though these differ in total RNA content by a factor of more than a thousand.

TABLE 4.2. Content of Ribosomal and Maternal Messenger RNA's in Mature Sea Urchin and Amphibian Eggs

Species	Total RNA (ng per egg)	Amount of ribosomal RNA (ng per egg)[a]	Number of ribosomes	Amount of maternal messenger RNA (ng per egg)[b]	ng maternal messenger RNA / ng ribosomal RNA
Strongylocentrotus purpuratus	3.3	2.65	~10^9	0.05–0.1	0.018–0.036
Xenopus laevis	4000	3600	~10^{12}	40–70	0.011–0.019

[a] About 80% of the total sea urchin egg RNA and 90% of the total *Xenopus* egg RNA are ribosomal.
[b] See text for sources of values listed.

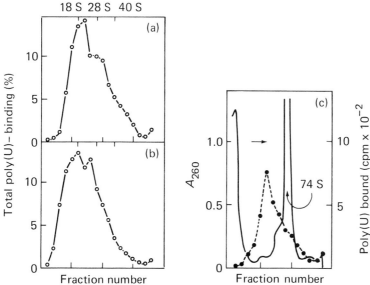

Fig. 4.2. Sucrose gradient analysis of poly(A)RNA from *Xenopus* oocytes. (a) Size distribution of poly(A)RNA extracted from mature oocytes and measured by binding of ^{3}H-poly(U) in gradient fractions. The gradient is calibrated by the positions of ribosomal RNA's (40 S, 28 S, 18 S) which are not shown. (b) Size distribution of *Xenopus* tissue culture cell poly(A)RNA in ribonucleoprotein particles, measured as above, after RNA extraction from each gradient fraction. (c) A homogenate of the eggs was prepared and the postribosomal supernatant displayed on the gradient. The 74 S peak represents the monosomes. Dotted line represents binding of radioactive poly(U), and solid line the optical density profile. From M. Rosbash and P. J. Ford (1974). *J. Mol. Biol.* **85**, 87.

Less extensive evidence exists for mammalian species. Schultz (1975) has reported that the unfertilized rabbit egg also contains poly(A)RNA. The poly(A) tracts of these eggs represent about 0.25% of the total RNA, suggesting that the content of poly(A)RNA is of the order of 2–3% of the total RNA if the usual lengths of 3'-poly(A) tracts and messenger RNA's are assumed. No cell-free translational data on mammalian embryo maternal messenger RNA's yet exist, but it is known that interference with RNA synthesis by α-amanitin fails to affect early cleavage or uptake of radioactive precursor into protein in mammalian eggs (Manes, 1973; Nadijcka and Hillman, 1974; Schultz, 1975). By analogy with the sea urchin and *Xenopus*, therefore, it seems probable that mammalian eggs rely largely on maternal message as development is initiated. The data of Schultz also suggest that the fraction of mammalian egg RNA present as maternal message is likely to be similar to the fraction of sea urchin and amphibian egg RNA which is maternal message (Table 4.2).

POSTRIBOSOMAL PARTICLES CONTAINING MATERNAL MESSENGER RNA

For some years it has been believed that maternal messenger RNA is localized in ribonucleoprotein particles migrating in velocity sedimentation analyses more slowly than do monosomes. Such postribosomal particles are the probable location of maternal messenger RNA's in *Xenopus* eggs, as illustrated in Fig. 4.2c (Rosbash and Ford, 1974). Sedimentation of these particles is unchanged by ethylenediamine tetraacetate (EDTA) treatment which, of course, strongly affects ribosomes and ribosomal subunits. The proposal that such particles contain maternal messenger RNA dates back to early experiments of Spirin (reviewed by Spirin, 1966), Spirin and Nemer (1965), and Nemer and Infante (1965). Spirin and his collaborators claimed that newly synthesized messenger RNA of sea urchin and fish embryos is associated with proteins in subribosomal particles. These particles were fixed and then isolated in cesium chloride gradients. They possess specific densities distinct from those of ribosomal subunits, and in contrast to ribosomes their protein content varied from 57–75%. RNA's stored in these particles ranged up to about 20 S in size, and these RNA's were able to stimulate crude cell-free protein synthesis systems. Such putative message bearing ribonucleoprotein particles were named "informosomes." Unfortunately, this term was immediately generalized to include virtually any ribonucleoprotein particle containing nonribosomal RNA's, whether viral, nuclear, or cytoplasmic (Spirin, 1966), and today it is little used. Nonetheless, the experiments on early sea urchin and fish embryo "informosomes" are to be regarded as the direct antecedents of modern studies on the particulate storage form of maternal messenger RNA.

There are now several direct demonstrations of postribosomal particles which contain maternal message. Gross *et al.* (1973a) homogenized sea urchin eggs under mild conditions and tested various centrifugal fractions for messenger activity in the mouse ascites cell-free protein synthesis system. In Fig. 4.3 several of the key experiments from this study are reproduced. All stimulatory activity is associated with particles sedimenting more slowly than do monoribosomes. At least under certain homogenization conditions much of the stored poly(A)RNA of these eggs is also present in such particles (Slater *et al.*, 1973). The size of the template active RNA's in the postribosomal particles was measured by Gross *et al.*, as shown in Fig. 4.3b and c. Here the RNA of two of the fractions indicated in Fig. 4.3a were extracted and tested in the Krebs system. Fraction "2" contains larger RNA's and makes larger proteins. Fraction "4" contains predominantly histone messenger RNA. This is shown in Fig. 4.3d, where

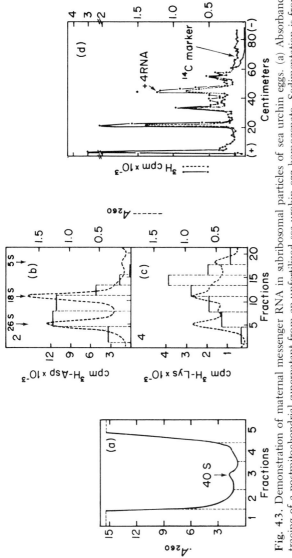

Fig. 4.3. Demonstration of maternal messenger RNA in subribosomal particles of sea urchin eggs. (a) Absorbance tracing of a postmitochondrial supernatant from an unfertilized sea urchin egg homogenate. Sedimentation is from right to left, i.e., the monoribosomes are included in fraction 1. (b) and (c) display on sucrose gradients the RNA extracted from fractions "2" (b) and "4" (c). Fractions from these RNA gradients were tested in the cell-free Krebs ascites protein synthesis system. The RNA was pooled as indicated and concentrated by ethanol precipitation. Equal aliquots of each indicated fraction were tested. The height of the bars indicates counts of ^{3}H-asparagine (b) or ^{3}H-lysine (c) incorporated in protein due to translation of the added RNA per fraction tested. The dashed lines indicate the A_{260} profile of the extracted RNA in the gradient. (d) ^{3}H-Phenylalanine-labeled proteins translated in the cell-free system from RNA of fraction 4 were digested with trypsin and the resultant peptides analyzed by paper electrophoresis (solid line). The dotted line shows the distribution of coelectrophoresed tryptic peptides of authentic histone labeled with ^{14}C-phenylalanine. From K. W. Gross, M. Jacobs-Lorena, C. Baglioni, and P. R. Gross (1973a). *Proc. Natl. Acad. Sci. U.S.A.* 70, 2614.

tryptic peptides of the proteins translated in the cell-free system are compared to those of authentic histones. Skoultchi and Gross (1973) also showed by sequence homology experiments that histone messenger RNA is present in these subribosomal particles. RNA extracted from them competes effectively with the hybridization to DNA of authentic histone messenger RNA. More than 95% of the total histone messenger RNA of the unfertilized egg could be detected in this subcellular fraction, as measured by the competition method. Unfertilized sea urchin egg RNA was independently shown to compete with the hybridization of histone messenger RNA by Farquhar and McCarthy (1973). Similar competition hybridization studies were reported by Lifton and Kedes (1976), who fractionated histone messenger RNA's extracted from unfertilized oocytes and demonstrated the presence of maternal messages for all major classes of histones.

Postribosomal particles have also been isolated from the eggs of the surf clam *Spisula solidissima* and these apparently contain maternal message for at least histone I. These experiments were reported by Gabrielli and Baglioni (1975). As did Gross *et al.*, these workers translated RNA obtained from the postribosomal particles of the unfertilized egg in cell-free systems (both ascites and wheat germ). They conclude, on the basis of electrophoretic mobility in two gel systems, that histone I is among the translated products.

In conclusion, the experiments reviewed here unequivocally demonstrate the existence of maternal message. In the case of the histone message studies, a specific component of the maternal messenger RNA stockpile is identified for the first time. Furthermore, the maternal messenger RNA can be recovered in subribosomal ribonucleoprotein particles, just as was proposed a decade ago by Spirin and others. However, an important note of caution is required here, for the sedimentation behavior on which the size estimates of the message bearing particles are based can be expected to vary according to ionic strength and other conditions. In addition the particles may have been broken loose from larger aggregates or otherwise affected during homogenization. The real dimensions, or more importantly, the intracellular localization of the maternal messenger bearing particles thus remains unknown.

Fertilization and the Utilization of Maternal Message in Sea Urchin Embryos

Within minutes after fertilization, sea urchin eggs begin to increase their rates of protein synthesis. Analysis of this phenomenon provided much of the original inferential evidence for maternal messenger RNA, since the increase in synthesis rate was found to occur even in activated enucleated egg fragments (Denny and Tyler, 1964). As a result of a vast amount of research, a quantitative picture of the rapid mobilization of maternal messenger RNA following fertilization has been constructed. We now briefly review the most convincing evidence, much of which is relatively recent. It should be noted, however, that the process by which protein synthesis rate is increased after fertilization was already correctly understood on a qualitative level some years ago, as can be seen in various reviews of earlier work (see, e.g., Spirin, 1966; Gross, 1967; Tyler, 1967; or Davidson, 1968).

Increase in protein synthesis following fertilization of sea urchin eggs is among the best documented molecular aspects of development. It was first detected many years ago by Hoberman *et al.* (1952) who studied the incorporation of deuterium into sea urchin egg proteins, and by Hultin (1952), who measured the incorporation of ^{15}N-amino acids. Nakano and Monroy (1958) injected ^{35}S-methionine into the coelom prior to induced ovulation, thus preloading the eggs with the radioactive amino acid. Their data suggested that a real protein synthesis rate change occurs on fertilization, rather than simply a change in permeability and pool specific activity. According to Rinaldi and Parente (1976), the rate of protein synthesis is about the same in unfertilized eggs and in ovarian oocytes. Hultin (1961b) was the first to show that cell-free preparations of sea urchin eggs display enhanced incorporation of labeled amino acids after fertilization. Though it begins within 5–10 minutes of sperm penetration (Epel, 1967) or even less (Timourian and Watchmaker, 1970), the increase in protein synthesis is not the earliest response of the egg. Other changes which lie outside the scope of this discussion and cannot be reviewed here, start within seconds of penetration. These include cortical granule breakdown, release of cortical granule contents, activation of various enzymes, and alteration of respiration rate (see, e.g., Tyler, 1963; Monroy, 1965; Epel, 1967, 1975; Carroll and Epel, 1975; Vacquier, 1975). Within 2 hours after fertilization the protein synthesis rate has dramatically increased by a factor of at least fifteen (Epel, 1967; Piatigorsky and Tyler, 1970; Humphreys, 1969, 1971).

INCREASE IN POLYSOMES FOLLOWING FERTILIZATION

Two main facts relevant to the mechanism of the increase of protein synthesis after fertilization had been established by the late 1960's. First, it was evident, as reviewed above, that the unfertilized egg possesses all the requisite components to support protein synthesis. Second, and more significantly, a number of investigations had shown that the fraction of ribosomes present in polysomes increases at the same time as does total protein synthesis rate after fertilization. This was first reported by Monroy and Tyler (1963) and then by Stafford et al. (1964), Malkin et al. (1964), Infante and Nemer (1967), Cohen and Iverson (1967), Mano (1971a), and Denny and Reback (1970) as well as others. The nature of the processes responsible for polysome increase remained in question, as this could occur by several alternative mechanisms. For example, the number of polysomes present in the unfertilized egg could be limited either by the amount of available messenger RNA or by specific deficiencies in the capability of the egg to carry out protein synthesis. In the latter case a quantitative lack of one of the necessary initiation factors or an inhibitor of protein synthesis in unfertilized eggs might be responsible. The literature contains many such proposals. These include the existence of a protein inhibiting messenger RNA and tRNA binding to ribosomes (Metafora et al., 1971), the presence of inactive polysomes (Piatigorsky, 1968), and the existence prior to fertilization of an inhibitor of chain initiation (MacKintosh and Bell, 1969). Some of these proposals could not be excluded by the cell-free protein synthesis experiments reviewed above, particularly proposed mechanisms involving deficiencies in the initiation systems of unfertilized eggs. This issue has now been largely settled by direct measurements of translational efficiency before and after fertilization.

Translational efficiency is defined as the number of polypeptide chains produced per polysomal messenger RNA molecule per unit time. Translational efficiency is directly proportional to the number of initiations per unit time and is inversely proportional to the length of time it takes to complete a protein. If limitations of the translational system, rather than sequestration of messenger RNA, are the cause of the very low rates of protein synthesis measured before fertilization, translational efficiency must be lower in unfertilized eggs. This effect should be easily noticeable, since as noted above the rate of protein synthesis increases fifteenfold after fertilization. Humphreys (1969) and MacKintosh and Bell (1969) measured the amount of time an average amino acid remains in a nascent polysomal protein, and in both studies this was found to vary less than a factor of two before and after fertilization. After the pool has equilibrated and a steady state condition obtained, the average time the amino acid is present in nascent protein, T, can be calculated simply as

$$T = C \frac{1}{dP/dt}$$

Here C is the quantity of radioactive amino acid present in nascent protein at steady state, and dP/dt is the rate at which radioactive amino acids appear in released, or completed, protein chains (i.e., the rate of flow of amino acids through the pool). If the specific activity of the precursor pool is known, C can be expressed in pmoles of radioactive amino acid in nascent protein, and dP/dt is in pmoles min^{-1}; alternatively if C is simply the quantity of radioactivity [counts per minute (cpm) in the amino acid], dP/dt is the rate of flow of radioactivity through the pool. According to Humphreys' measurement the average amino acid is present in nascent protein about 1.16 minutes in postfertilization polysomes and about 0.61 minutes in unfertilized egg polysomes. Thus the polysomal apparatus of the fertilized egg actually translates at about one-half the rate of the unfertilized egg polysomes. The main conclusion is that increase in the rate of translocation after fertilization cannot be the cause of the increase in protein synthesis rate. Furthermore, Humphreys (1969) showed that the size distribution of both the polysomes and the newly synthesized proteins are similar, at least within a factor or two, before and after fertilization. Therefore, the difference in synthesis rate cannot be due to a significant distinction in the number of ribosomes attaching to each messenger RNA. It follows that change in the rate of initiation per message is not the explanation of the difference in protein synthesis rate between fertilized and unfertilized eggs. The results of these experiments show that changes in translational efficiency cannot account for the rapid increase in protein synthesis rate. The increase in number of polysomes after fertilization therefore requires a large increase in the amount of messenger RNA available for polysome assembly. Since the increase in polysomes is not due to new RNA synthesis, these results mean that fertilization causes the activation of maternal messenger RNA from an inactive or sequestered state, for use by the ready translational apparatus of the egg.

Subsequent measurements (Humphreys, 1971) showed that the fraction of ribosomes in polysomes increases linearly after fertilization for about 2 hours. At this point 20% of total ribosomes are in polysomal structures, which is approximately 30 times the fraction of ribosomes in polysomes in the unfertilized egg (MacKintosh and Bell, 1969; Denny and Reback, 1970; Humphreys, 1971). The polysomal content of the unfertilized egg may vary according to conditions, however, as shown by MacKintosh and Bell (1969). The linear increase in polysome number over the first 2 hours is reflected in the linear increase in synthesis rate during this period (Fry and Gross, 1970a). Combined with the twofold decrease in translocation rate noted above, the thirtyfold increase in polysome content provides a quantitative explanation for the overall fifteenfold change in protein synthesis rate. Later in development the fraction of ribosomes as polysomes increases further to 50% in the blastula stage (see, e.g., Infante and Nemer, 1967), and other measurements (Galau *et al.*, 1974) suggest values as high as 60% in the gastrula. As noted earlier, Berg and Mertes (1970) showed that the overall rate of protein synthesis changes little from the mesenchyme blastula stage on.

From measurements of polysome content Humphreys (1971) estimated the mass of messenger RNA being translated in early sea urchin embryos. Humphreys assumed that about 2% of the polysomal RNA is messenger RNA, leading to the calculation that about 6 pg of messenger RNA would be present at any given time on the polysomes of 2–6-hour *Lytechinus* embryos. As noted in Table 4.1, however, recent measurements show that in *Strongylocentrotus* the fraction of polysomal RNA which is message is 4% (in gastrulae). If this value is used instead, the polysomal message content in the 2–6-hour embryos would be about 12 pg. This is no more than one-fourth the amount of maternal messenger RNA stored in the egg (Table 4.2). Since the turnover rate of maternal message once on the polysomes is unknown, the actual flow of maternal messenger RNA into the polysomes cannot be calculated (see Chapter 5 for a discussion of RNA turnover and synthesis rates). Other measurements show that the accumulation of *newly transcribed* messenger RNA in the polysomes can account for less than 10–15% of the total polysomal message in these early cleavage stage embryos (Humphreys, 1971; and see Chapter 5). This analysis fully confirms the conclusion drawn by most earlier workers, namely, that fertilization results in the mobilization of maternal messenger RNA.

MECHANISMS OF MATERNAL MESSENGER RNA MOBILIZATION

The means by which the maternal messenger RNA is activated after fertilization remain unknown. Several authors have suggested that the message is released by protease digestion of a protein "coat," and there is some evidence that increase in protease activity indeed accompanies fertilization (Lundblad, 1955; Mano, 1966; Mano and Nagano, 1970). Since the maternal message is now known to be stored in ribonucleoprotein particles (see above), this remains a viable hypothesis. Mano and Nagano (1970) found that a cell-free sea urchin ribosome system could be stimulated by subcellular fractions from unfertilized eggs after brief trypsin treatment. Experiments of this kind were initially reported by Monroy *et al.* (1965). However, no direct evidence for protease release of maternal message in the egg yet exists. The RNA in maternal message-containing particles has not been shown to be activated by protease treatment, and a requirement for a protease is thus still to be demonstrated. Another possibility is a change in the messenger RNA itself. For example, polyadenylation of maternal message could be involved. An experiment reported by Mescher and Humphreys (1974) renders this particular explanation highly unlikely. These authors completely blocked postfertilization adenylation with the adenosine analogue cordycepin (3-deoxyadenosine) and found that the usual sharp rise in protein synthesis following fertilization nonetheless occurred on schedule. However they may operate, the mechanisms by which maternal messages are activated at fertilization must be partially specific. We have already seen (Chapter 3) that particular sets of maternal messages are translated at certain embryonic stages. The activation mechanism is probably related to the internal location of the messenger RNA ribonucleoprotein particles. Thus a complete explanation will probably require further understanding of maternal message localization in a cytological as well as a molecular sense.

An additional aspect of variation in maternal messenger RNA translation after fertilization deserves note, and this is the transient change in rate of translation at first cleavage. First cleavage metaphase occurs 40–80 minutes after fertilization, depending on the species. A brief pause in the overall linear increase in protein synthesis rate has been observed by both Mano (1970, 1971b) and Fry and Gross (1970a) at first cleavage metaphase. The pause lasts until anaphase when the synthesis rate increase is resumed. Both the pause and resumption of increase in synthesis rate occur in the presence of actinomycin. Furthermore, the same phenomenon is seen in parthenogenically activated and enucleated half-eggs (Fry and Gross, 1970a). Some evidence suggests that soluble protein factors affecting translation control this fluctuation (Mano, 1971b,c). These observations provide a further warning of the complexity of the control systems which govern the utilization of stored maternal message.

Changes in Protein Synthesis Rate during Early Development in Other Organisms

VARIATIONS IN THE EFFECT OF FERTILIZATION ON PROTEIN SYNTHESIS RATE

Rapid increase in protein synthesis rate and in polysome content similar to that seen in sea urchin eggs after fertilization has been observed in a few other marine organisms, but is by no means the rule. To date only a relatively small number of species have been studied in any detail. Among those in which a more than twofold increase in polysome content occurs within a few hours of fertilization are two lamellibranch molluscs: the surf clam, *Spisula solidissima* (Firtel and Monroy, 1970), and the coot clam,

Mulinia lateralis (Kidder, 1972b). In *Spisula*, the polysome content increases from about 10% during maturation (i.e., germinal vesicle breakdown, polar body extrusion, and pronuclear formation), to over 25% at 5 hours. By the trochophore larva stage at 20 hours, 50% of the ribosomes are present in polysomes. In this organism fertilization initiates the events of maturation, in contrast to the sea urchin, in which the eggs have already completed maturation when they are fertilized. The relative rate of protein synthesis rises sharply after fertilization (Bell and Reeder, 1967; Firtel and Monroy, 1970) along with the polysome content. Another marine egg displaying a rapid rise in relative protein synthesis rate following fertilization is that of the polychaete annelid, *Sabellaria alveolata* (Guerrier and Freyssinet, 1974). Apparent increases up to fivefold in protein synthesis rate within 2 hours were measured. However, more gradual rises in protein synthesis rate are observed in various other marine eggs after fertilization. Relative protein synthesis rate in embryos of the marine mud snail, *Ilyanassa*, increases less than twofold during the first day of development (Collier and Schwartz, 1969) and continues to increase slowly through gastrulation (day 3) to a final level only about three times that at fertilization. Here again change in polysome profile parallels change in protein synthesis rates. About a twofold increase in the amount of polysomes occurs between the unfertilized egg and the 8-cell cleavage stage, and a threefold increase compared to the unfertilized eggs is observed by the gastrula stage (Mirkes, 1972). Other species whose embryos display similar patterns include *Urechis caupo* (Gould, 1969), where relative protein synthesis rate increases only about twofold after fertilization, the loach (teleost) *Misgurnus fossilis* (Krigsgaber and Neyfakh, 1972), and the rabbit (Manes and Daniel, 1969). The main interest here, other than descriptive, is in understanding the distinction between organisms in which protein synthesis is rapidly accelerated by fertilization and those in which it rises only gradually. After the cleavage period the same process appears to occur in all species. The level of protein synthesis established by the blastula stage remains nearly constant or increases only slightly throughout early development until postgastrular organogenesis. The main differences thus pertain to changes in protein synthesis rate very early in embryogenesis.

Detailed study of protein synthesis rates before and after fertilization in *Xenopus* and *Rana pipiens* eggs has provided some general insight into this problem. Fertilization has only a small effect on the rate of protein synthesis in these eggs, increasing it by less than 1.5-fold (Ecker and Smith, 1968; Smith and Ecker, 1969a). Instead a rate increase occurs about the time of germinal vesicle breakdown, early in the process of maturation. This was first detected by Smith *et al.* (1966) and can be regarded as a response to the hormonal stimulation which provokes ovulation (Ecker, 1972; see Smith, 1975, for review). The severalfold increase in synthesis rate which occurs at this time in *Rana pipiens* oocytes normally antecedes fertilization by as much as 24 hours. Since oocytes enucleated before germinal vesicle breakdown display the very same changes in rate (Smith and Ecker, 1969a; Ecker, 1972), the new protein synthesis must be coded by maternal message. Similar events occur in *Xenopus*. During the maturation period the polysome content of *Xenopus* eggs also increases severalfold to include about 3% of the ribosomes (Woodland, 1974). It then rises slowly to about 15% by 10 hours postfertilization (5000 cell blastula), whereafter it remains constant until after gastrulation. However, this change is gradual and is not an immediate response to fertilization as in the sea urchin. These events are illustrated in Fig. 4.4, which is excerpted from Woodland's (1974) report. Two conclusions can be drawn from the protein synthesis rate changes reviewed here. The first is that increase in polysome content always underlies the earliest developmental changes in synthesis rate. In both the sea urchin and frog it is clear that this process depends on mobilization of maternal messenger RNA. That is, at the beginning of development the translational apparatus is less utilized than later, and the assembly of polysomes requires the activation

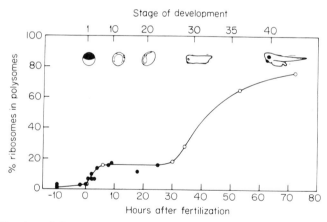

Fig. 4.4. Fraction of ribosomes in polysomes during development of *Xenopus laevis* embryos. The time of development (lower abscissa) applies to embryos developing at 23°C, though a lower temperature was actually used in the experiments. The upper abscissa shows the stage of the embryos, according to Nieuwkoop and Faber (1956). For prefertilization stages the following time scale is used: Oogenesis is taken to end 10 hours before fertilization; 2 hours before fertilization the oocyte "matures" to form an egg, the diagnostic characteristic being the breakdown of the oocyte nuclear membrane. The value for this point was obtained by incubating normal oocytes with progesterone at 5 μg/ml *in vitro*. From H. R. Woodland (1974). *Dev. Biol.* **40**, 90.

or release of maternal message. Second, in cases where protein synthesis rate does not increase sharply at fertilization, this is simply because that process, including mobilization of maternal message, has occurred earlier, sometime between ovulation and fertilization. As stated by Smith and Ecker (1970) in reference to amphibian eggs, "from this point of view, fertilization or artificial activation are simply points superimposed on a continuous biosynthetic process which was initiated by the hormonal induction of maturation."

RELATIVE PROTEIN SYNTHESIS RATES IN MAMMALIAN EMBRYOS AND IN HYDRATED *ARTEMIA* EMBRYOS

Direct measurements of absolute protein synthesis rate do not yet exist for any mammalian embryos. Several relative measurements have been made on preimplantation mouse embryos at various stages of development. These studies have measured amino acid incorporation rates in embryos cultured *in vitro* and in some cases uptake rates as well (Monesi and Salfi, 1967; Monesi *et al.*, 1970; Tasca and Hillman, 1970; Brinster, 1971; Epstein and Smith, 1973). Interpretation of the incorporation data is complicated by the fact that the ability of the embryo to take up exogenous amino acids increases sharply as the embryos develop (Tasca and Hillman, 1970; Epstein and Smith, 1973). When incorporation is considered relative to uptake, it appears that no detectable change in protein synthesis rate occurs at fertilization (Epstein and Smith, 1973), and only in the third day of development, in the early blastocyst stage, is an increase in protein synthesis rate observed. The amount of this increase is estimated by Epstein and Smith (1973) at threefold to ninefold, comparing the 3 day blastocyst to the 8- to 16-cell (2-day) embryo. The number of cells is increased to a roughly proportional extent, since the early blastocyst has 40–60 cells, and a possible interpretation is that the new embryonic nuclei are providing some or all of the RNA's needed to support this change in protein synthesis. During the first 2 days of development when the protein synthesis rate does not change significantly while the cell number increases from 1 to 16 cells, maternal messenger RNA's and other components are probably being used (see Chapter 2). However, as noted in Chapter 3, newly synthesized messenger RNA's probably also play a significant role during the later cleavage stages in mouse embryos. Rabbit embryos appear to show a pattern of protein synthesis which in terms of developmental stage is similar to that of mouse embryos (Karp *et al.*, 1974). The amount of incorporation into protein relative to the acid-soluble pool remains constant through cleavage (3 days, 16 cells), but then rises about fourfold by 4.5 days (128 cells), just before blastocyst formation.

Mobilization of stored messenger RNA may be evident in a completely different context during embryogenesis in *Artemia salina*, the brine shrimp. In this organism an encysted dessicated gastrula can be produced under certain conditions. All detectable metabolism then ceases. Upon rehydration, this cryptobiotic state is relieved, and development proceeds (Clegg, 1967). Resumption of metabolism is associated with a continuous increase in protein synthesis rate (Golub and Clegg, 1968; Clegg and Golub, 1969; Hultin and Morris, 1968), and underlying this is a parallel increase in polysome content. Polysome assembly is first detectable within 3 minutes in gastrulae prehydrated at 0°C and then raised to 30°C, or within 30 minutes without prehydration (Clegg and Golub, 1969). Messenger RNA is stored in dormant *Artemia* cysts (Nilsson and Hultin, 1974) and can be extracted from homogenates. The extracted RNA, which sediments as a cytoplasmic particulate fraction together with mitochondria, strongly stimulates protein synthesis in an *E. coli* cell-free system. The stored messenger RNA, or both new and stored message, is apparently required to load the new polysomal structures after rehydration. The main molecular features of this system are clearly reminiscent of early cleavage embryos in species such as the sea urchin.

Synthesis and Inheritance of Some Specific Embryonic Proteins

In this section we consider four specific embryonic proteins, or sets of proteins. These are the histones, the RNA polymerases, the DNA polymerases, and microtubule proteins.

NUCLEOSOMES AND THE REQUIREMENT FOR HISTONES DURING EARLY DEVELOPMENT

Several aspects of histone molecular biology are relatively well known, and cannot be reviewed at length here, except insofar as they pertain to expression of the histone genes in early development. Histones are the highly basic proteins complexed to DNA in all animal chromatin, and exist in five main species. Four of these, histones IIb_1 (or f2a2), IIb_2 (or f2b), III (or f3), and IV (or f2a1) occur in stoichiometric proportions to each other. In native chromatin these histones are complexed in globular structures about 100 Å in diameter containing two molecules of each of the four histones (Kornberg, 1974; Kornberg and Thomas, 1974; Noll, 1974; Olins and Olins, 1974; van Holde *et al.*, 1974; Oudet *et al.*, 1975). The DNA double helix is wrapped around these particles, which are termed "nucleosomes" or "nu bodies." Nucleosomes occur adjacent to each other, or nearly so, throughout most of the chromatin. The result is a "packing ratio", i.e., ratio of the length of DNA in chromatin to the length of the chromatin, of about 6.8. Histone I (or f1) is located externally to the nucleosomes. While histone I is also ubiquitous, no stoichiometric relation between DNA content and histone I or between histone I and the other histones has been discovered. Except for histone I the amino acid sequences of the histones appear to have been remarkably invariant throughout evolution (e.g., de Lange *et al.*, 1968, 1969a,b, 1973; Iwai *et al.*, 1970; Yeoman *et al.*, 1972; Bailey and Dixon, 1973; Hooper *et al.*, 1973).

Nucleosomes are clearly an ubiquitous and a basic feature of the organization of animal chromatin. A major problem faced by embryonic systems is therefore the provision of sufficient histones to engage the rapidly expanding DNA complement of the embryo in nucleosomes. Analyses of embryo chromatin show that large amounts of histone I are also required. Figure 4.5 illustrates the change in cell number with time in embryos of *Strongylocentrotus purpuratus* (Hinegardner, 1967). On the right-hand ordinate the DNA content of the embryos is also given, calibrated on the basis of the genome size of this organism, 1.78 pg of DNA per diploid cell (Hinegardner, 1974). Cell number increases exponentially between 1 and 10 hours of development, resulting in an increase in nuclear DNA content of about two orders of magnitude. Since the mass ratio of histone to DNA in a nucleosome is a constant (about 1.16 by mass), the relative requirement for the nucleosomal histones as a function of time is also represented by Fig. 4.5.

HISTONE GENE CLUSTERS IN THE SEA URCHIN

The structural genes for the histones are present in multiple copies in all the genomes which have been studied. In the sea urchin there are several hundred copies of these genes (Kedes and Birnstiel, 1971; Weinberg *et al.*, 1972; Skoultchi and Gross, 1973), in *Xenopus* about 30 (Jacob *et al.*, 1976), and in humans more than 20 (Wilson *et al.*, 1974). These values were obtained by hybridizing histone messenger RNA with excess DNA. The rate at which histone messenger RNA–DNA duplex forms indicates that the histone genes are present in the numbers of copies cited. Two separate studies have now shown that the histone genes are organized as a tandemly repeating cluster of all five structural gene sequences, together with spacer DNA. These studies were made possible by the purification of the DNA containing the histone genes, accomplished by isopycnic centrifugation in actinomycin–cesium gradients, combined with hybridization and other procedures (Birnstiel *et al.*, 1974; Shutt and Kedes, 1974). Kedes *et*

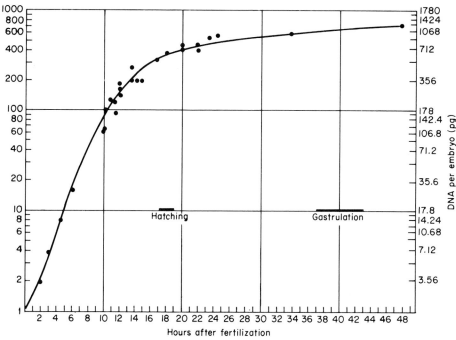

Fig. 4.5. Cell division and nuclear DNA content as a function of time in *Strongylocentrotus purpuratus* embryos. Embryos were grown at 15°C and the number of cells counted in squashes prepared with EDTA in order to form a monolayer of blastomeres on the slide (left ordinate). The right-hand ordinate gives DNA content in pg DNA per embryo, calculated from the cell number on the basis that each diploid cell contains 1.78 pg (Hinegardner, 1974). From R. T. Hinegardner (1967). *In* "Methods in Developmental Biology." (F. H. Wilt and N. K. Wessells, eds.), p. 139. Crowell, New York.

al. (1975a,b) constructed hybrid DNA molecules *in vitro*, consisting of the bacterial plasmid PSC-101 bearing a tetracycline resistance factor and fragments of sea urchin DNA. These were used to transfect tetracycline-sensitive bacteria. Plasmids were recovered from the tetracycline-resistant transfectants, and those containing histone genes were isolated and replicated. Each such clone represents a single region of the repetitive histone gene set, rather than a mixture of all regions as in the original histone DNA preparation. Histone DNA fragments obtained by the clonal procedure were challenged with messenger RNA's for the various histones. This experiment showed that the individual clonal DNA fragments each contain sequences coding for the several specific messenger RNA's tested. Schaffner *et al.* (1976) arrived at the conclusion that the genes for the different histones are clustered by a completely different route. Their approach was to treat purified sea urchin histone DNA with two different restriction endonucleases, mapping the fragments obtained by means of their overlaps. The restriction enzyme EcoRI cuts the histone gene set only once per repeating unit, releasing fragments of about 6000 nucleotides. Again these could be shown to contain sequences hybridizing with all the histone messenger RNA's, including histone I. The histones are small proteins, and the messenger RNA's for each are only on the order of 400 nucleotides in length. Almost half of the repeating histone gene cluster appears to be spacer DNA. Schaffner et al. (1976) have also cloned sea urchin histone DNA and in an elegant series of experiments showed that the order of the histone structural genes is IV-IIb$_2$-III-IIb$_1$-I. These genes are all transcribed from the same strand and in the same direction, from left to right as written (Gross et al., 1976b).

Early embryos must contain either a large reserve of preformed histones or must synthesize histones very actively. From the preceding discussion at least the four nucleosome histones must be provided together, since they occur stoichiometrically. The clustering of their genes and their uniform orientation (Gross et al., 1976b) could imply a common and coordinate transcription level regulatory system. The complement of histones present in sea urchin embryo nuclei is basically similar to that of mammalian chromatin in electrophoretic pattern, amino acid composition, and other chemical properties (see, e.g., Vorobyev et al., 1969; Thaler et al., 1970; Benttinen and Comb, 1971; Hill et al., 1971; Ozaki, 1971; Easton and Chalkley, 1972; Seale and Aronson, 1973b). Several variants of each major class of histones are synthesized, and about 11 individual species are known in sea urchin embryos (Levy et al., 1975; Cohen et al., 1975; Lifton and Kedes, 1976). Active synthesis of histones has been known for some years to occur in early sea urchin embryos, and to be correlated with DNA synthesis (see Fig. 4.5 for the course of DNA synthesis in sea urchin embryos). Histone synthesis is activated along with total protein synthesis immediately after fertilization in sea urchin embryos, and newly synthesized histones can also be recovered from the chromatin in very early cleavage embryos (Ruderman and Gross, 1974), some prior failures to identify histones at such stages notwithstanding. In fact histones are present even in unfertilized egg nuclei (Evans and Ozaki, 1973). Throughout blastulation histones continue to be synthesized at a high rate (Thaler et al., 1970; Seale and Aronson, 1973a; Ruderman and Gross, 1974). When the rate of DNA synthesis decreases (Fig. 4.5), the rate of histone synthesis does likewise (e.g., Seale and Aronson, 1973a), and when DNA synthesis in early embryos is blocked by hydroxyurea, histone synthesis also falls sharply (Kedes et al., 1969; Ruderman and Gross, 1974). Three particularly interesting aspects of histone synthesis in sea urchin embryos require consideration here. These are the rate of histone synthesis compared to the rate of synthesis of other proteins; the existence of stage-specific changes in the patterns of histones synthesized; and the relative roles of maternal messenger RNA and newly transcribed messenger RNA in supporting embryonic histone synthesis.

QUANTITATIVE ASPECTS OF HISTONE SYNTHESIS DURING SEA URCHIN EMBRYOGENESIS

Quantitative data on the absolute amount of histones synthesized have been presented by Seale and Aronson (1973a) and Moav and Nemer (1971). In midblastula 25–40% of total protein synthesis is histone synthesis. These measurements were foreshadowed by the radioautographic observation that a large fraction of the proteins synthesized in these embryos accumulate in the blastomere nuclei (Kedes *et al.*, 1969). Of these nuclear proteins, a large proportion are nonhistone chromatin proteins, but when histone synthesis is at its peak rate, about half are histones (Seale and Aronson, 1973a). The maximum rate is attained in the 100- to 200-cell blastula. By gastrula stage only about 10% of total protein synthesis is accounted for by histones, an absolute rate of histone synthesis less than 7% of the maximum measured (cf. the decrease in DNA synthesis rate at this stage in Fig. 4.5). The maximum rate is 1–2 pg min^{-1} of histones. The absolute rate of DNA synthesis in embryos of this stage is estimated at a similar value, about 1.5 pg min^{-1} (Moav and Nemer, 1971; see Fig. 4.5). Therefore the absolute rate of histone synthesis is tightly linked to the rate of DNA synthesis. This is also the implication of the temporal correlation between cell division rate and histone synthesis rate and the effects of hydroxyurea noted above. These data argue against the presence of a large stockpile of preformed histones, at least one which persists to the blastula stage. Nor can detectable quantities of the five histones be extracted from the cytoplasm of unfertilized sea urchin eggs, according to Benttinen and Comb (1971).

Histones are synthesized on relatively small polysomes which form a recognizable and distinct peak in sucrose gradients. These were first recognized as the site of histone synthesis by virtue of the fact that the ratio of lysine to tryptophan is very high in the nascent proteins of these polysomes (Nemer and Lindsay, 1969; Kedes *et al.*, 1969). An important point is that both newly synthesized and maternal histone messenger RNA appear to be present on the light polysomes of cleavage and blastula stage embryos. We have already reviewed experiments which demonstrate maternal message for the histones in the postribosomal fraction of unfertilized sea urchin eggs (Gross *et al.*, 1973a; Farquhar and McCarthy, 1973; Skoultchi and Gross, 1973; Lifton and Kedes, 1976; see Fig. 4.3).

When isolated, much of the messenger RNA on the light polysomes sediments around 9 S, and it is now clear that most of the mass of the 9 S messenger RNA is histone message. Gross et al. (1973b) added 9 S message to the cell-free Krebs ascites translation system and showed that tryptic peptides of the translation products are similar to those of authentic histones. Very little background representing additional protein synthesis is observed. The total 9 S RNA from late cleavage stage sea urchin embryos has been resolved into several subfractions by slab gel electrophoresis (Grunstein et al., 1974; Levy et al., 1975). When translated individually in vitro each is found to be highly enriched in templates for specific histones. The light polysomes contain newly synthesized (i.e., labeled) 9 S RNA, and this too consists predominantly of histone message sequences. Thus, Skoultchi and Gross (1973) showed that radioactive 9 S RNA includes the same sequences as bulk 9 S RNA in hybridization competition experiments. Hybridization of the labeled species to DNA is decreased 80% in the presence of excess total light polysome RNA. At the ratios used this experiment would not have worked (i.e., the unlabeled RNA would not have been in excess) were much of the mass of message in the light polysomes some other species of messenger RNA. Furthermore, labeled 9 S RNA separates into bands which migrate coincidentally with the individual histone messages studied by in vitro transcription (Levy et al., 1975; Gross et al., 1976a). As of this writing a partial sequence and fingerprint analyses exist for one of these newly synthesized messenger RNA fractions, that which codes for histone IV (Grunstein and Schedl, 1976; Grunstein et al., 1976). This RNA is about 400 nucleotides long, compared to a necessary codogenic sequence length of 306 nucleotides. Poly(A) tracts are absent, a known characteristic of histone messenger RNA's in other material. The oligonucleotides which have been sequenced are consistent with the codogenic sequence for histone IV, as inferred from the amino acid sequence of this protein. Transcription of new histone messenger RNA sequences has been observed in isolated nuclei derived from cleavage stage sea urchin embryos but not from later stages (Shutt and Kedes, 1974). The assay used for the histone message sequences in these experiments was hybridization to purified histone DNA and competition with this hybridization by 9 S messenger RNA.

The mass of histone messenger RNA which must be present in these embryos is significant. Over half of the polysomes are in the light class at blastula stage (Moav and Nemer, 1971), and most of these synthesize histones. Since at least 25% of amino acid incorporation into protein at midblastula is incorporation into histones, this should also be the approximate percent of the mass of the polysomal message which is histone messenger. The polysomes contain about 60 pg of total messenger RNA in midblastula (see Chapter 5 for this calculation). Thus 15 pg of the blastula message may be histone messenger RNA. This amount of histone message is consistent with that obtained by calculating the amount needed to support the synthesis of 1–2 pg min^{-1} of histone. Without knowledge of histone messenger RNA turnover, it cannot be calculated whether the quantity of histone message in the cleavage stage is too great to be supplied totally by new synthesis, i.e., whether reliance on maternal message is obligatory. At the 40-cell cleavage-stage histone synthesis is only about 5% of total protein synthesis (Seale and Aronson, 1973a), and this is within the percentage of total polysomal message which is newly synthesized, considering all species of message (Humphreys, 1971).

The proportion of histone message which is maternal, as opposed to newly synthesized, has been investigated in actinomycin experiments (Kedes *et al.*, 1969; Thaler *et al.*, 1970; Johnson and Hnilica, 1971; Ruderman and Gross, 1974). This inhibitor blocks almost all incorporation of labeled nucleosides into the 9 S messenger RNA fraction. However, all five histone species continue to be synthesized, and the rapidly expanding embryo chromatin complement still contains histones, though in somewhat reduced amounts. During early cleavage approximately two-thirds of histone synthesis is actinomycin resistant, while by the blastula stage this value drops to less than one-third (Ruderman and Gross, 1974). Presumably this is not an artifact due to changing sensitivity to actinomycin, but it is not possible to be certain. On the basis of the relative amounts of unfertilized egg RNA and blastula polysomal RNA needed to compete with the hybridization of labeled histone messenger RNA, both Skoultchi and Gross (1973) and Farquhar and McCarthy (1973) concluded that the amount of histone messenger RNA in eggs is about one-fourth of that in light polysomes at midblastula. Thus we may estimate roughly that the amount of maternal histone messenger RNA stores in the egg is in the range of 4 pg, or as much as 4–8% of the total maternal messenger RNA stockpile (Table 4.2).

An important principle is revealed by these observations, one which may be broadly generalizable (Chapter 6); this is that the histones are translated from *both* maternal and newly synthesized embryo messenger RNA's in the early embryo.

QUALITATIVE ALTERATIONS IN HISTONE SYNTHESIS DURING EARLY DEVELOPMENT

During early sea urchin development several changes in the pattern of histone synthesis can be detected. These affect the proportions of individual histones synthesized, and may include modifications such as acetylation and phosphorylation. A stage-specific alteration noticed in several species is in the type of histone I synthesized. New varieties of this histone appear at specific stages, such as blastula or gastrula, depending on the species, and these differ in molecular weight by as much as 1000 daltons (Johnson and Hnilica, 1971; Cohen *et al.*, 1973; Seale and Aronson, 1973b; Ruderman and Gross, 1974; Ruderman *et al.*, 1974; Arceci *et al.*, 1976). Scheduled changes in other major histones are also known to occur during early sea urchin development, aside from those merely reflecting modification by phosphorylation or acetylation (Poccia and Hinegardner, 1975; Cohen *et al.*, 1975). Poccia and Hinegardner (1975) showed that the species of histone I initially present is simply diluted out during growth by the later appearing form, rather than being specifically degraded. Cell-free translation studies have identified the distinct messages for these histone I variants in embryo polysomes at the respective stages when they are synthesized (Ruderman *et al.*, 1974; Arceci *et al.*, 1976). Such changes in the messenger RNA population seem a likely result of transcriptional variation. The data reviewed earlier on post-transcriptional alterations in protein synthesis patterns make this a somewhat dangerous assumption. In the case of histone I variants, however, it appears that maternal messenger RNA exists only for that form of histone I which is synthesized before gastrulation. This was established by Arceci *et al.* (1976), who showed that the *in vitro* translation products of total oocyte RNA include only the pregastrular form of histone I, while pluteus RNA contains only the message for the postgastrular form. The structural genes for the postgastrular form are evidently transcribed only during embryogenesis, from the gastrular stage on. It remains to be seen whether maternal messages for all the other forms of histone synthesized up to the pluteus stage are inherited by the embryo, or alternatively, only for those forms which appear initially. It is clear from the work of Lifton and Kedes (1976) cited above that messenger RNA for all five major classes of histone are stored in the mature oocyte. If the embryos are fed at the pluteus stage, thus enabling them to develop further (Hinegardner, 1969), the pattern of histones soon comes to resemble that generally seen in other adult animal tissues, such as calf thymus (Poccia and Hinegardner, 1975).

QUANTITATIVE ASPECTS OF HISTONE SYNTHESIS IN *XENOPUS*

It is of interest to compare the course of histone synthesis in *Xenopus* embryos with that observed in sea urchins. To facilitate this comparison several of the quantitative measurements and estimates reviewed in the last several paragraphs are summarized in Table 4.3, where they can be compared to the equivalent parameters for amphibian embryos. In the

TABLE 4.3. Comparison of Histone Synthesis Parameters in Sea Urchin and *Xenopus* Embryos[a]

Organism	Number of histone genes per haploid genome	Maximum rate of histone synthesis per embryo (pg min^{-1})	Stage of development at maximum synthesis rate	Maximum rate of DNA synthesis per embryo (pg min^{-1})	Approximate maximum amount of histone messenger RNA in translation per embryo (pg)	Fraction of total protein synthesis as histone synthesis at maximum stage (%)	Estimated amount of stored maternal histones per embryo[c] (ng)
Sea urchin[b]	400–1000	1–2	Midblastula	1.5	15	25–40	None
Xenopus	30	100	Late blastula	1500	200–400	3	190

[a] Sources of the estimates listed in this table are given in the text.

[b] Most of the studies referred to have been done on *Arbacia punctulata*, *Strongylocentrotus purpuratus*, and *Lytechinus pictus*.

[c] In other words, other than histones associated with the oocyte chromatin.

amphibian system the quantitative demands are very different. The
Xenopus embryo produces over 10^5 pg of new DNA within the first 10
hours of development (late blastula), compared to about 10^3 pg in the late
blastula of the sea urchin. The latter also requires about twice as long to
develop to this stage. Furthermore, as noted above, the genome of
Xenopus has less than one-tenth the number of histone genes as the sea
urchin genome. It seems unlikely *a priori* that new messenger RNA synthesis could supply the massive quantities of histone required by this embryo during cleavage and blastula formation. Rates of histone synthesis in
Xenopus embryos have been measured by Adamson and Woodland (1974,
1976) and Woodland and Adamson (1976). Though the identification of
histones in Xenopus embryos is not supported by the depth of information
which has accumulated with regard to sea urchin embryo histones, we
rely on the data of Adamson and Woodland for purposes of comparison.
Though this is the only available source, these measurements seem unlikely to be seriously in error. Histones are identified by their characteristic
acid solubility and electrophoretic mobility patterns and also by the coincidence in two-dimensional separations of tyrosine-containing tryptic
peptides with those of authentic histones. At maximum rate during the
blastula stage at least 100 pg min^{-1} of histones are being synthesized
(Adamson and Woodland, 1974), requiring at any one time 200– 400 pg of
histone messenger RNA. Note, however, that this is only about 0.3– 2.0%
of the probable maternal messenger RNA stockpiled in the Xenopus egg
(see Table 4.2). Histone synthesis thus represents at least 3% of total
protein synthesis [the total protein synthesis rate is 200 ng hr^{-1} (see
above)] during pregastrular development. In activated eggs, where the
total protein synthesis rate is about 40 ng hr^{-1}, histone synthesis represents
as much as 7.5– 10% of total protein synthesis (Adamson and Woodland,
1974, 1976).

There is direct evidence that this synthesis results from the translation
of maternal messenger RNA's. Adamson and Woodland (1976) showed
that more than 97% inhibition of RNA synthesis in matured oocytes with
actinomycin does not in the least affect the absolute rate of histone synthesis, nor does enucleation of the eggs. During the period of maturation
following treatment of ripe ovarian oocytes with progesterone, the rate of
histone synthesis increases as much as eightyfold, compared to about
twofold for total protein synthesis rate in Xenopus (see above). It is important to note that a large part of this increase occurs in enucleated eggs as
well (Adamson and Woodland, 1976). Direct demonstrations of histone
maternal messenger RNA in both Xenopus and Triturus oocytes were
accomplished by Ruderman and Pardue (1976). RNA was extracted from
ovaries and oocytes in this work and translated in the cell-free wheat germ

system. *In vitro* synthesis of the major histone classes from this template was identified on the basis of electrophoretic mobilities. An experiment of Ecker and Smith (1971) regarding the fate of the proteins synthesized from maternal message in *Rana* eggs is in accordance with this conclusion. Nuclei were injected into enucleated cleavage-stage embryos after newly synthesized embryo proteins had been labeled, and these proteins were found to accumulate rapidly in the injected nuclei. Probably a large fraction of these proteins are histones.

Unlike the sea urchin case, in *Xenopus* embryos histone synthesis rate is not correlated with DNA synthesis rate. Instead histone synthesis rate is already established at 35–65 pg min^{-1} in unfertilized, activated eggs and newly fertilized eggs, when the DNA content is only that of the single zygote nucleus, 6 pg. This rate of histone synthesis is already half the highest rate ever attained. Furthermore, during the first S phase leading to cleavage, this rate of histone synthesis remains constant rather than fluctuating coincidentally with DNA synthesis, according to Adamson and Woodland (1976). From the high rate of histone synthesis compared to DNA content, it follows that histones synthesized in cleavage are held for subsequent use when DNA synthesis rate outstrips histone synthesis rate. Adamson and Woodland calculated that in the late blastula 15 times more DNA mass is being synthesized per unit time than histone. However, the stoichiometry of chromatin requires a nearly equal mass of DNA and histone, and *Xenopus* embryo chromatin is known to contain all the usual ·histones at this stage (Destrée *et al.*, 1973; Byrd and Kasinsky, 1974; Adamson and Woodland, 1974). Even the quantity of histones accumulated in early cleavage is not sufficient to satisfy requirements beyond midblastula. About four times more histone is needed by late blastula than is synthesized in the whole period between fertilization and late blastula. It was inferred by Adamson and Woodland (1974) that the embryo inherits a large amount of preformed histone protein, as well as messenger RNA, from oogenesis, and this view was confirmed by subsequent work. Woodland and Adamson (1976) showed that histone synthesis takes place in ovarian oocytes of all stages, though at much lower rates than in activated eggs and early embryos. Full-sized ovarian oocytes synthesize histones at rates averaging about 0.7–1.7 pg min^{-1}, though this is dependent somewhat on individual females. It was also noted that the rate was higher if the follicle cells were removed. Synthesis of all histones except histone I was observed. Since oogenesis lasts for several months (see Chapter 8), this rate is sufficient to account for the accumulation of the necessary amount of histone, about 100–200 ng.

By iodinating proteins extracted from oocytes and studying their electrophoretic mobility and tryptic peptides, a large pool of histone stored in the mature *Xenopus* oocytes has been demonstrated. This again includes all the nucleosomal histones but not histone I. The source of the latter remains a mystery, since little newly synthesized histone I has so far been detected either in activated eggs or early embryos, probably for technical reasons. Woodland and Adamson (1976) concluded as a rough estimate that about 190 ng of total histone is stored in the mature oocyte, and this value is consistent with the deficit in the quantity of histone synthesized during early development.

A summary of histone synthesis rates and estimated histone messenger RNA content for *Xenopus* embryos is found in Table 4.3. The data we have reviewed show that *Xenopus* embryos require histone at such a rapid rate that storage of these proteins from previous periods of synthesis is required. These periods are ovarian oogenesis, maturation, and early cleavage. Rate of histone synthesis is stepped up as much as fiftyfold between the ovarian oocyte and fertilization, comparing the rate of synthesis in ovarian oocytes, $0.7-1.7$ pg min^{-1}, with that in activated eggs, which is as high as 65 pg min^{-1}. Over the same period total protein synthesis increases only about twofold. Most of the newly synthesized histone made after fertilization must be encoded on maternal message. The most general conclusions from comparing the amphibian and echinoderm systems (Table 4.3) are that the sea urchin embryo relies on both histone message synthesis and maternal histone message, while the *Xenopus* embryo relies on maternal message and stored histones. Thus in both systems a double source of these essential proteins exists.

DNA POLYMERASES

DNA polymerase displays a simpler pattern of accumulation, in both sea urchin and *Xenopus* embryos. DNA polymerase activity is usually assayed as DNA primer-dependent incorporation of ^{3}H-thymidine into double-stranded DNA. This activity is found in sea urchin egg homogenates and has been shown to remain constant between fertilization and gastrula stages (Loeb *et al.*, 1969; Fansler and Loeb, 1969). According to Loeb and Fansler (1970), the polymerase protein turns over at the same rate as does total protein (see first section of this chapter). The embryo, or fertilized egg, contains a large amount of polymerase, about 5×10^8 molecules in S. *purpuratus* eggs, or enough for one polymerase molecule to bind every 1600 nucleotide pairs of DNA in the hatching blastula (Loeb, 1970). While the quantity of enzyme does not change during development, its localization in the embryonic cells does. Thus Loeb and Fansler (1970), Fansler and Loeb (1972), and Mazia (1966) demonstrated that the polymerase shifts between the nucleus and the cytoplasm in each cell cycle.

DNA polymerase accumulated during oogenesis is also present in large quantities in *Xenopus* eggs. Both nuclei (Gurdon, 1968) and double-stranded DNA (Gurdon *et al.*, 1969) injected into mature eggs serve as primers for DNA synthesis. Several different polymerases are present (Benbow *et al.*, 1975; Ford *et al.*, 1975), and a complex stage-specific pattern of appearance of these variants has been described (Grippo and Lo Scavo, 1972; Benbow *et al.*, 1975). A major point is that the total amount of polymerase activity for double-stranded DNA primer remains essentially constant from fertilization through gastrulation, i.e., through the production of over 10^5 cells (Benbow *et al.*, 1975). At neurula stage a new form of polymerase appears, and the total activity then increases further. Ovarian oocytes manufacture DNA polymerase continuously, and by ovulation they already contain a third of the total embryo polymerase activity. During the maturation period the remaining polymerase activity appears. This may represent a newly synthesized protein, since its characteristics are distinct from those of the ovarian enzyme. Thus at fertilization the *Xenopus* egg contains an enormous store of presynthesized DNA polymerase. Benbow *et al.* (1975) and Ford and Woodland (1975) stress that control systems other than simply the quantity of polymerase must operate to regulate DNA synthesis rates. These may include cyclic nuclear localization of the polymerases, as in the sea urchin, but other factors operate as well. While the DNA polymerase activity can be extracted and displayed in homogenates (Benbow *et al.*, 1975), only a small fraction of this activity is available *in situ* when measured by injection of primer DNA into mature oocytes and eggs. Benbow and Ford (1975) argued from this that control factors must be present to regulate DNA synthesis, and these authors may have demonstrated such a factor in a cell-free DNA synthesis system. Nuclei were isolated from adult *Xenopus* liver, a tissue displaying little DNA synthesis, and were incubated with cytoplasmic homogenates in the presence of deoxynucleoside triphosphates. When the cytoplasm is obtained from oocytes, little synthesis by the endogenous liver DNA polymerases is observed, but cytoplasm from fertilized eggs or early embryos supports active DNA synthesis in the liver nuclei. This could be assayed in electron micrographs by the appearance of initiation loops in the liver DNA, as well as by deoxynucleotide incorporation. Preliminary experiments indicate that the responsible control factor is a protein. Since all the usual requirements for DNA synthesis, such as ligases, nucleases, precursors, and polymerases, are present even in oocytes (e.g., Gurdon and Laskey, 1970; Woodland and Pestell, 1972; Ford *et al.*, 1975), it is considered likely that the control protein is some form of initiation factor (Benbow and Ford, 1975).

As discussed earlier in this section, the rate of DNA synthesis is very high in early embryos. The postulate that this rate is controlled by initiation factors seems a reasonable one in view of the presence of large amounts of preformed polymerase, and also because the number of initiation points for DNA synthesis seems to be extremely high in rapidly dividing embryonic tissues. DNA replication in animal cells occurs by means of bidirectional replication forks which move outward from the point of origin, forming closed replication loops or "eye forms" (e.g., Huberman and Riggs, 1968; Callan, 1972; Wolstenholme, 1973; Kriegstein and Hogness, 1974; Blumenthal *et al.*, 1974). Apparent replication loops from *Drosophila* embryos are shown in Fig. 4.6a (Kriegstein and Hogness, 1974). This remarkable photograph shows such loops in a single DNA molecule about 119,000 nucleotide pairs in length. As required for the forks terminating each replication loop, the two branches of each fork are homologous in sequence. This is illustrated in Fig. 4.6b, a partial denaturation map of one replication loop. Interfork distances in embryonic cells of *Drosophila* have been measured both by electron microscopy and by radioautography (Wolstenholme, 1973; Kriegstein and Hogness, 1974; Blumenthal *et al.*, 1974). It is found that the forks are more closely spaced in very rapidly dividing cells than in other cells. In *Drosophila* cleavage-stage embryos the nuclei divide as rapidly as every 9.6 minutes, and S phase is less than 3.6 minutes long (Rabinowitz, 1941). The mean interfork distance is in the range of 9000–11,000 nucleotide pairs. Lee and Pavan (1974) have studied another dipteran insect, *Cochliomyia hominivorax* (the screwworm) in which DNA replication is even more rapid. Replication loops in this genome are shown in Fig. 4.6c. Here the mean interfork distance is only about 6900 nucleotide pairs. In *Drosophila* tissue culture cells where S phase lasts 600 minutes, the rate of fork movement is close to that in embryo cells, about 2600 nucleotide pairs min^{-1}. However, the distances between replication sites are much greater, as the DNA of these cells displays mean interfork distance classes averaging 28,000 and 57,000 nucleotide pairs (Blumenthal *et al.*, 1974). In *Xenopus* tissue culture cells the distances separating areas of DNA replication are over 100,000 nucleotide pairs, and the distance between points of origin averages about 180,000 nucleotide pairs (Callan, 1972). The average fork movement rate, about 4500 nucleotides min^{-1} is similar to that in *Drosophila* embryos. These data are consistent with the view that DNA replication rates are controlled by the frequency of initiation sites, rather than by the rate of fork movement.

RNA POLYMERASES

The RNA polymerases of early embryonic cells have also been quantitatively investigated. These enzymes occur in *Xenopus* embryo cells in the same three chromatographic and functional classes as found in other animal cell types (Roeder, 1974b). Included are polymerases responsible for ribosomal RNA synthesis (class I polymerases) and the α-amanitin-sensitive polymerases responsible for heterogeneous nuclear RNA synthesis (class II). The same is true of sea urchin embryos (Roeder and Rutter, 1970). In both organisms the amount of activity of each class remains almost constant from fertilization through gastrulation (Roeder and Rutter, 1970; Roeder *et al*., 1970; Kohl *et al*., 1973; Roeder, 1974a). All three forms of polymerase are present in the mature *Xenopus* oocyte germinal vesicle in the amounts also found in the fertilized egg except for class II which increases slightly by fertilization (Roeder, 1974a). Huge quantities of polymerase are also located in the germinal vesicle of *Rana pipiens* oocytes, according to Wassarman *et al*. (1972). These authors found that the polymerase activity of a single germinal vesicle equals that of a tadpole, which contains 400,000 cells, and a similarly dramatic equivalence exists between the polymerase content of the fertilized *Xenopus* egg and that of a 60,000 cell gastrula. Beyond this stage the RNA polymerase content of *Xenopus* embryos increases, just as does the DNA polymerase content (Roeder, 1974a).

These data indicate that at fertilization sea urchin and *Xenopus* eggs contain all the RNA and DNA polymerases they will require far into development. Only in the postgastrular period is the quantity of either class of enzyme augmented. The amounts of both DNA and RNA polymerase which are accumulated during oogenesis are equivalent to the contents of many thousands of somatic cells. Thus, as far as is known at present, no quantitatively significant role is played by synthesis of any of these polymerases following fertilization, though *de novo* synthesis probably contributes to the determination of the maximum level of transcription after gastrulation.

MICROTUBULE PROTEINS

As a final example we now consider the synthesis and accumulation of microtubule protein. Radioautograph experiments showed some years ago that newly synthesized protein accumulates in the mitotic apparatus of cleaving sea urchin eggs (Gross and Cousineau, 1963b; Stafford and Iverson, 1964). Amino acid label incorporated into isolated mitotic structures, from a preparation of Stafford and Iverson (1964) and D. W. Stafford (unpublished) is shown in Fig. 4.7. Protein synthesis is required in order for cleavage to take place and protein synthesis inhibitors prevent the formation of the mitotic apparatus. The first cleavage spindle is an exception in that it still forms if the protein synthesis inhibitor is added 30 minutes or less before cleavage (Hultin, 1961b; Wilt *et al*., 1967). Much of

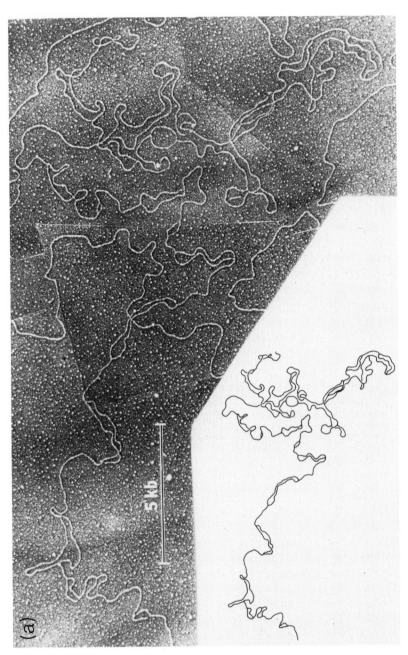

Fig. 4.6. Replication forms in embryo cell DNA. (a) Fragment of replicating chromosomal DNA from cleaving *Drosophila* embryo nuclei. The fragment shown contains 23 "eye forms" in a length of 119,000 nucleotide pairs. Cleavage nuclei were lysed in a solution containing EDTA and detergents, and the lysate was layered on a CsCl gradient. It was centrifuged at 20°C for 72 hours at 40,000 rpm in an ultracentrifuge. Fractions containing DNA were pooled, dialyzed, and spread for electron microscopy by the formamide technique of Davis *et al.* (1971), using 40 and 10% formamide in the hyperphase and hypophase, respectively. 1 kb = 1000 nucleotide pairs. From H. J. Kriegstein and D. S. Hogness (1974), *Proc. Natl. Acad. Sci. U.S.A.* **71**, 135.

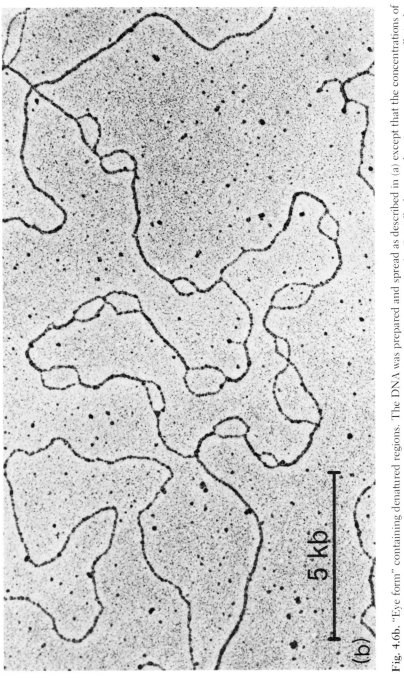

(b)

Fig. 4.6b. "Eye form" containing denatured regions. The DNA was prepared and spread as described in (a) except that the concentrations of formamide in the hyperphase and hypophase were increased to 80 and 50%, respectively. Partially denatured regions are seen as small single-stranded loops. The distance of each consecutive such loop from the fork at either end is about the same on both sides of the "eye form," indicating sequence homology. 1 kb = 1000 nucleotide pairs. From H. J. Kriegstein and D. S. Hogness (1974). *Proc. Natl. Acad. Sci. U.S.A.* **71**, 135.

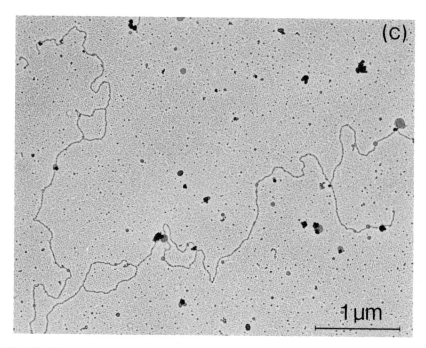

Fig. 4.6c. Electron micrograph of replicating DNA molecule from *Cochliomyia hominivorax* embryos. From C. S. Lee, and C. Pavan (1974). *Chromosoma* **47**, 429.

the label incorporated into the cleavage spindles is now known to be in newly synthesized microtubule proteins (Cohen and Rebhun, 1970; Bibring and Baxandall, 1971; Meeker and Iverson, 1971). These proteins are also present in embryonic cilia (Stephens, 1972).

Microtubule proteins have been highly purified from the soluble phase of sea urchin eggs by use of specific high affinity ligands such as vinblastine sulfate and colchicine (Raff *et al.*, 1971; Raff and Kaumeyer, 1973). These studies show that sea urchin eggs and embryos contain a large quantity of microtubule protein subunits not polymerized into supramolecular structures. About 120 pg of microtubule proteins exist in the *Arbacia* egg. This is equivalent to approximately 0.37% of the total protein of the egg (32 ng) or about 1.1% of the nonyolk soluble protein (Raff and Kaumeyer, 1973). Furthermore, the total amount of microtubule protein remains virtually constant throughout early development, i.e., at the level present when fertilization occurs. In *Xenopus* a similar situation appears to exist, as a relatively vast microtubule protein pool amounting to 1% of the total egg

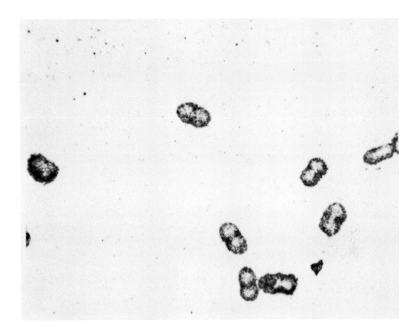

Fig. 4.7. Radioautograph of isolated mitotic structures of cleavage stage sea urchin eggs. The eggs had been incubated in seawater containing ^{14}C-leucine. Courtesy of Dr. D. W. Stafford.

protein is present in mature oocytes (Smith and Ecker, 1969b; Pestell, 1975). The microtubule protein pool of these eggs also remains roughly constant throughout early development. Similar data exist for *Spisula* (Burnside *et al.*, 1973).

It is interesting that despite the presence of a large pool of microtubule protein in sea urchin eggs, these proteins are very actively synthesized (Meeker and Iverson, 1971; Raff *et al.*, 1971; Hirama and Mano, 1974). The fraction of total protein synthesis represented by microtubule proteins is about 0.5–1.5% in late cleavage *Strongylocentrotus* eggs (Raff and Kaumeyer, 1973), or 6 pg hr^{-1} (see Table 4.1). At this rate the microtubule protein pool would be turned over every 20 hours, since the total quantity of microtubule protein remains constant. Calculations based on the measured rate of microtubule protein synthesis (in another species, *Arbacia*) suggest a synthesis rate of about 16 pg hr^{-1} (Raff and Kaumeyer, 1973). These data indicate that though a significant fraction of total protein synthesis is devoted to microtubule proteins, the size of the microtubule protein pool is large relative to the amount produced per hour by

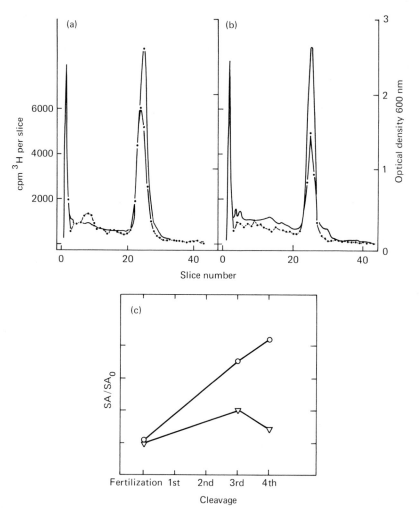

Fig. 4.8. Labeling of microtubule proteins of *Lytechinus pictus* embryos in the presence and absence of high molecular weight RNA synthesis. Microtubule proteins were precipitated with vinblastine, and displayed in a gel electrophoresis system. Solid line indicates the optical density trace of the vinblastine precipitate. Dashed line indicates radioactivity in the vinblastine precipitate. (a) Control embryos labeled at 16-cell stage with ^{3}H-leucine. (b) Embryos cultured in the presence of 250 μg/ml of actinomycin D continuously from 60 minutes before fertilization. Embryos were labeled at the 16-cell stage the same as controls. From R. A. Raff, G. Greenhouse, K. W. Gross, and P. R. Gross (1971). *J. Cell Biol.* **50**, 516. (c) Changes in relative specific activity of microtubule proteins and total soluble proteins during early cleavage of *Arbacia*, in the presence of actinomycin. Specific activity (SA) of microtubule proteins at various cleavage stages is related to the specific activity of mi-

synthesis. The question now arises as to whether synthesis of microtubule protein depends on newly transcribed or maternal messenger RNA.

Raff *et al*. (1971, 1972) and Raff and Kaumeyer (1973) showed that sea urchin eggs treated with actinomycin continue to synthesize tubulin at exactly the same rates as do controls. This is illustrated in Fig. 4.8 which displays ^{3}H-leucine incorporation into microtubule proteins in the absence (Fig. 4.8a) and presence (Fig. 4.8b) of actinomycin (Raff *et al*., 1971). Enucleated eggs activated parthenogenically also synthesize microtubule proteins at control rates (Raff *et al*.,1972). Furthermore, neither the microtubule protein pool size, nor the relative increase in microtubule protein synthesis rate after fertilization (Fig. 4.8c) is affected by actinomycin treatment. A very similar result has been obtained by Raff *et al*. (1975) with *Ilyanassa* embryos. These embryos also display an increasing rate of microtubule protein synthesis after fertilization, first detectable on day 2 of development (early gastrula). Actinomycin-treated *Ilyanassa* embryos carry on microtubule protein synthesis for up to 36 hours, qualitatively and quantitatively unaffected by the complete abolition of RNA synthesis. The synthesis of microtubule proteins in these embryos thus appears to depend exclusively on maternal messenger RNA. This is particularly clear in the sea urchin case, where it has been demonstrated that the amount of microtubule protein synthesis is significant, and where the actinomycin experiments are supported by observations on microtubule protein synthesis in enucleated egg cytoplasm.

The provision of microtubule protein in sea urchin embryos is thus accomplished by means similar to those by which histones are provided in *Xenopus* embryos. That is, a large amount of the proteins themselves are stored in the egg, having been synthesized during oogenesis, but active synthesis on maternal messages also takes place. It is interesting to note that in both cases which we have examined where a specific protein is synthesized on embryo polyribosomes, its source is not unique. Either there are both maternal and embryo messengers (histones in the sea urchin), or maternal message plus stored maternal protein (histones in *Xenopus*, and microtubule proteins in sea urchins).

---

crotubule protein labeled in the 30 minute interval starting at fertilization (SA$_0$). Specific activities (SA) of 150,000 *g* soluble proteins (cpm/mg protein) are also related to (SA$_0$) for soluble proteins. Actinomycin D was present at 20 μg/ml. Microtubule proteins (O) and total soluble proteins (∇). From R. A. Raff, H. V. Colot, S. E. Selvig, and P. R. Gross (1972). *Nature (London)* **235**, 211.

5

Transcription in Early Embryos

Classes of RNA synthesized in amphibian embryos include heterogeneous nuclear RNA's, messenger RNA's, ribosomal RNA's, and
transfer and 5 S RNA's. A simple set of equations for analyzing the
synthesis and turnover kinetics of an RNA species with constant precursor specific activity is provided. From the quantities of labeled
RNA of various classes accumulated, plus other data, estimates are
obtained of the steady state quantities of nuclear and messenger RNA
in *Xenopus* embryos (Table 5.1), and for the synthesis rate of the
heterogeneous nuclear RNA. At blastula stage the relative rate of
synthesis of nuclear RNA increases. Analysis of the rate of accumulation of newly synthesized ribosomal RNA shows that even in the
postgastrular stages when ribosomal RNA synthesis appears prominent, the ribosomal genes are being transcribed at a very low rate.
This rate is less than 1% of the calculated maximum rate. If the
ribosomal genes were transcribed in pregastrular embryos at the same
rate, ribosomal RNA synthesis would have been undetectable. Therefore, it is questionable whether regulation of ribosomal RNA synthesis
occurs at all during early development in amphibian embryos. Direct
measurements of absolute synthesis and decay rates exist for the heterogeneous nuclear RNA of sea urchin embryos. The molecular
characteristics of this RNA are reviewed, and it is noted that several
classes exist, distinguished by their poly(A) content. The functional
meaning of these classes is unknown. The absolute synthesis rate of
sea urchin embryo nuclear RNA (Table 5.2) declines severalfold between the cleavage and the pluteus stage. The heterogeneous nuclear

RNA turns over rapidly, with a $t_{1/2}$ of about 10 to 20 minutes. Studies of messenger RNA synthesis and decay rates are also reviewed, and equations are presented for obtaining these parameters when the precursor specific activity changes with time. The steady state quantity of messenger RNA in sea urchin gastrulae, as calculated from the synthesis and decay kinetics of labeled messenger RNA, is close to the total polysomal messenger RNA content. Therefore, by this stage, most polysomal message is newly synthesized. Most of the messenger RNA appears to turn over with a $t_{1/2}$ of about 5.3 hours. The rate of synthesis accounts for only 4–7% of the instantaneous rate for total heterogeneous nuclear RNA. These data are collated in Table 5.3. In the sea urchin embryo, about half of the messenger RNA is polyadenylated and half is not. At least some distinctions exist in the sets of sequences included in these two messenger RNA categories. Mitochondrial RNA synthesis is quantitatively important in cleavage stage sea urchins but not in later embryos. Until feeding, the ribosomal RNA genes of sea urchin embryos are transcribed at an extremely low level. On feeding, the rate of their transcription increases sharply. No clear evidence for regulation of ribosomal RNA synthesis prior to feeding exists for these embryos. RNA synthesis in mammalian embryos is also discussed. In rabbit and mouse embryos, ribosomal RNA synthesis can be observed as early as the 4- to 8-cell stage, and heterogeneous nuclear RNA synthesis is reported in the earliest embryos. Unlike echinoderm and amphibian eggs, net growth of the embryo begins very early in development, and this is correlated with the precocious ribosomal RNA synthesis. Observations on RNA synthesis in a variety of other embryos are collated in Table 5.4. These are mainly qualitative, but lead to the general conclusion that transcription of both heterogeneous nuclear RNA and messenger RNA always begin very early, during cleavage. This is paradoxical since, as shown in prior chapters, neither morphogenetic events nor patterns of protein synthesis seem to depend on embryo genome action during the pregastrular period.

This and the following chapter are concerned with transcriptional activity in the genomes of embryonic cells. Significant measurements exist for several animals regarding rates of embryonic RNA synthesis and turnover, and the classes of RNA made. These matters are reviewed here, and we reserve for Chapter 6 discussion of sequence homology experiments which shed light on the number of different RNA sequences and changes in RNA populations occurring in early embryos. In this chapter consideration is confined to broad classes of RNA, namely, the ribosomal RNA's, transfer RNA's, heterogeneous nuclear RNA's, and messenger RNA's. Several distinct patterns of RNA synthesis are evident, and it is most convenient to treat this subject according to taxon.

RNA Synthesis in Amphibian Embryos

ACCUMULATION OF LABELED RNA SPECIES

At present the best studied amphibian species are *Xenopus laevis* and *Rana pipiens*. *Xenopus*, like most other amphibians, sheds an egg which is relatively impermeable to all RNA precursors except $^{14}CO_2$ (Cohen, 1954; Flickinger, 1954). A labeled precursor can be administered by two other routes as well. One method, initially described by Kutsky (1950) for *Rana* eggs and studied further by Grant (1958), consists of injecting ^{32}P-orthophosphate into the peritoneal cavity of the gravid female. In this way the eggs are loaded with this precursor before becoming impermeable as they traverse the oviduct. In the second group of procedures the external permeability barrier is attacked by direct physical means, viz., cutting the embryo open, dissociating the cells, or microinjecting the isotope into the embryo. RNA synthesis was shown to occur during cleavage by Decroly *et al.* (1964) and Brown and Littna (1964), following earlier explorations on *Rana pipiens* eggs such as that of Grant (1958). The latter microinjected [2-^{14}C]glycine into the eggs, while Decroly *et al.* (1964) and Brown and Littna (1964) used the method of injecting ^{32}P-orthophosphate into the maternal body cavity. The latter experiments were the first to show that even unfertilized eggs and 2- to 8-cell embryos synthesize heterogeneously sedimenting RNA.

Sucrose gradient analyses of the RNA synthesized at cleavage and later stages in *Xenopus* embryos are reproduced in Fig. 5.1 (Brown and Littna, 1964). Since the embryos in this study were labeled from the ^{32}P-precursor pool present from the beginning of ovulation, the radioactive RNA includes all stable RNA species accumulated from the time of ovulation up to each stage. It can be seen in Fig. 5.1 that a much larger amount of heterogeneously sedimenting labeled RNA is present in blastula stage embryos than in cleavage, and even more is present during gastrulation. The following discussion shows that most of this RNA is nuclear, and that it turns over rapidly. Therefore, part of the apparent increase in the quantity of labeled, heterogeneously sedimenting RNA is due simply to the sharp

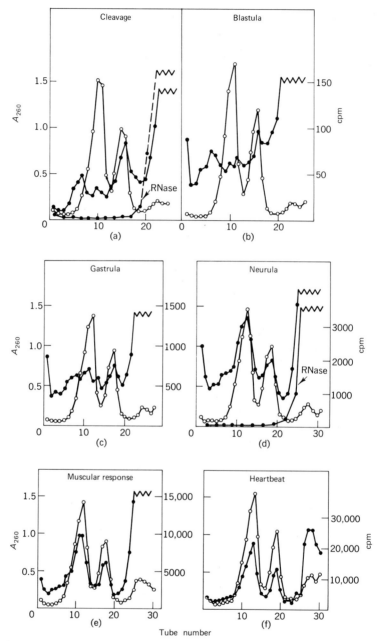

Fig. 5.1. Sedimentation pattern of RNA from sibling *Xenopus* embryos at different developmental stages. Each density gradient centrifugation was performed on RNA isolated from 150 embryos which had been labeled with ³²P-phosphate during ovulation (see text).

increase in the number of synthesis sites (i.e., nuclei) as development proceeds, rather than to accumulation of labeled species per cell. However, as detailed below, the overall change in the quantity of labeled heterogeneous RNA which is seen in Fig. 5.1 probably reflects a change in rate of synthesis per nucleus. In addition, some fraction of the radioactivity probably represents the accumulation of messenger RNA's. Another change in RNA synthesis pattern seen in Fig. 5.1 is the apparent onset of ribosomal RNA synthesis at the gastrula stage. Synthesis of this class of RNA's cannot be convincingly observed earlier but becomes dominant after neurulation. As noted in Chapter 4 the ribosomal RNA of early *Xenopus* embryos is maternal, and thus the absorbance peaks (open circles) in Fig. 5.1 remain essentially constant in size throughout the experiment. We now discuss each of these points beginning with the rate of heterogeneous nuclear RNA synthesis.

The properties of most of the newly synthesized RNA's of pregastrular *Xenopus* embryos are typical of the heterogeneous nuclear RNA's of other animal cell types. In Fig. 5.2 radioautographs of *Xenopus* embryo cells exposed to ³H-uridine for 1 hour are shown (Bachvarova and Davidson, 1966). At the 5000-cell blastula stage the embryos were bisected manually to allow immediate penetration of the labeled precursor. The grains remaining after various washes to remove acid-soluble precursor represent newly synthesized RNA (see RNase control, Fig. 5.2e and f). The nuclear location of the vast majority of grains is particularly evident in Fig. 5.2b and d. Wallace (1966) also observed in radioautographs that newly synthesized nonribosomal RNA of *Xenopus* embryos is predominantly nuclear. Aside from its sedimentation behavior (Brown and Littna, 1964; Mariano and Schram-Doumont, 1965; Brown and Gurdon, 1966; Bachvarova *et al.*, 1966) and nuclear location (see also Mariano and Schram-Doumont, 1965), pulse labeled RNA of these embryos displays other characteristics of heterogeneous nuclear RNA. Thus, it hybridizes with some repetitive DNA sequences (Brown and Gurdon, 1966; Denis, 1966), it has a more or less DNA-like base composition (Brown and Littna, 1964; Bachvarova *et al.*, 1966; Brown and Gurdon, 1966), and it turns over very rapidly (Mariano and Schram-Doumont, 1965; Brown and Gurdon, 1966; Shiokawa and Yamana, 1968). Rapid turnover is, of course, indicated by

---

Sedimentation is from right to left. RNase controls are shown in (a) and (d). The stages and hours after fertilization of each group of embryos are (a) early cleavage, stages 2 to 7 of Nieuwkoop and Faber (1956), 1.5 to 3 hours; (b) midblastula, stages 8 to 9, 5 to 6 hours; (c) gastrula, stages 10 to 11, 28 hours; (d) neurula, stages 13 to 14, 34 hours; (e) muscular response, stages 25 to 26, 54 hours; (f) heartbeat, stages 33 to 34, 74 hours. Open circles represent absorbance at 260 nm; filled circles, radioactivity. From D. D. Brown, and E. Littna (1964). *J. Mol. Biol.* 8, 669.

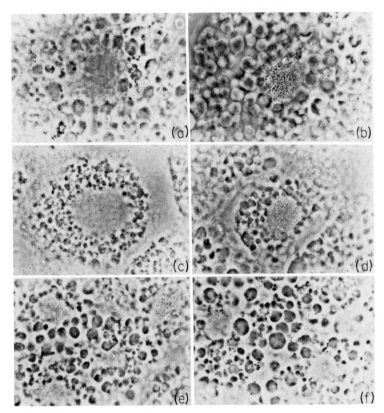

Fig. 5.2. Radioautographs of RNA synthesis in cells of *Xenopus* blastulae. Intranuclear labeling is demonstrated at various stages. In each case dorsal halves of embryos were immersed in a solution containing ^{3}H-uridine for 1 hour. The bisected embryo halves are completely permeable to the precursor at all stages. (a) Presumptive endodermal cell, stage 7 (Nieuwkoop and Faber, 1956); (b) endodermal cell, stage $8\frac{1}{2}$; (c) equatorial cell, stage 8; (d) equatorial cell, stage $8\frac{1}{2}$; (e) inner equatorial cell, stage 9; (f) same as (e), except treated with ribonuclease. From R. Bachvarova and E. H. Davidson (1966). *J. Exp. Zool.* **163**, 285.

the fact that this RNA class is preferentially labeled in short periods of time, as shown in detail in what follows. It is to be noted that some of the studies of Brown and Gurdon (1966) cited here were carried out on *o nu* embryos (see Chapter 4 and below for a description of these anucleolate mutants). Possible confusion between rapidly turning over heterogeneous RNA and ribosomal precursor RNA with similar kinetic properties can thus be excluded.

Heterogeneous and ribosomal ^{32}P-RNA's accumulate with the kinetics

shown in Fig. 5.3 when the precursor pool is labeled by injection of
[32]P-orthophosphate into the maternal body cavity. The quantities of RNA
shown on the ordinate of Fig. 5.3a are calculated from the incorporated
radioactivity and from the measured specific activity of the α-phosphates
in the nucleoside triphosphate precursor pool. During development the
specific activity of these precursors changes, but only by a factor of about
four, so that errors due to this source are not expected to be large. Furthermore, calculations based on these pool specific activities yield a value
for the quantity of new DNA accumulated which is within 20% of the true
values measured (directly) by Dawid (1965). The course of DNA accumulation is also shown in Fig. 5.3a. A steep rise in the amount of heterogeneous ^{32}P-RNA accumulated in the embryos occurs just before gastrulation (stage 10), according to Fig. 5.3a, and by gastrulation the amount of
this class of newly synthesized RNA is about 50 ng per embryo. The
amount of accumulated RNA is expressed as a ratio to the amount of
DNA at various stages in Fig. 5.3b. Here it can be observed that for a long
period, between gastrulation and swimming tadpole (stages 10–26), the
per cell content of heterogeneous ^{32}P-RNA remains approximately the
same. By this time the total RNA content per cell has decreased to a
constant value (Bristow and Deuchar, 1964). The plateau in heterogeneous ^{32}P-RNA/DNA ratios seen in Fig. 5.3b reflects the steady state
concentration of newly synthesized heterogeneous nuclear RNA, which
turns over rapidly, and also of messenger RNA. Unfortunately, direct
measurements of the quantities and synthesis rates of heterogeneous nuclear RNA and messenger RNA are not available for *Xenopus* embryos.
We must therefore rely on rough estimates for some of these parameters,
based on the data shown in Fig. 5.3 plus other information. These estimates are discussed in the following paragraphs and some of them are
summarized in Table 5.1.

RNA SYNTHESIS AND TURNOVER KINETICS FOR THE CASE
OF CONSTANT PRECURSOR SPECIFIC ACTIVITY

We first consider the general problem of analyzing the synthesis kinetics
of a population of RNA molecules which turns over rapidly in relation to
the time scale of the observations. This problem has been considered by
many authors (see, e.g., Kafatos and Gelinas, 1974). The approach taken
below is useful not only in the present context but also in later sections of
this book. We assume that label is introduced, that the precursor pool
equilibrates rapidly, and that the specific activity of the precursor pool
remains essentially constant throughout the labeling period. The rate of

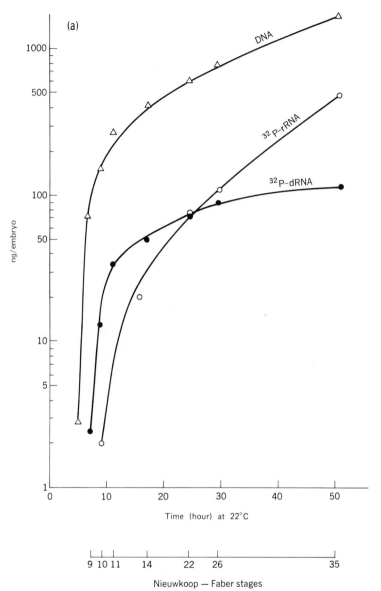

Fig. 5.3. Accumulation of newly synthesized RNA's in *Xenopus* embryos. (a) Amounts of heterogeneous high molecular weight ^{32}P-RNA ("dRNA," i.e., DNA-like in base composition) and ribosomal ^{32}P-RNA ("rRNA") present in the embryo at each stage of development, calculated from the total ^{32}P incorporated in each RNA species. Since specific activities of the α-phosphates of the nucleoside triphosphate precursors were determined for each sample of embryos, the total radioactivity in each class of RNA could be converted to nanograms of RNA. Values for DNA are taken from Dawid (1965). See Fig. 5.1 for correlation between developmental stage and Nieuwkoop–Faber stage numbers. From D. D. Brown and E. Littna (1966a). *J. Mol. Biol.* **20**, 81.

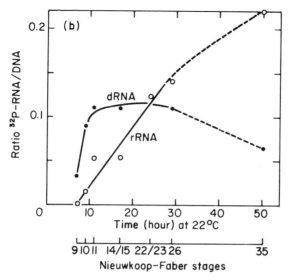

Fig. 5.3b. Amounts of heterogeneous ^{32}P-RNA and ribosomal ^{32}P-RNA relative to DNA in the embryo at each stage of development. Ratios were calculated from the data shown in (a). From D. D. Brown and E. Littna (1966a). *J. Mol. Biol.* 20, 81.

change in the quantity of the RNA which bears labeled nucleotides is given by

$$\frac{dC}{dt} = k_s - k_d C \tag{5.1}$$

where k_s is the RNA synthesis rate, in units of mass × time^{-1} (in the present calculations, pg min^{-1}); k_d is the first-order decay constant, in units of time^{-1} (here min^{-1}); and C is the mass of RNA bearing label in the cell or embryo at time t (here in pg). C can be considered the quantity of RNA synthesized since some earlier time and still present when the observation is made. The radioactivity incorporated into the RNA is converted to molar or mass units by reference to the specific activity of the nucleoside triphosphate precursor pool. Since the decay of the RNA is assumed to be first-order in equation (5.1) the half-time of the decay process, $t_{1/2}$ is related to k_d by

$$k_d = \frac{\ln 2}{t_{1/2}} \tag{5.2}$$

The amount of labeled RNA present t minutes after the introduction of a label is given by the solution to equation (5.1):

$$C = \frac{k_s}{k_d}(1 - e^{-k_d t}) \tag{5.3}$$

when the starting quantity of labeled RNA, is zero. This expression describes the accumulation of newly synthesized (i.e., labeled) RNA as a function of time. The steady state quantity of the RNA is obtained simply from equation (5.1) or (5.2).
At steady state,

$$C_\infty = \frac{k_s}{k_d} \tag{5.4}$$

SYNTHESIS RATES AND STEADY STATE QUANTITIES OF THE HETEROGENEOUS RNA'S OF *XENOPUS* EMBRYOS

We now attempt to estimate the steady state quantities of heterogeneous nuclear RNA and messenger RNA in *Xenopus* embryos. In Fig. 5.3b it can be seen that the amount of total heterogeneous ^{32}P-RNA accumulated per cell at and after stage 10 is about 0.11 times the amount of DNA, or about 0.68 pg. This quantity is not atypical; for example, mouse L cells, which are nearly tetraploid, contain about 1.8 pg of heterogeneous nuclear RNA (Brandhorst and McConkey, 1974). The total accumulated high molecular weight heterogeneous ^{32}P-RNA of *Xenopus* embryos includes both nuclear and messenger RNA. (We assume here that other kinds of heterogeneous RNA are insignificant.) The quantity of total polysomal messenger RNA can be estimated independently from measurements of polysome content (Woodland, 1974) as shown in Table 5.1 for stage 14 (neurula) and stage 26 (tailbud) embryos. Labeled non-ribosomal RNA is associated with embryo ribosomes as early as cleavage (Brown and Littna, 1964). By using *o nu* embryos, Brown and Gurdon (1966) were able to ensure that this newly synthesized polysomal RNA is nonribosomal, and Gurdon and Ford (1967) also showed directly that the polysomes of *o nu* tadpoles are loaded with newly synthesized message. Thus a reasonable, though by no means certain, assumption is that all the polysomal message is to be included in the accumulated newly synthesized heterogeneous RNA of Table 5.1. According to the data shown there, stage 14–26 *Xenopus* embryo cells should contain 0.25–0.34 pg of polysomal messenger RNA, or 22–29 ng per embryo. It is interesting that this amount is significantly less than the probable amount of messenger RNA stored in the unfertilized egg. As discussed in Chapter 4 (see Table 4.2) the content of poly(A)RNA in *Xenopus* oocytes is in the range of 40–70 ng. The *minimum* steady state content of heterogeneous nuclear RNA per cell can now be estimated as the total heterogeneous RNA minus all the messenger. This is about 0.5 pg per nucleus. The nuclear RNA is known to turn over rapidly, as noted above. If we assume a rate typical of animal cells, e.g., a half-life of 20–40 minutes or k_d in the

TABLE 5.1. Approximate Quantities of Heterogeneous Nuclear and Messenger RNA in *Xenopus laevis* Embryos at Stages 14 (Neurula) and 26 (Tailbud)

Stage	DNA content per embryo[a] (ng)	No. of cells[b]	Heterogeneous ^{32}P-RNA accumulated[c]		Total RNA[d]		Polysomal messenger RNA[e]		Steady state quantity of heterogeneous nuclear RNA per cell[f] (pg)
			Per embryo (ng)	Per cell (pg)	Per embryo (ng)	Per cell (pg)	Per embryo (ng)	Per cell (pg)	
14	410	6.5×10^4	50	0.77	4000	62	22	0.34	0.43
26	750	1.2×10^5	90	0.75	4000	33	29	0.25	0.50

[a] Dawid (1965).
[b] Calculated on the basis of 6.3 pg per diploid cell (Dawid, 1965).
[c] Brown and Littna (1966a).
[d] Brown and Littna (1964).
[e] Calculated on the basis that 90% of the total RNA is ribosomal; 15% of the ribosomes are in polysomes at stage 14 and 20% at stage 26 (Woodland, 1974). Calculation based on 4% of the polysomal RNA as messenger RNA.
[f] Calculated on the assumption that ^{32}P-RNA consists only of heterogeneous high molecular weight nuclear RNA, all of which turns over rapidly, plus messenger RNA, all of which is synthesized by the embryo.

range 0.017–0.035 min^{-1}, application of equation (5.4) would suggest an approximate synthesis rate for heterogeneous nuclear RNA of about 0.01–0.02 pg min^{-1} per cell. Not even tentative estimates such as these can as yet be made for the messenger RNA of Xenopus embryos, since the kinetics of turnover and accumulation have not been measured.

Brown and Littna (1966a) reported that a considerable amount of heterogeneous ^{32}P-RNA is accumulated during the maturation period, i.e., up to about 6 ng per egg. This is an appreciable fraction of the amount of such RNA present in the embryo even at the neurula stage (50 ng; see Table 5.1) and an even larger fraction earlier, when the accumulated pool of heterogeneous ^{32}P-RNA is only one-tenth as great (Fig. 5.3a). Even if all the heterogeneous RNA synthesized in the maturation period were stable, the single oocyte nucleus would have to produce this RNA at 2–3 orders of magnitude times the rate estimated above for the neurula cell nucleus in order to accumulate 6 ng in the 10-hour maturation period; this is highly unlikely. Smith (1975) and Webb et al. (1975) pointed out that during much of the maturation period the chromosomes are in a condensed metaphase state. Such chromosomes are unlikely to be actively synthesizing RNA. A more probable conclusion (Smith, 1975) is that the heterogeneous ^{32}P-RNA accumulating during the maturation period consists mainly of mitochondrial transcripts.

APPARENT CHANGE IN THE RELATIVE RATE OF NUCLEAR RNA SYNTHESIS AT BLASTULA STAGE

Among the interesting aspects of the patterns of heterogeneous RNA synthesis in amphibian embryos is the sudden increase in transcriptional activity which takes place at the late blastula stage. This feature seems to be peculiar to amphibians, or at least is not widely observed in other groups. Figure 5.3b shows that at blastula stage the quantity of newly synthesized heterogeneous RNA per cell increases sharply. Such increases in RNA synthesis from preloaded precursor pools were also observed by earlier workers (e.g., Grant, 1958). Bachvarova and Davidson (1966) bisected embryos to circumvent the permeability barrier, and showed that the amount of incorporation of ^{3}H-uridine into high molecular weight RNA during a 1-hour labeling period increases sharply after stage 8 (early blastula). The newly synthesized RNA is of the typical heterogeneous nuclear type. At least a twentyfold relative increase in the amount of labeled RNA per nucleus can be demonstrated as the embryos progress from stage 8 to stage $8\frac{1}{2}$, i.e., within a period of about 1 hour. This is seen clearly in radioautographs, as well as in experiments in which the RNA is extracted and the radioactivity incorporated is measured. In Fig. 5.2

(Bachvarova and Davidson, 1966) radioautographs are displayed which compare nuclear RNA synthesis in cells of various regions of the dorsal half of the embryo before and after stage 8. On the left (Fig. 5.2a and c) are endodermal and equatorial cells from stage 7 blastulae, while on the right (Fig. 5.2b and d) the same cell types labeled in the same way are shown at stages $8\frac{1}{2}$ and 8, respectively. The dorsal half-embryos used in these experiments take up isotopes into their acid-soluble pools as rapidly as do totally dissociated single embryo cells throughout the period studied, and no significant changes in the radioactivity of the total and soluble pools which could explain the increased incorporation into RNA take place over this period. However, no direct precursor pool specific activity measurements were made in this study, and only a relative estimate of the magnitude of the increase in absolute synthesis rate for heterogeneous nuclear RNA can be derived.

Bachvarova *et al.* (1966) also noticed that tRNA synthesis is activated at the same stage of development. Sudden increases in absolute rates of synthesis of tRNA were measured by Brown and Littna (1966b), Gurdon (1967), and Woodland and Gurdon (1968). As before the data of Brown and Littna were obtained by using the ^{32}P-orthophosphate preloading procedure, while in the experiments of Gurdon labeled nucleosides were injected directly into the embryos. All of these studies agree that the quantity of newly synthesized tRNA per cell increases significantly between blastula and midgastrula stages. Since tRNA is almost certainly synthesized from the same intranuclear precursor pool as is heterogeneous nuclear RNA, this finding reinforces the evidence for large relative increases in the per nucleus synthesis rates of heterogeneous RNA species.

RIBOSOMAL RNA SYNTHESIS IN *XENOPUS* EMBRYOS

Figure 5.1 and 5.3 indicate that clearly distinguishable synthesis of ribosomal ^{32}P-RNA does not occur until gastrula stage, after which the content of ^{32}P-ribosomal RNA rises continuously (Brown and Littna, 1964, 1966a). The amount of ribosomal RNA synthesized per cell increases throughout the period in which heterogeneous RNA content remains constant (Fig. 5.3b). This general pattern was also reported by Gurdon (1967), Woodland and Gurdon (1968), and Knowland (1970). From Fig. 5.3a the total amount of ribosomal RNA synthesized during early development can be seen to amount to only a small fraction of the maternal ribosomal RNA. At stage 26 about 100 ng of labeled ribosomal RNA has accumulated. This, however, is not more than 3% of the total (maternal) ribosomal RNA complement (Table 4.2). Ribosomal RNA turns over only very slowly in *Xenopus* embryos, displaying a half-life of about 3.5 days (Chase and Dawid, 1972). On a per nucleus basis the rate of ribosomal RNA synthesis suggested by these data is relatively low, even taking the turnover into account. Thus the small amount of ribosomal ^{32}P-RNA present at stage 26 is the product of 10^4 to 10^5 cells. At stage 26 each cell contains about 0.85 pg of ribosomal ^{32}P-RNA, and this has accumulated at a nearly constant rate per cell for 20 hours (Fig. 5.3b). Use of equation (5.1) shows that the ribosomal RNA synthesis rate is only about 7.3×10^{-4} pg min^{-1} per nucleus during this period, or less than 200 molecules of ribosomal RNA min^{-1} per (diploid) cell! This is less than 1% of the rate of ribosomal RNA synthesis which could be supported by 940 genes per diploid genome (Brown and Weber, 1968), assuming a transcription rate of 15 nucleotides sec^{-1}. This rate was measured for *Xenopus* oocytes (D. M. Anderson and L. D. Smith, personal communication). The question now arises as to whether ribosomal RNA synthesis at the low rate of 7.3×10^{-4} pg min^{-1} per cell could have been detected in earlier embryos, with only 10^1 to 10^3 nuclei, in any of the experiments reviewed here. The significance of this issue is that failure to detect ribosomal RNA synthesis has been used as an argument that ribosomal genes are regulated during embryogenesis. That is, it has been supposed that these genes are not transcribed in amphibian embryos prior to gastrulation, whereupon they are activated.

   A simple calculation from the data of Brown and Littna (1966a) shows that if ribosomal RNA were synthesized at 7.3×10^{-4} pg min^{-1} per cell in an early embryo, say the 1000-cell stage, it could never have been observed at the precursor specific activities reported. The same argument can be applied to earlier studies (e.g., Brown and Littna, 1964; Gurdon and Brown, 1965). In other investigations Brown and Gurdon (1966), Gurdon (1967), Woodland and Gurdon (1968), and Knowland (1970) compared the amount of radioactivity in ribosomal RNA's to that in other nucleic acid species, including DNA. No incorporation into ribosomal RNA prior to gastrulation could be detected. Again, however, this is what would be expected even were the rate of ribosomal RNA synthesis in each nucleus the same before and after gastrulation. Thus in comparing radioactivity incorporated in ribosomal RNA to that incorporated in DNA, it is necessary to take into account that during blastulation the rate of DNA synthesis (cf. Chapter 4) is several orders of magnitude higher than is the rate of ribosomal RNA synthesis after gastrulation. In this situation ribosomal RNA synthesis could not be detected. These arguments, of course, cannot show that ribosomal RNA synthesis is *not* regulated, since the same results would have been obtained if it is. Other experiments relevant to this question have been carried out on disaggregated amphibian embryo cells. This line of investigation was begun by

Yamana and Shiokawa (1966) and Shiokawa and Yamana (1967a), who found that embryo cells dissociated by the procedures of Morrill and Kostellow (1965) take up isotopes and synthesize RNA for many hours. Shiokawa and Yamana reported that cells taken from embryos at stages before ribosomal RNA synthesis is supposed to be activated, e.g., blastula, fail to synthesize ribosomal RNA *in vitro*. Cells from later embryos, e.g., neurulae, synthesize easily detectable amounts of ribosomal RNA. Furthermore, medium obtained from blastula cultures inhibited neurula cells from synthesizing ribosomal RNA. Synthesis of tRNA and DNA were reported to be unaffected by this medium (Yamana and Shiokawa, 1966; Shiokawa and Yamana, 1967b; Wada *et al.*, 1968). Ribosomal RNA synthesis in the inhibited cultures is not abolished, but rather is diminished severalfold relative to a low molecular weight RNA (probably tRNA). Subsequent attempts by Landesman and Gross (1968) and by Hill and McConkey (1972) to repeat this set of observations failed. Landesman and Gross (1969) showed that dissociated embryo cells of postgastrular stages synthesize a rapidly turning over 40 S ribosomal precursor, but could not reproduce the observation of an early ribosomal RNA synthesis inhibitor. Unfortunately, in the dissociated cell experiments of Shiokawa and others in which earlier and later embryonic stages were compared, the number of cells present in the blastular cultures was generally 1–2 orders of magnitude lower than that in the neurular cell cultures, since the number of embryos used for each stage remained about the same. Thus the lack of evident ribosomal RNA synthesis in the blastular cell cultures is not simple to interpret. An interesting postscript to this story has been provided by Laskey *et al.* (1973), who were able to confirm the earlier claims of Shiokawa and Yamana. They reported that a perchloric acid extract from blastulae causes a decrease in the ratio of ribosomal to transfer RNA synthesis in dissociated neurula cells by a factor of 3–4. Little evidence as to the real specificity of this inhibitory factor exists; for example its effect on heterogeneous nuclear RNA synthesis is unknown. In any case, the inhibitor in question is clearly not a particularly potent one, and so far it must be said that this line of experimentation has not provided critical evidence to support the concept that ribosomal RNA transcription is regulated in early embryogenesis.

Ribosomal RNA synthesis becomes detectable in different regions of the embryo at slightly different stages. Woodland and Gurdon (1968) showed that in the gastrula–neurula period (stages 12–18) ribosomal RNA synthesis cannot be detected in endoderm cells, while it is easy to detect in other cells when the embryos are labeled by direct injection of precursors. Ribosomal RNA was identified in this work by its elution position from a MAK (methylated serum albumen–Kieselguhr) column after correcting for radioactive heterogeneous RNA eluting at the same position. *o nu* control embryos were used, since these synthesize no ribosomal RNA while carrying out a normal level of heterogeneous RNA synthesis (Brown and Gurdon, 1964, 1966; Gurdon and Brown, 1965; Woodland and Gurdon, 1968; Knowland and Miller, 1970; Miller and Knowland, 1970). The endoderm contains only 3–4% of the total embryo nuclei at these stages, but Woodland and Gurdon were able to detect DNA and tRNA synthesis in endoderm cells. Relative to the radioactivity incorporated in these nucleic acid species, the incorporation of precursor into ribosomal RNA was only about one-tenth as active as in the rest of the embryo. This difference disappears by the tailbud stage. Subsequently, Miller (1972) confirmed that endoderm cells in midgastrulae (stage 11) synthesize very little ribosomal RNA relative to ectodermal cells. Synthesis of ribosomal precursor could also be demonstrated in ectoderm cells by using a [methyl-^{3}H]methionine label, while any such synthesis in endoderm cells occurs at too low a rate to be detected. In none of these studies are direct measurements of ribosomal RNA synthesis rate available. The data are expressed relative to incorporation into tRNA and DNA, and quantitative conclusions are difficult to draw, since differences also exist in the rates of DNA and tRNA synthesis. Thus tRNA synthesis seems to be higher in gastrular endoderm cells, relative to DNA synthesis (Woodland and Gurdon, 1968; Miller, 1972), while DNA synthesis rate is more or less proportional to DNA content (Woodland and Gurdon, 1968). A conservative view is that a transient severalfold difference in the per nucleus ribosomal RNA synthesis rate probably exists between endoderm and other cell types during gastrulation. This difference has yet to be quantitated, and the degree of its specificity remains unknown.

In pregastrular amphibian embryos typical nucleoli are absent, and instead intranuclear aggregations of fibrous electron-dense material are observed (e.g., Hay and Gurdon, 1967). Radioautograph experiments show that RNA synthesis occurs in the vicinity of these bodies as well as in non-nucleolar regions (Karasaki, 1965). A correlation exists between the stage at which definitive nucleoli appear, midgastrula, and the biochemical observations which place the "onset" of ribosomal RNA synthesis at this stage, e.g., Fig. 5.1 (reviewed by Brown, 1966). Cytological observations also correlate well with the regional studies reviewed above. Thus Woodland and Gurdon (1968) report that definitive nucleoli appear in only 20% of endoderm cells at the neurula stage, but in 80% of the nuclei of other cell types. It is somewhat dangerous, however, to equate the appearance of definitive nucleoli with ribosomal RNA synthesis. In very rapidly dividing cells nucleolar elements may fail to coalesce into the definitive nucleolar structures typically observed. This may be due simply to absence of a sufficiently long interphase period. Emerson and Humphreys (1971) showed that the multiple "immature" nucleoli of cleaving sea urchin embryos, which are similar to those of pregastrular amphibian embryos, could be induced to coalesce and form normal looking nucleoli by interfering with DNA synthesis with fluorodeoxyuridine.

To summarize, we conclude that transcription of ribosomal RNA genes may be regulated in amphibian embryos as is conventionally believed, but a convincing demonstration of such regulation is difficult to make and is still lacking. The best evidence so far is the apparent disparity between ribosomal RNA synthesis in endoderm as opposed to ectoderm cells at the gastrula stage. However, the main point that emerges is a different one. This is that the rate of ribosomal RNA synthesis in embryos even at the tailbud stage is apparently very low, less than about 200 molecules min^{-1} per cell, for the whole diploid set of 940 ribosomal genes. Ribosomal RNA synthesized by the embryo accounts for only about 2.5% of the egg RNA by tailbud stage (1.2×10^5 cells), and its synthesis occurs at less than 1% of the probable maximum rate. Most ribosomal genes are therefore inactive most of the time throughout early embryogenesis, both before and after gastrulation.

Ribosomal protein synthesis is apparently coordinated with the synthesis of ribosomal RNA. Hallberg and Brown (1969) studied synthesis of these proteins in o nu tadpoles which lack the ribosomal genes and produce no ribosomal RNA. Ribosomal protein synthesis occurred at less than 4% of the control rates in these embryos. Nor are ribosomal proteins produced in early cleavage. This finding is interesting because, while the number of nuclei is relatively low in early cleavage embryos, the protein synthesis apparatus is almost as active as at later stages (see Chapter 4). Ribosomal proteins may thus be encoded by new messenger RNA's and not by maternal message. In contrast, the synthesis of 5 S ribosomal RNA is not coordinated with that of the large ribosomal RNA's (Miller, 1973). The 5 S RNA genes are located near the telomeres of most or all chromosomes in Xenopus (Pardue et al., 1973) rather than at the nucleolar organizer regions. Miller (·1973) showed that these genes are expressed at normal levels in the completely anucleolate o nu homozygote as well as in partially anucleolate mutant embryos.

MITOCHONDRIAL RNA SYNTHESIS IN XENOPUS EMBRYOS

The mature oocytes of Xenopus and Rana pipiens contain 250–300 times more mitochondrial DNA than the chromosomal DNA content of the germinal vesicle, which is around 12.5 pg, the 4C value (Dawid, 1965, 1972). The amount of mitochondrial DNA is 3.1–3.8 ng in Xenopus oocytes and 4.5 ng in *Rana* oocytes (Dawid, 1972; Chase and Dawid, 1972), and this is sufficient to provide about 1.8×10^8 mitochondrial genomes. The mass of mitochondrial DNA per embryo does not change throughout early development, and mitochondrial DNA synthesis is not detected until tadpole stages (Chase, 1970; Chase and Dawid, 1972). Approximately 1% of the total RNA of the oocyte is mitochondrial RNA. Synthesis of mitochondrial RNA's occurs at only a low rate in pregastrular embryos (Chase and Dawid, 1972; Young and Zimmerman, 1973). At stages 3–10 (cleavage and blastula) the rate of mitochondrial ribosomal RNA synthesis is less than 0.5 pg min^{-1} per embryo, according to Chase and Dawid (1972). Webb *et al.* (1975) found a similar or slightly lower rate of mitochondrial ribosomal RNA synthesis in mature ovarian oocytes. It follows that the events associated with ovulation and fertilization do not greatly stimulate mitochondrial RNA synthesis. During gastrulation the synthesis of mitochondrial ribosomal RNA becomes more active, and in the tailbud stage this synthesis attains a rate of about 3 pg min^{-1} per embryo (Chase and Dawid, 1972). About 15 species of tRNA as well as some other heterogeneous RNA species are synthesized. By this time, however, there are two orders of magnitude more nuclear DNA than mitochondrial DNA in the embryo. Furthermore, the rate of stable RNA synthesis in mitochondria is low relative to the amount of mitochondrial DNA present (Dawid, 1972). Comparison with the amounts of RNA synthesized by the embryo indicates that at no time after early cleavage does mitochondrial transcription contribute in a quantitatively significant way to overall RNA synthesis. In this, as in other respects, transcriptional patterns in the amphibian embryo differ strikingly from those found in other kinds of embryo. An example is the sea urchin embryo, our next subject of consideration.

Transcriptional Rates and Patterns in Sea Urchin Embryos

THE HETEROGENEOUS NUCLEAR RNA OF SEA URCHIN EMBRYOS

A far greater number of measurements have been carried out on sea urchin embryos than on any other form, and a fairly complete picture of overall synthesis patterns can be deduced. Throughout much of early development the main products of transcription are the heterogeneous nuclear RNA's. Nuclear RNA synthesis occurs even in mature oocytes prior to fertilization (Levner, 1974). Many investigators noted easily labeled heterogeneously sized RNA's in sea urchin embryos beginning with the earliest cleavage stages. This RNA was found to be more or less DNA-like in base composition and to be' predominantly nuclear in location (see, e.g., Wilt, 1963, 1964; Comb *et al.*, 1965; Gross *et al.*, 1965a; Aronson and Wilt, 1969; Wilt *et al.*, 1969; Emerson and Humphreys, 1970; Aronson *et al.*, 1972; Hogan and Gross, 1972). The nuclear location probably depends on binding to intranuclear constituents, since even in demembranated nuclei greater than 90% of this class of RNA remains associated with ribonucleoprotein particles (Aronson and Wilt, 1969; Aronson *et al.*, 1972; Wilt *et al.*, 1973). The main features of sea urchin embryo heterogeneous nuclear RNA are its large molecular size, its high rates of synthesis and turnover, the presence of poly(A) tracts of various lengths, and its interspersed sequence organization and extremely high sequence complexity. Consideration of the latter two subjects is deferred to Chapter 6, and we now review evidence relating to the synthesis rates and general properties of this class of embryonic RNA's.

Like other animal cell heterogeneous nuclear RNA's, those of sea urchin embryos sediment at very high velocities in nondenaturing sucrose gradients. In many of the references cited above aqueous sedimentation velocities are reported for these RNA's in the range 50 S to 100 S (see also Brandhorst and Humphreys, 1972; Peltz, 1973; Sconzo *et al.*, 1974; Dubroff and Nemer, 1975). These RNA's also migrate as a class of very large molecules in nondenaturing gel electrophoresis systems (e.g., Peltz, 1973; Giudice *et al.*, 1974). However, several careful size studies have been carried out under denaturing conditions or after formylation, when RNA molecules behave essentially as random coils (Boetdker, 1968). These studies show that sea urchin embryo nuclear RNA's are of much more moderate dimensions than previously thought. Peltz (1973) found that the mean size of formylated heterogeneous nuclear RNA in sea urchin embryos is about 45 S, though some components sediment more rapidly. In nonaqueous denaturing conditions both intrastrand base pairing and intermolecular aggregation are suppressed, and because of the greater effect of charge repulsion in these conditions, RNA molecules behave as even more extended forms than in aqueous formaldehyde media (see, e.g., Ricard and Salser, 1975). The apparent size of sea urchin nuclear RNA molecules measured in denaturing velocity sedimentation gradients is slightly less than in aqueous formaldehyde gradients. Kung (1974) and Dubroff and Nemer (1975) found that in 99% dimethyl sulfoxide or 70% formamide gradients most of the nuclear RNA sediments as a relatively narrow peak, the mode value of which is about 8000–9000 nucleotides by reference to ribosomal RNA markers. Similar results were obtained in 4% agarose acrylamide gels run in 99% formamide (Kung, 1974). In contrast to what is reported for mammalian cells, the nuclear RNA of sea urchin embryos appear to be no more than 2–4 times as large as the messenger RNA. A broad range of messenger RNA sizes is present, the number average of which, as measured under either denaturing or aqueous conditions, is at least 2000 nucleotides (Kung, 1974; Nemer, 1975; Nemer et al., 1974, 1975; see also references in Chapter 4). Of this, up to 200 nucleotides could be poly(A) tracts. It is hard to exclude the possibility that the nuclear RNA lengths observed in these studies are the result of strand scission during extraction, but reconstruction experiments in which trace quantities of RNA of known size are added to the extraction mixtures reveal no detectable degradation (Kung, 1974). Furthermore, some very large cytoplasmic RNA's, which may be messengers, have been noted by several investigators. These approach the size of the nuclear RNA's (Brandhorst and Humphreys, 1972; Kung, 1974; Guidice et al., 1972, 1974; Rinaldi et al., 1974; Sconzo et al., 1974). To prove the genuinely cytoplasmic location of heterogeneous RNA's of the size of sea urchin nuclear RNA Giudice et al. (1974) manually dissected sea urchin eggs into nucleate and non-nucleate halves after labeling. The same very large RNA's could be found in the purely cytoplasmic non-nucleate preparation as in whole embryo cytoplasm. This rules out the possibility that such large cytoplasmic RNA's originate as an artifact of nuclear leakage during cell fractionation. It should be noted that the actual sizes of the RNA's reported by Giudice and his associates are probably too large, according to the measurements of Nemer et al., Kung, and others cited above. However, the existence of some cytoplasmic RNA's similar in length to nuclear RNA's is consistent with the overall low ratio of mean nuclear RNA to mean messenger RNA length. Thus the paradox posed by the greater than tenfold ratio of nuclear to messenger RNA lengths reported for some mammalian systems may not be applicable to transcriptional processes in the sea urchin. On the other hand, sequences of the same enormous length as are transcribed into giant nuclear RNA's in mammalian cells could be transcribed from single initiation sites in sea urchin cells, except that the transcripts would undergo strand scission even while chain elongation is proceeding. This issue cannot yet be resolved.

Several classes of heterogeneous sea urchin nuclear RNA have been distinguished by Dubroff and Nemer (1975) on the basis of poly(A) content. About 70% of labeled nuclear RNA at the blastula stage contains no poly(A) tracts of length sufficient to promote binding either to oligo(dT) cellulose or to poly(U) filters. The poly(U) filters, however, trap two classes of nuclear RNA. About 15% of the nuclear RNA contains internal tracts consisting of about 25 adenylic acid residues. The remaining 15% of the nuclear RNA contains 3'-poly(A) tracts averaging 175 nucleotides in length, similar to those found on polysomal messenger RNA. The latter class of nuclear RNA's is also of slightly smaller size, with a modal sedimentation velocity in denaturing gradients of 31 S compared to 36 S and 37 S for the nonadenylated and short poly(A) nuclear RNA fractions. The interrelations of these nuclear RNA classes, their fate, and their function remain unknown. The presence of base-paired regions is a feature of sea urchin nuclear RNA's as of other animal cell nuclear RNA's (e.g., Stern and Friedman, 1970; Harel and Montagnier, 1971; Jelinek and Darnell, 1972; Jelinek *et al.*, 1974). These regions probably account for some of the large differences in sedimentation behavior observed when normal, and denatured or formaldehyde-treated, nuclear RNA molecules are compared. They consist in part of hairpinlike structures or "foldbacks" formed from nearly homologous sequences of reversed orientation. Kronenberg and Humphreys (1972) found such double-stranded regions in the heterogeneous nuclear RNA's of sea urchin embryos, and showed that they include about 0.2% of the RNA nucleotides.

A variety of low molecular weight nuclear RNA's is also synthesized in sea urchin embryo nuclei. These include tRNA's, 5 S RNA's, and other homogeneous species of unknown function. About nine such homogeneous species were detected by Hogan and Gross (1972), all less than a few hundred nucleotides in length. The synthesis of several apparently homogeneous species of RNA in this size range was also observed in sea urchin blastulae by Frederiksen *et al.* (1973).

SYNTHESIS AND DECAY KINETICS OF HETEROGENEOUS NUCLEAR RNA IN SEA URCHIN EMBRYOS

We now consider the rates at which high molecular weight heterogeneous nuclear RNA is synthesized in the embryo blastomeres. Several studies of the behavior of the nucleotide pools in sea urchin embryos have been carried out (Kijima and Wilt, 1969; Aronson and Wilt, 1969; Wilt *et al.*, 1969; Wilt, 1970; Emerson and Humphreys, 1970; Brandhorst and Humphreys, 1971, 1972; Wu and Wilt, 1974; Galau *et al.*, 1976a). At relatively low external nucleoside concentrations the specific activity of some of these pools reaches a maximum value within $\frac{1}{2}$ hour, and in some conditions much less, and it then remains essentially constant for a period of many hours. This is shown for the ATP (Brandhorst and Humphreys, 1971) and UTP (Wilt *et al.*, 1969; Aronson and Wilt, 1969) pools in Fig. 5.4a–c. This renders the treatment given in equations (5.1) to (5.4) easy to apply. The GTP pool (Fig. 5.4d) behaves differently in blastulae and gastrulae at similar exogenous nucleoside concentrations, i.e., around $10^{-7} M$ (Kijima and Wilt, 1969; Galau *et al.*, 1976a). The specific activity of the GTP pool decreases rapidly to a steady state level, after reaching an

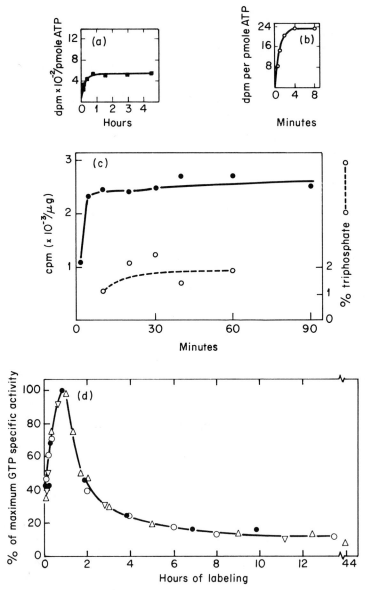

Fig. 5.4. Kinetic behavior of nucleoside triphosphate pools in sea urchin embryos. (a) Specific activity of ATP pool, in dpm × 10^{-2}/pmole ATP in a 2.5% suspension of blastulae labeled continuously with ^{3}H-adenosine (1×10^{-7} M; 28 Ci/mmole). (b) Rapid approach to saturation of ATP pool specific activity (in dpm/pmole ATP) in an experiment in which the embryos were labeled as above for 3 minutes, at 4×10^{-7} M in 10% suspension. They were then diluted. (a) and (b) from B. P. Brandhorst and T. Humphreys (1971). *Biochemistry* 10,

early maximum, and as will be seen from the following discussion, the kinetics of this decrease probably reflect the synthesis of cellular RNA's. The GTP pool would be expected to behave differently at earlier stages of development, since there are then many fewer nuclei, and therefore less RNA synthesis occurring per embryo, while the overall pool size remains about the same. Thus Kijima and Wilt (1969) observed that the specific activity of the acid-soluble pool in the presence of about 4×10^{-7} M exogenous ^{3}H-guanosine declines only slightly from its initial maximum value in cleavage-stage embryos, while at later stages the pool behaves similarly to that illustrated in Figure 5.4d. The GTP pool evidently turns over more rapidly than the ATP pool. If the exogenous nucleoside concentration is greatly increased, e.g., to 10^{-3} M, the pools expand considerably (Grainger and Wilt, 1976). More exogenous nucleoside can then be incorporated in the pool, thus facilitating labeling with radioactive or density isotopes. Expansions on the order of twofold were measured by Grainger and Wilt (1976) for the ATP pool at these high concentrations of exogenous precursor.

By measuring the kinetics of approach to steady state specific activity, absolute synthesis and turnover rates have been obtained for the heterogeneous RNA of stages between fertilization and pluteus. Incorporation and decay kinetics for the total rapidly labeled RNA of sea urchin embryos are illustrated in Fig. 5.5. This figure shows a density label experiment of Grainger and Wilt (1976). Here pool expansion was used to achieve significant incorporation of ^{15}N- and ^{13}C-labeled nucleosides into newly synthesized total RNA, and the newly synthesized RNA was separated from previously extant cellular RNA's by isopycnic centrifugation. The precursor mixture also included radioactively labeled nucleosides. The absolute

877. Copyright by the American Chemical Society. (c) Specific activity (left ordinate) of nucleotides in a 1% suspension of blastulae labeled with ^{3}H-uridine (4×10^{-8} M; 26.6 Ci/ mmole). Specific activity values are given as cpm $\times$ $10^{-3}/\mu$g of total acid-soluble nucleosides and nucleotides. About 2% of the total labeled nucleotides are UTP (right ordinate). From F. H. Wilt, A. I. Aronson, and J. Wartiovaara (1969). *In* "Problems in Biology: RNA in Development" (E. W. Hanly, ed.), pp. 331. Univer. Utah Press, Salt Lake City; and A. I. Aronson and F. H. Wilt (1969). *Proc. Natl. Acad. Sci. U.S.A.* **62**, 186. (d) Specific activity of GTP in late blastulae and early gastrulae labeled at 3×10^4 ml^{-1}, approximately 1.5%, with ^{3}H-guanosine (5×10^{-7} M; 0.5–2.6 Ci/mmole). Several different experiments are indicated by the various symbols, and for ease of presentation these are normalized to the maximum specific activity value, i.e., at 50 minutes. Data are then expressed as percent of maximum specific activity (ordinate). The actual specific activity obtained was directly correlated with the external specific activity. For example, when this was 2 Ci/mmole, the 50 minute (peak) GTP specific activity was about 2×10^{11} dpm/mmole GTP, or about 4% of the specific activity of the medium. From Galau, G. A., Lipson, E. D., Britten, R. J., and Davidson, E. H. (1976a). *Cell*, in press.

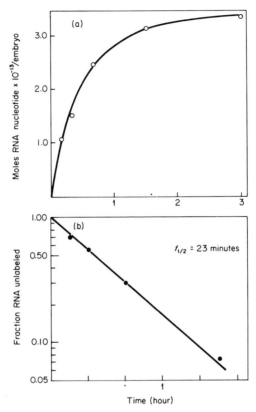

Fig. 5.5. Incorporation and decay kinetics of total heterogeneous RNA in sea urchin mesenchyme blastulae. Most of the label is present in nuclear RNA. The newly synthesized RNA was labeled with [15]N- and ^{13}C-containing nucleosides. The medium also contained all four ^{3}H-nucleosides. Newly synthesized RNA was separated from other cellular RNA's by isopycnic centrifugation in cesium formate gradients. (a) Molar accumulation of newly synthesized RNA in S. *purpuratus* embryos over the first 3 hours of labeling. The quantities of nucleotides incorporated were calculated from the measured pool specific activities. The curve is of the form given by equation (5.3). At the plateau or steady state level approached by the initial phase of the incorporation curve, the embryo contains about 0.3 pmole of heterogeneous RNA, or approximately 100 pg. Since there are about 600 cells, this represents about 0.17 pg of heterogeneous nuclear RNA per cell (see independent kinetic calculation referred to in text). (b) Decay kinetics of total heterogeneous RNA, based on the data shown in (a). The slope of this line gives the first-order decay rate constant k_d which is here evaluated at 0.03 min^{-1}. The curve is of the form specified in equation (5.5), and the half-time ($t_{1/2}$) of 23 minutes is related to k_d as specified in equation (5.2). From Grainger, R. M. and Wilt, F. H. (1976). *J. Mol. Biol.* **104**, 589.

quantity of the isolated, newly synthesized RNA could thus be calculated from the amount of label incorporated and the specific activity of the measured precursor pool at each time point. The function shown in Figure 5.5a is the molar accumulation curve described by equation (5.3). By the use of equation (5.3) an instantaneous rate of total heterogeneous RNA synthesis can be calculated of about 3 pg min^{-1} per embryo. A number of essentially similar molar accumulation analyses have been carried out on sea urchin embryos at different stages by various workers, though Grainger and Wilt (1976) are the only ones to have utilized density labeled precursors to isolate the newly synthesized RNA. Most of the available measurements are listed in Table 5.2. The compilation of data in this table demonstrates that the various measurements are satisfactorily coherent for each stage, irrespective of variations in procedure and the specific precursors used.

An unexpected conclusion to be drawn from the total heterogeneous RNA synthesis rates shown in Table 5.2 is that the rate per nucleus *declines* slightly during early development. This was first pointed out by Kijima and Wilt (1969) and has been verified by other workers, particularly Brandhorst and Humphreys (1971) (see references in Table 5.2). The amount of this decline is roughly threefold between cleavage and the postgastrular stages.

Most of the total labeled heterogeneous RNA at short labeling times is nuclear in early embryos [note that the rate of total RNA synthesis measured by Grainger and Wilt (1976) is about the same as that measured for isolated nuclear RNA's by Brandhorst and Humphreys (1972)]. Therefore the turnover rate observed in the experiments shown in Fig. 5.5 provides a good approximation of the decay kinetics for nuclear RNA. Figure 5.5b illustrates a decay analysis of the data in Fig. 5.5a. The curve plotted here is given by the expression

$$\ln \frac{A_{max} - A}{A_{max}} = -k_d t \qquad (5.5)$$

where A is the quantity of newly incorporated nucleotides in the RNA at time t, relative to the maximum amount of incorporated nucleotides, A_{max}. A_{max} could be the steady state amount of newly incorporated nucleotides or the amount at the beginning of a chase. Equation (5.5) can be derived from equation (5.3) or from simple first-order decay kinetics. The expression $[(A_{max} - A)/A_{max}]$ is the fraction of heterogeneous RNA remaining unlabeled, and $(-k_d)$ is the slope. The value of k_d from the experiment shown is 0.03 min^{-1}. The half-life of the heterogeneous nuclear RNA in mesenchyme blastulae thus appears to be about 23 minutes [equation (5.2)]. Brandhorst and Humphreys (1971, 1972) calculated a rate of heterogeneous nuclear RNA decay which is somewhat higher than this, reporting for this class of RNA a half-life of less than 10 minutes. Their study was based on molar accumulation curves for total heterogeneous RNA using labeled adenosine as the precursor. Within the same

TABLE 5.2. Synthesis Rates for Heterogeneous RNA's in Sea Urchin Embryos

Stage (cell no.)	Species	Precursor	Rate of synthesis of total heterogeneous RNA (pg min⁻¹)	
			Per cell	Per embryo
Cleavage				
Mature oocyte (1)[a]	Strongylocentrotus purpuratus	U	1.5×10^{-5} [a]	1.5×10^{-5} [a]
32–64	Strongylocentrotus purpuratus	U	1.0×10^{-2} [b]	0.48[b]
16–32	Strongylocentrotus purpuratus	G	2.4×10^{-2} [c]	0.57[c]
40	Lytechinus pictus	A	7.9×10^{-3} [d]	0.31[d]
Early blastula				
350	Strongylocentrotus purpuratus	U	6.7×10^{-3} [b]	2.3[b]
140	Lytechinus pictus	A	1.5×10^{-2} [e]	2.2[e]
300	Lytechinus pictus	A	8.6×10^{-3} [d]	2.6[d]
Mesenchyme blastula				
600	Strongylocentrotus purpuratus	U	4.5×10^{-3} [b]	2.7[b]
470	Lytechinus pictus	A	9.7×10^{-3} [e]	4.6[e]
Not stated	Strongylocentrotus purpuratus	A,U,G,C	5.1×10^{-3} [f]	3.1[f]
Gastrula				
670	Strongylocentrotus purpuratus	U	3.5×10^{-3} [b]	2.3[b]
600	Lytechinus pictus	A	2.2×10^{-3} [d]	1.3[d]
Prism				
800	Strongylocentrotus purpuratus	U	3.9×10^{-3} [b]	2.8[b]
Pluteus				
1340	Lytechinus pictus	A	3.1×10^{-3} [e]	4.1[e]

[a] Levner (1974). In the sea urchin the maturation divisions are completed prior to fertilization and nuclear RNA synthesis in the spawned oocyte occurs in the haploid pronucleus.

[b] Roeder and Rutter (1970). Where more than one observation for a given stage is reported, the data are averaged. Cell number is taken from Hinegardner (1967) (see Fig. 4.5). It is assumed that 25% of the RNA nucleotides are uridylic acid.

[c] Calculated from data of Wilt (1970).

[d] Calculated from data listed by Wu and Wilt (1974).

[e] Data from Brandhorst and Humphreys (1971, 1972).

[f] Calculated from Grainger and Wilt (1976) who give a rate of 9.3×10^{-15} moles nucleotide min⁻¹ per embryo. The blastulae are taken to contain 600 cells. Data are shown in Fig. 5.5a, and details are given in the caption to this figure and in the text.

labeling period their data are not very different from those shown in Fig. 5.5a, but on the basis of several longer time points, at 4–6 hours after the start of labeling, they resolved the overall curves into two kinetic components. The component with the more rapid turnover was later associated with nuclear RNA, and the component with the slower turnover with polysomal message in cell fractionation experiments (Brandhorst and Humphreys, 1972). However, if the lower decay rates measured subsequently in direct studies of polysomal message turnover (see below) are imposed on their data, the faster component is found to turn over with kinetics similar to those reported by Grainger and Wilt. The possibility remains that the *Lytechinus* embryos studied by Brandhorst and Humphreys (1971, 1972) differ in their nuclear RNA decay rates from the *Strongylocentrotus* embryos used for the experiments of Fig. 5.5, or that the observed decay rates were affected by some experimental variable such as how the embryos were handled, or the use of adenosine as a precursor.

For the purposes of the following calculations we provisionally accept a $t_{1/2}$ of 20 minutes for the rapidly turning over heterogeneous nuclear RNA of sea urchin embryos. Very similar rates have been measured for the heterogeneous nuclear RNA of other animal cells, e.g., mouse L cells (Brandhorst and McConkey, 1974; see review in Davidson and Britten, 1973). We assume that only a minor proportion of the newly synthesized total heterogeneous RNA in each time interval is messenger RNA, an assumption which is supported by calculations given below. Thus by inserting in equation (5.4) the synthesis rates shown in Table 5.2 and the nuclear RNA decay rates just discussed, a kinetic estimate of the average steady state content of heterogeneous nuclear RNA can be obtained. This turns out to be about 0.17 pg per nucleus for mesenchyme blastulae, using the data of Grainger and Wilt [i.e., $k_s = 5 \times 10^{-1}$ pg min^{-1} per cell (Table 5.2) and $k_d = 0.03$ min^{-1}] or one tenth the mass of DNA per nucleus. The same value can also be obtained graphically from the steady state (plateau) quantity of newly synthesized RNA in Fig. 5.5a. It is interesting that the data reviewed earlier for *Xenopus* embryos also suggests a steady state nuclear RNA content close to 10% of that of the nuclear RNA mass.

SYNTHESIS AND DECAY KINETICS FOR A CASE OF CHANGING PRECURSOR SPECIFIC ACTIVITY: CALCULATION OF MESSENGER RNA SYNTHESIS RATES

The amount of messenger RNA in sea urchin embryos can be estimated independently of kinetic measurements from the fraction of ribosomes present in polyribosomes. This calculation is given for the blastula–gastrula stage in Table 5.3. The most likely value for *Strongylocentrotus*

TABLE 5.3. Messenger RNA Synthesis and Decay Rates in Embryos of *Strongylocentrotus purpuratus* during the Blastula–Gastrula Period

---

Polysomal messenger RNA per embryo (pg):
   59–86[a]
   70[b]

Half-life: $t_{1/2}$ (min) k_d (min^{-1})[e]
                345[c] 2.0×10^{-3}
                320[d] 2.16×10^{-3}

Estimated rates of polysomal messenger RNA synthesis
   (pg min^{-1}):
   1.2×10^{-1} to 1.8×10^{-1} [f]
   1.35×10^{-1} [g]

Fraction of heterogeneous nuclear RNA synthesis as synthesis of
   messenger RNA[h]:
   0.044–0.067

---

[a] The fraction of polysomal RNA in messenger RNA can be calculated independently of kinetic measurements from the data of Nemer *et al.* (1974). These authors report that 1–1.5% of the total blastula cytoplasmic RNA is polyadenylated, but this is only 45% of the total messenger RNA. Thus, 2.2–3.3% of the cytoplasmic RNA, or 3.7–5.5% of the polysomal RNA, is messenger RNA (assuming 60% of the cytoplasmic RNA is polysomal; see below). The calculation assumes 3.3 ng RNA per egg (Whiteley, 1949) and 60% of the ribosomes as polysomes (Galau *et al.*, 1974).

[b] From the steady state value approached by the molar accumulation curve in Fig. 5.6b, assuming a polysomal RNA content calculated as in footnote *a*. See legend to Fig. 5.6.

[c] Galau *et al.* (1976a). This measurement is shown in Fig. 5.6c.

[d] Nemer *et al.* (1975). This measurement was obtained by monitoring the decline in polysomal messenger RNA radioactivity during a cold nucleoside chase. ^{3}H-Uridine was the label.

[e] Calculated by application of equation (5.2) from $t_{1/2}$ values.

[f] Calculated by use of equation (5.1) for the steady state situation on the assumption that all of the messenger RNA on the polysomes is newly synthesized. If some of it is maternal, the synthesis rate would be lower. Thus $k_s = (2.1 \times 10^{-3}$ min^{-1})(59 pg, as calculated in footnote *a*) $= 1.2 \times 10^{-1}$ to $(2.1 \times 10^{-3}$ min^{-1})(87 pg) $= 1.8 \times 10^{-1}$ pg min^{-1} per embryo.

[g] Calculated as described in text and in Fig. 5.6 from incorporation of radioactive precursor into messenger RNA, using the precursor specific activity data shown in Fig. 5.4d [i.e., by use of equation (5.7)]. This value is judged the most secure.

[h] The rate of total heterogeneous RNA synthesis used is an average value taken from Table 5.2 for *S. purpuratus* at mesenchyme blastula and gastrula stages, i.e., 2.7 pg min^{-1} per embryo.

purpuratus blastulae and gastrulae is thus estimated at 59–86 pg. The polysome content is somewhat lower at earlier stages. For instance, Infante and Nemer (1967) measured about 30% of ribosomes in polysomes at midcleavage, which would suggest a value for total polysomal message of around 35 pg per embryo. While calculations of message content on this simple basis could be influenced by underloading of polysomes (Nemer *et al.*, 1975) or other irregularities, errors due to such sources are not likely to exceed 50% of the listed values and certainly cannot affect the order of magnitude.

Messenger RNA turnover rates have been measured in two ways. One method, shown in Fig. 5.6 (Galau *et al.*, 1976a), is similar to that illustrated for total heterogeneous RNA in Fig. 5.5. Here the embryos are labeled with exogenous ³H-guanosine and the molar content of newly incorporated guanylic acid residues in the messenger RNA is calculated from the radioactivity of the polysomal message and the GTP pool specific activity. Since, as shown in Fig. 5.4d, the specific activity of the GTP presursor pool follows a complex function with time, the curve shown in Fig. 5.6b must be calculated in a different manner. We begin in this case with a differential equation analogous to equation (5.1), but describing the rate of change in the radioactive label in the RNA, R [rather than in the mass of RNA bearing label, C in equation (5.1)]:

$$\frac{dR}{dt} = k_s S - k_d R \tag{5.6}$$

Here S is the specific activity of the precursor pool at any given time, t, and k_s and k_d are the rates of synthesis and decay, respectively, as before. If k_s is in units of pg min^{-1}, S is in units of cpm pg^{-1}. Labeled messenger RNA does not appear in the cytoplasm until about 20 minutes (Fig. 5.6). Inserting a lag time, L, to take this fact into account, the solution to equation (5.6) is for our case

$$R(t) = k_s \int_L^t e^{-k_d(t-t')} S(t') \, dt' \tag{5.7}$$

In the case of the experiment shown in Fig. 5.6a and b, the value of S was evaluated at each time t' between L and t from the data in Fig. 5.4d. This is a generally useful procedure since it does not depend on a precursor pool specific activity which remains constant with time. k_s and k_d can then be extracted by least-squares methods, using the observed radioactivity (R) of the messenger RNA, after subtracting radioactivity present in ribosomal RNA. The curves in Fig. 5.6b were generated with equation (5.3), using the values of k_s and k_d derived in this manner. A useful kinetic treatment of the flow of radioactive precursors into macromolecules was

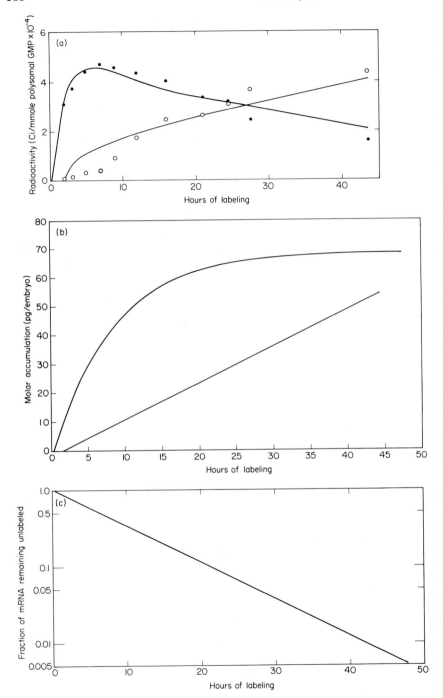

presented many years ago by Roberts *et al.* (1955) and many similar treatments are now in the literature.

Figure 5.6b shows the molar accumulation curve for messenger RNA prepared from isolated polysomes, and in Fig. 5.6c the decay curve based on the same set of data is illustrated. Here it is seen that the total labeled message turns over with a half-time of about 5.8 hours. A very similar value, 5.3 hours, was obtained by Nemer *et al.* (1975), who relied on a completely different procedure. In their experiments the embryos were exposed to labeled uridine at a concentration of about $2 \times 10^{-6}\,M$, and after 2 minutes washed with $10^{-3}\,M$ unlabeled uridine and then chased for some hours with unlabeled $10^{-4}\,M$ uridine. Incorporation into stable RNA species soon stops, showing the chase to be reasonably effective, and the decay of polysomal messenger RNA specific activity is then measured. The same 5.3-hour $t_{1/2}$ was observed by Nemer *et al.* (1975) for both poly(A)-containing and poly(A)-lacking message populations.

These data permit two estimates of the messenger RNA synthesis rate. Equation (5.1) can be used to calculate the synthesis rate from the steady state message content (assuming all the messenger RNA turns over), and the messenger RNA decay constant can be derived from either the chase experiment of Nemer *et al.* (1975) or the experiment shown in Fig. 5.6. Alternatively, this rate can be derived, as described above, from the label-

---

Fig. 5.6. Synthesis and decay kinetics for messenger RNA in *Strongylocentrotus purpuratus* at the mesenchyme blastula–gastrula stage (600-cell stage). (a) Accumulation of ³H-guanosine in polysomal messenger RNA (closed circles) and in ribosomal RNA (open circles). Messenger RNA was released from isolated polysomes by the specific polysomal disaggregating agent, puromycin. Ribosomal RNA was isolated and displayed in velocity sedimentation gradients containing formaldehyde. Similar ribosomal RNA labeling rates were obtained when the ribosomal RNA was isolated from purified ribosomal subunits. The GTP pool specific activity changes as a function of time as shown in Fig. 5.2d, which is taken from the same series of experiments. The curves shown were calculated from these precursor specific activity data, together with the amount of radioactivity measured in the total polysomal RNA, and in the purified ribosomal RNA of the polysomes. These curves represent the function $R(t)$, generated by use of equation (5.7) (see text for symbols and discussion). The data were fit by least-squares methods, and from this analysis the best values of k_s and k_d were obtained. It was assumed that k_d for ribosomal RNA is effectively zero. The values of these parameters were: for messenger RNA, $k_d = 2 \times 10^{-3}\,\mathrm{min}^{-1}$; $k_s = 1.35 \times 10^{-1}\,\mathrm{pg\,min}^{-1}$ per embryo; for ribosomal RNA, $k_s = 2.16 \times 10^{-2}\,\mathrm{pg\,min}^{-1}$. From these values the steady state quantity of polysomal messenger RNA is 70 pg [equation (5.2)] and the percent of total polysomal RNA which is messenger RNA is 4.4%. (b) The k_s and k_d values obtained from several experiments such as that in (a) agreed closely, and the average values were used to generate the molar accumulation curves shown, by application of equation (5.3). (c) Decay kinetics of polysomal messenger RNA, from the same set of data. The curve shown is of the same form as that illustrated in Fig. 5.5b, and is thus described by equation (5.5). Half-life for the messenger RNA is 5.8 hours. (a) and (b) from Galau, G. A., Britten, R. J., and Davidson, E. H. (1976a). *Cell*, in press.

ing kinetics of the messenger RNA. These calculations are partially independent, since the steady state estimate is based on the polysome content and the decay constant, while the kinetic estimate depends on measurements of the absolute specific activity of the GTP precursor pool. Figure 5.6 also provides an additional measure of the content of newly synthesized polysomal message at steady state, i.e., the plateau value. The molar accumulation curve shown in Fig. 5.6b approaches a steady state value of 70 pg of polysomal message per embryo, in good agreement with the estimates based on total polysome content. The results of all of the kinetic calculations, and their ancillary parameters, are displayed in Table 5.3. From them we conclude that in the 600-cell sea urchin embryo the rate of messenger RNA synthesis is likely to be about $0.12-0.18$ pg min^{-1}, or only 4-7% of the total heterogeneous RNA synthesis rates shown in Table 5.2. Most of the remaining synthesis is heterogeneous nuclear RNA.

The messenger RNA synthesis rate estimates shown in Table 5.3 disagree by a factor of 2 to 3 with two other calculations now in the literature. Brandhorst and Humphreys (1971, 1972) concluded that 15% of the total heterogeneous RNA synthesis per unit time is messenger RNA synthesis, but this result requires a nuclear RNA turnover rate which is higher than that measured by Grainger and Wilt (1976), as noted above, and also a more rapidly turning over messenger RNA component with a half-life of about 60-75 minutes. However, the data shown in Fig. 5.6 fit significantly less well with two kinetic components, one of which turns over with a half-life of 60-75 minutes, than with a single kinetic component whose half-life is 5-6 hours. Wu and Wilt (1974) measured the rate of accumulation of total cytoplasmic poly(A)RNA and also calculated a presumptive cytoplasmic poly(A)RNA synthesis rate amounting to 10-15% of the total heterogeneous RNA synthesis rate. However, much of the poly(A) included in this analysis was on molecules which were too small to represent bona fide poly(A) messenger RNA. In any case the higher rates reported by these authors cannot be easily reconciled with the turnover kinetics measured by Nemer *et al.* (1975) and Galau *et al.* (1976a), given the reasonably firm nonkinetic estimates of total polysomal message content shown in Table 5.3.

While it is clear that early in cleavage most messenger RNA must be maternal, there is no direct information as to how far into development maternal messages persist. The best evidence remains that reviewed in earlier chapters. There, it will be recalled, we found that a changing set of maternally programmed protein synthesis patterns persist even into the blastula period. Furthermore, little convincing evidence for the influence of new transcription on the biological course of events prior to postgastrular organogenesis could be adduced. The possibility thus remains open that maternal messenger RNA remains present through much of early development. Specific measurements exist only for very early cleavage. As reviewed in Chapter 4, the data of Humphreys (1971) show that during the first 2 hours after fertilization more than 85–90% of the polysomal message is maternal. Nonetheless, it is clear that some newly synthesized messenger RNA is being translated even at these earliest stages. Many reports indicate the presence of newly synthesized messenger RNA in the embryo polysomes throughout early development (e.g., Infante and Nemer, 1967; Kedes and Gross, 1969; Humphreys, 1971; Brandhorst and Humphreys, 1972). The kinetic data of Table 5.3 show that the percent of newly synthesized, as opposed to maternal, messenger RNA in the polysomes must be large, if not 100%, by the gastrula stage. This conclusion follows from the convergence of the kinetic estimate of the steady state content of newly synthesized messenger RNA with the other estimates of total polysomal message in Table 5.3. However, there is an important additional point to be made. Though maternal message may play a quantitatively minor role after cleavage, the possibility cannot be discounted that it remains qualitatively important, perhaps all the way up to the feeding pluteus stage.

CLASSES OF MESSENGER RNA IN SEA URCHIN EMBRYOS

Three general classes of newly synthesized messenger RNA have been distinguished in sea urchin embryos. These are the histone messenger RNA's, quantitatively most significant during the midblastula phase (see Chapter 4); nonhistone messenger RNA's lacking 3'-poly(A) sequences; and nonhistone messenger RNA's containing poly(A). From early in development some newly synthesized cytoplasmic RNA's contain poly(A) (Slater and Slater, 1974; Wu and Wilt, 1974; Nemer *et al.*, 1974, 1975; Nemer, 1975). These are of message size, and can be extracted from polysomes. However, it is now clear from the work of Nemer *et al.* (1974, 1975), Fromson and Duchastel (1975), and Fromson and Verma (1976) that an important fraction of nonhistone polysomal messenger RNA in sea urchin embryos is not polyadenylated. This message fraction has the same size and the same synthesis and turnover kinetics as does the poly(A) messenger RNA (Nemer *et al.*, 1975). Poly(A) messenger RNA enters the cytoplasm after a lag of 15–30 minutes (Wu and Wilt, 1974; Nemer *et al.*, 1975), and this lag is the same as that measured for non-poly(A), and for total, messenger RNA (Aronson, 1972; Brandhorst and Humphreys, 1972).

Fromson and Duchastel (1975) and Fromson and Verma (1976) also report that in the cell-free wheat germ system the non-poly(A) messenger RNA is translated at least as well as is poly(A)RNA.

Figure 5.7 (Nemer *et al.*, 1975) summarizes the percent of radioactive precursor incorporated into the three classes of messenger RNA after 60 minutes of labeling. The proportions shown may not be directly equivalent to fractions of mass, since labeling and turnover kinetics for histone messenger RNA at the earlier stages are probably different from those of the nonhistone messages; that is, the fraction of total radioactivity as histone message seems likely to be significantly higher than the fraction of total messenger RNA nucleotides in histone messenger RNA (cf. discussion of the amount of histone message in Chapter 4). However, since non-poly(A) and poly(A) message display similar turnover and entry kinetics, the proportion of the total radioactivity in each class should also represent their mass proportions. Figure 5.5 shows that during blastulation poly(A) messenger RNA rises to around 50% of the total, and this proportion is maintained thereafter. Unfortunately, the molecular function of poly(A) tracts is as yet unknown, and therefore the meaning of the increase in poly(A) message content during these early stages cannot be assessed.

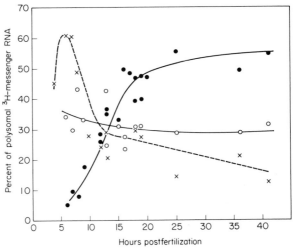

Fig. 5.7. Percent of ^{3}H-uridine radioactivity in polysomal messenger RNA which lacks poly(A), contains the 3'-poly(A) sequence, and is presumptive histone messenger RNA. Labeling was for 60 minutes. The presumptive histone messenger RNA (- -×- -) is calculated simply as non-poly(A) message about 9 S in size. The classification of poly(A)-containing (—●—) and nonhistone poly(A)-lacking (—○—) messages is based on binding to oligo(dT)–cellulose columns. Since no detergent was included in the lysis or other cell fraction buffers, only messenger RNA's present on free polysomes are monitored. From M. Nemer (1975). *Cell* **6**, 599.

According to Nemer *et al.* (1974) and Fromson and Verma (1976), however, the non-poly(A) and poly(A) sets of message represent at least partially distinct sequence populations. In the experiments of Nemer *et al.* (1974) cDNA transcribed from the poly(A)-containing messages is shown to be unable to react with the non-poly(A) messenger RNA, though it reacts well with the poly(A) messenger RNA. It follows from this observation that Fig. 5.7 describes the transcription of three distinct sets of structural genes during early development.

MITOCHONDRIAL RNA SYNTHESIS IN SEA URCHIN EMBRYOS

We have not yet discussed another class of nonribosomal RNA synthesized in early sea urchin embryos, viz., mitochondrial RNA. Synthesis of mitochondrial RNA's occurs in enucleated egg cytoplasm (Chamberlain, 1970; Craig, 1970; Craig and Piatigorsky, 1971). Mitochondrial transcription can be severely inhibited with ethidium bromide but not with actinomycin (Craig and Piatigorsky, 1971; Chamberlain and Metz, 1972). Mitochondrial ribosomal RNA's and tRNA's as well as other RNA's of unknown function, probably messenger RNA's, have been observed (Chamberlain, 1970; Chamberlain and Metz, 1972; Devlin, 1976). Most of this RNA remains bound in the mitochondria (Hartmann *et al.*, 1971). The crucial evidence that it is indeed mitochondrial comes from RNA–DNA hybridization experiments. These have convincingly demonstrated hybridization of newly synthesized cytoplasmic RNA with mitochondrial DNA (Hartmann and Comb, 1969; Craig, 1970; Hartmann *et al.*, 1971). Chamberlain (1970) pointed out that the sedimentation profile of RNA extracted from enucleated egg fragments and from whole embryos during the first few cleavages is strikingly similar, and suggested that much of the newly synthesized RNA in these embryos is actually of cytoplasmic origin (i.e., mitochondrial). Later experiments showed that 50% of the radioactivity incorporated over a 1-hour labeling period at the beginning of cleavage is associated with mitochondria in isopycnic density gradients (Chamberlain and Metz, 1972), and similar results were reported by others (e.g., Hartmann *et al.*, 1971). The specific activity of the intramitochondrial precursor pools is not known, relative to that of the total egg or nuclear precursor pools. Nor have the turnover and absolute incorporation rates of mitochondrial RNA been extensively measured. Thus the general contribution of mitochondrial RNA synthesis cannot be determined adequately from the available data. One observation (Humphreys, 1973) indicates a (mainly) mitochondrial RNA synthesis rate in the fertilized egg of about 0.2 pg min^{-1}. Comparison of this value to the total per embryo and per nucleus heterogeneous RNA synthesis rates shown in Table 5.2 confirms that about 50% of the total synthesis per embryo is mitochondrial at the cleavage stage. However, the mitochondrial contribution is major only early in development, when few nuclei yet exist, and at later stages it becomes much less important. If the 0.2 pg min^{-1} per embryo rate persists, it would represent less than 10% of the total heterogeneous RNA synthesis after cleavage (Table 5.2.). Neither the relative rate nor the nature of mitochondrial transcription seems to change during development, according to Devlin (1976). This author identified 8 distinct species of poly(A)RNA synthesized by mitochondria in enucleated egg fragments, and showed that together with the mitochondrial tRNA's and ribosomal RNA's, these species account for 96% of the mitochondrial genome (assuming single strand transcription). The 8 putative messenger RNA species continue to be synthesized in the same proportion at all stages, and Devlin concluded that throughout embryogenesis the entire mitochondrial genome continues to be transcribed at a constant rate.

SYNTHESIS OF RIBOSOMAL AND TRANSFER RNA IN SEA URCHIN EMBRYOS

It remains to discuss the synthesis of the discrete species of stable RNA, the ribosomal and transfer RNA's. Though these are easiest to identify, controversy still exists as to when the genes for these RNA species are activated. The problem, first pointed out by Emerson and Humphreys (1970), is the detection of a small amount of ribosomal or transfer RNA synthesis in the presence of much more incorporation into heterogeneous RNA's in the early embryos. The corresponding difficulty was discussed above, in connection with amphibian embryos. A good example of this problem is to be found in tRNA synthesis. Early workers (Glišin and Glišin, 1964; Gross et al., 1965a; see also Hynes et al., 1972) reported that tRNA synthesis does not begin until the mesenchyme blastula stage, and indeed a distinct 4 S RNA peak cannot be seen in gradients displaying newly synthesized RNA until this stage. However, O'Melia and Villee (1972) showed that internally labeled tRNA can be extracted even from cleaving sea urchin embryos if appropriate purification procedures are used. The synthesis of 5 S RNA also occurs during cleavage, and the view that the genes for these RNA species are not active during cleavage is now inadmissible.

Ribosomal RNA synthesis is clearly evident in sedimentation and compositional analyses of newly synthesized RNA only after the late blastula or early gastrula stages (Nemer, 1963; Comb et al., 1965; Nemer and Infante, 1967b; Giudice and Mutolo, 1967; Sconzo et al., 1970a). This is true as well for the RNA synthesized by disaggregated isolated embryo cells (Sconzo *et al.*, 1970b) and for isolated cell nuclei extracted at different stages (Hogan and Gross, 1972). In the postgastrular period the high G + C 28 S and 18 S ribosomal RNA's represent a significant fraction of accumulated labeled RNA. In 5 hours of labeling at the pluteus stage, for instance, newly synthesized 28 S and 18 S ribosomal RNA's constitute about one-fourth of the mass of the total newly synthesized RNA, the remainder being mainly heterogeneous RNA's (Emerson and Humphreys, 1970). Two factors quantitatively affect the visibility of ribosomal RNA synthesis at earlier stages. These are the smaller number of nuclei and the more active synthesis of the heterogeneous RNA's (see Table 5.2 and discussion above). Far more newly synthesized heterogeneous RNA is accumulated in the large nuclei of cleavage-stage embryos than in later stages (Emerson and Humphreys, 1970). Assuming that the same rate of ribosomal RNA synthesis as measured in the pluteus nuclei were to obtain throughout earlier development, Emerson and Humphreys concluded that ribosomal RNA would constitute less than 10% of the radioactivity sedimenting around 28 S at the blastula stage. Ribosomal RNA synthesis at the pluteus rate would thus be virtually undetectable in cleavage-stage embryos without extensive purification. Emerson and Humphreys (1970, 1971) used methylated albumin–Kieselguhr (MAK) column fractionation to extract a labeled high G+C RNA from blastulae which chromatographs in the position expected for ribosomal RNA, but this RNA fraction was not further characterized or identified specifically as ribosomal RNA.

As in the literature on amphibians, an argument often used to support the concept that ribosomal genes are activated only in the late blastula stage is that this is when definitive nucleoli first appear (e.g., Millonig, 1966; Karasaki, 1968). However, as noted earlier, Emerson and Humphreys' (1971) experiments showed that this may simply be the result of a rate of cell division which is too high to allow normal nucleolar morphogenesis.

The evidence reviewed provides no completely satisfactory resolution of the question whether ribosomal RNA synthesis is regulated during early sea urchin development. This issue may be an empty one in any case, since even the rate of ribosomal RNA synthesis measured in gastrula and pluteus larvae is so low as to appear more like a leaky state of repression. Emerson and Humphreys (1970) report a rate of ribosomal RNA synthesis in the pluteus larvae of about 1.2×10^{-3} pg hr^{-1} per nucleus. The measurements of Galau *et al.* (1976a) indicate a synthesis rate of about 1.7×10^{-3} to 2.8×10^{-3} pg hr^{-1} per nucleus in the blastula to gastrula period (see Fig. 5.6). These values represent only about 300–700 molecules of ribosomal RNA synthesized per hour per cell. Since there are reported to be about 260 copies of ribosomal RNA genes per haploid genome (Patterson and Stafford, 1971), less than 2 molecules hr^{-1} are being transcribed per gene. As in the case of amphibian embryos, the most significant statement to be made about ribosomal RNA synthesis in early sea urchin embryos at all stages is thus that it is almost totally repressed. The significance of the view taken here has been emphasized by Humphreys (1973) in experiments on ribosomal RNA synthesis in fed and unfed plutei. The results of some of these experiments are summarized in Fig. 5.8. Both the total mass and the synthesis rate of ribosomal RNA are increased greatly by feeding. By 4 days after feeding the ribosomal RNA content of the embryo has increased by a factor of four, while the trickle of ribosomal RNA synthesis in the unfed pluteus fails even to maintain the starting level. The main regulatory event affecting ribosomal RNA synthesis during embryogenesis, if not the only one, thus occurs only at feeding.

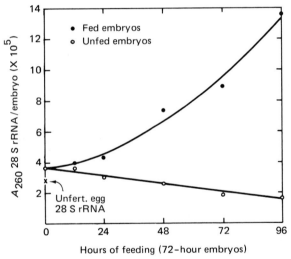

Fig. 5.8. 28 S ribosomal RNA in eggs and embryos either starved or fed at pluteus stage. *Lytechinus pictus* eggs were fertilized and allowed to develop at a concentration of 100 eggs/ml seawater. When a functional gut had differentiated (72–90 hours), they were fed 20000 *Rhodomonas* (Hinegardner, 1969) per embryo. The seawater was changed every 12 hours and an equal number of algae was added again. Samples were taken at points indicated (closed and open circles), the embryos washed, and their RNA extracted with about 95% yield. The RNA was centrifuged on a sucrose gradient and the A_{260} in 28 S RNA determined. From T. Humphreys (1973). *In* "Developmental Regulation, Aspects of Cell Differentiation" (S. J. Coward, ed.), p. 1. Academic Press, New York.

Transcription in Mammalian Embryos

Relatively little information exists regarding transcription in mammalian embryos compared to that available for amphibian and echinoderm embryos. The difficulty of obtaining sufficient material for experimentation is the primary cause of this situation, but other factors also are involved. These include the relatively large quantity of extraembryonic trophoblast tissue, which makes it difficult to study the embryo per se after very early stages. In any case, only preimplantation embryos have been investigated at the molecular level. Most studies have focused on rabbit and mouse embryos. The rabbit egg is about eight times the volume of the mouse egg and cleaves much more rapidly. Thus at 2 days after fertilization the rabbit egg contains 16 cells, at 3 days 128 cells, at 4 days over 1000 cells (Daniel, 1964). The mouse egg first cleaves toward the end of the second day after ovulation, has only about 10–30 cells on day 3, and attains 100 cells only on day 4 (Ellem and Gwatkin, 1968; Olds *et al.*, 1973).

RIBOSOMAL RNA SYNTHESIS IN EARLY MAMMALIAN EMBRYOS

In both rabbit and mouse embryos, in contrast to echinoderm and amphibian embryos, the total RNA content remains essentially constant only for the very earliest period of development. Manes (1969) found about 20 ng of total RNA in the fertilized rabbit ovum, and this remains the approximate bulk RNA content for the first day after fertilization. After this the RNA content begins to increase, and by day 3 (100-cell blastocyst stage) it has doubled. By day 4 the embryo contains 120 ng of total RNA, and by day 6 it contains 2790 ng. New bulk RNA synthesis, i.e., ribosomal RNA synthesis, is thus very important from early in development. Sedimentation analyses of newly synthesized RNA have demonstrated ribosomal RNA synthesis in rabbit embryos as early as 2 days (16 cells), and certainly by 3 days (Manes, 1971; Schultz, 1973). In the mouse embryo ribosomal RNA synthesis begins earlier in developmental time, but at about the same point in real time. The mouse ovum contains 0.5 ng of total RNA (Olds *et al.*, 1973), and no more than this is observed until 3–4 days postovulation (>30 cells). At 90 hours a mean RNA content of about 1.4 ng is reported. In this embryo ribosomal RNA synthesis is clearly detectable even at the 4-cell stage (Ellem and Gwatkin, 1968; Woodland and Graham, 1969; Pikó, 1970). In mouse embryos the onset of detectable ribosomal RNA synthesis is correlated with a sharp increase in polymerase I activity (Moore, 1975; Versteegh et al., 1975). This increase occurs after the 4- to 8-cell state and is accompanied by a decrease in the activity of form II polymerase. These polymerases are distinguished by the sensitivity of form II to α-amanitin, which does not affect form I. The total polymerase activity remains constant until the blastocyst period, after which it increases rapidly (Versteegh et al., 1975).

To summarize, as early as cleavage in the mouse embryo and the 16-cell morula in the rabbit embryo, ribosomal RNA synthesis results in an early increase in ribosomal RNA content, i.e., net growth. Preimplantation mammalian embryos constantly absorb nutrients from the fallopian and subsequently the uterine environments. In sharp contrast to the embryos of the organisms discussed earlier, which undergo net growth only after extensive differentiation, organogenesis, and feeding, mammalian embryos are expanding systems almost from the beginning.

SYNTHESIS OF HETEROGENEOUS RNA'S IN MAMMALIAN EMBRYOS

Nuclear RNA synthesis begins at the earliest stages, according to radioautographic observations (e.g., Mintz, 1964; Hillman and Tasca, 1969; Bernstein and Mukherjee, 1972; Karp et al., 1973). The newly synthesized RNA is very heterogeneous in size, and at least in nondenaturing gels and gradients much of it appears large (Woodland and Graham, 1969; Manes, 1971; Knowland and Graham, 1972; Schultz et al., 1973b). A gel electrophoresis pattern of newly synthesized RNA extracted from 2 cell mouse embryos is shown in Fig. 5.9 (Knowland and Graham, 1972).

Unfortunately, few data on absolute synthesis and turnover rates exist for mammalian embryos. Measurements of nucleoside and nucleotide pool behavior by Epstein and Daentl (1971) and Daentl and Epstein (1971, 1973) in mouse embryos illustrate some of the difficulties. Uptake of nucleosides appears to depend in part on exogenous conditions, such as the particular precursor used, the embryonic stage, the precursor concentration, and the presence of other nucleosides. Both pool expansion and compartmentalization are claimed to occur. No absolute rate calculations based on incorporation data were possible for stages prior to the 3 day 30- to 60-cell (blastocyst). At this stage significant pool expansion could be obtained with exogenous uridine, and on the assumption that the internal pool approaches the specific activity of the external medium, Daentl and Epstein (1971) arrived at an RNA synthesis rate equivalent to about 0.25 pg min^{-1} of RNA per embryo, or about 6×10^{-3} pg min^{-1} per cell. A similar value was roughly estimated by Epstein and Daentl (1971) from

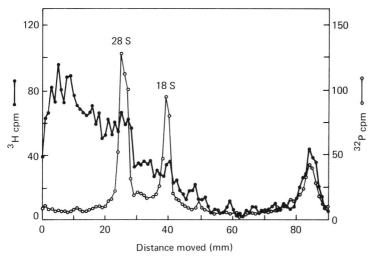

Fig. 5.9. Gel electrophoresis profile of newly synthesized RNA from 2-cell mouse embryos. RNA was extracted from 1000 2-cell embryos which had been labeled *in vitro* for 11–16 hours with 100 μCi/ml ^{3}H-uridine (33.8 Ci/mmole). Closed circles represent migration of the labeled RNA in 2.7% SDS polyacrylamide gel; open circles indicate the positions of ^{32}P-ribosomal and 4 S RNA markers. From J. Knowland and C. Graham, (1972). *J. Embryol. Exp. Morph.* **27**, 167.

adenosine incorporation experiments. Though based on some problematical assumptions, this rate is of the same order as that measured for early sea urchin embryo cells (Table 5.2). The mouse genome is about 4 times larger than the sea urchin genome, however. RNA synthesis appears to occur at about the same rate per cell over the 2- to 3-day period as the embryo increases from 8–16 cells to 30–60 cells (Epstein and Daentl, 1971). Hillman and Tasca (1973) also found that the ratio of RNA incorporation to precursor radioactivity increases only modestly per embryo between the 4-cell and morula stage, a result which is contrary to earlier conclusions of Ellem and Gwatkin (1968). The data of the latter authors, however, consisted solely of incorporation measurements, as pool radioactivity measurements were not then considered feasible. Subsequently, Hillman and Tasca (1973) showed that the rate of precursor uptake increases over the 4-cell to morula period. There does seem to be a sharp acceleration in the relative rate of total high molecular weight RNA synthesis at a much earlier time, during the 2- to 4-cell stage (Woodland and Graham, 1969). Expressed in terms of the ratio of radioactive precursor (^{3}H-uridine) incorporated in RNA to acid-soluble radioactivity, the apparent synthesis rate increases ninefold over this period.

TRANSFER RNA SYNTHESIS AND MESSENGER RNA
SYNTHESIS IN MAMMALIAN EMBRYOS

About the same time as ribosomal RNA synthesis becomes detectable in mouse embryos, 4 S RNA synthesis (presumably tRNA) does so as well. Like ribosomal RNA synthesis this also occurs at the 4-cell stage in the mouse. Ellem and Gwatkin (1968) showed incorporation of label into tRNA by using MAK column chromatography to distinguish this class of RNA's, and the same result was obtained by Woodland and Graham (1969), who relied on size determinations by gel filtration and sucrose gradients. The latter authors demonstrated labeled pseudouridylic acid in tRNA, and also found this RNA to be methylated in 8-cell embryos. These observations show that the incorporation of label into this RNA is not simply end group turnover, but rather represents *de novo* synthesis of tRNA.

One apparent difference between rabbit and mouse embryos is that tRNA synthesis begins at the 2-cell stage (Manes, 1971), prior to the appearance of detectable ribosomal RNA synthesis in rabbit embryos, while in the mouse, synthesis of both species of RNA becomes observable at the same time, i.e., the 4-cell stage. Of course, these differences could reflect changes in relative detectability rather than real changes in synthesis rate. Stage-specific changes in both the level of methylation of tRNA and the quantity of at least one set of tRNA's, the methionyl-tRNA's, have been reported in rabbit embryos. Manes and Sharma (1973) found that the level of methylation of total tRNA decreases sixfold at blastocyst formation (60–80 hours; 128–1024 cells). Clandinin and Schultz (1975) obtained the converse result in studies of the ability of methionyl-tRNA to accept methyl groups in an *in vitro* transfer system. Methionyl-tRNA extracted from blastocyst rabbit embryos was a tenfold better methyl acceptor than Met-tRNA extracted from 4- and 6-day embryos. The conclusion was drawn that the morula (2-day) stage embryo contains hypomethylated tRNA. Another striking change occurring at the time of blastocyst formation is the greater than tenfold increase in the quantity per embryo of both $tRNA_f^{Met}$ (about one third of the total Met-tRNA) and $tRNA^{Met}$. About 0.032 pmole of total tRNA is reported by Clandinin and Schultz (1975) in the 2-day embryo, and 0.377 pmole in the 4-day blastocyst. These changes are possibly significant with respect to the notable increase in total protein synthesis rate which occurs at blastocyst formation in the rabbit (see Chapter 4). Obvious qualitative changes in the pattern of protein synthesis do not occur at this stage, however (Chapter 3).

Additional observations concern messenger RNA synthesis in rabbit embryos. Polyribosomes have been extracted from morulae and blastocysts (Schultz, 1973), and newly synthesized RNA with the properties of messenger RNA is associated with these polysomes. This RNA is heterogeneous in sedimentation pattern, contains poly(A) tracts, and bands with polyribosomal particles in isopycnic gradients. As reviewed in Chapter 4, Schultz (1975) also showed that the unfertilized rabbit egg contains polyadenylated RNA, i.e., putative maternal messenger RNA, and the amount of this does not change during cleavage. Most (>70%) of the newly synthesized polysomal messenger RNA labeled with ^{3}H-uridine also contains poly(A), according to poly(U) filter binding studies (Schultz *et al.*, 1973b). The polysomal poly(A) turns over slowly, according to Schultz (1974), who reported complex accumulation kinetics which could be resolved into two components decaying with half-times of 7 and 18 hours. Unfortunately, direct measurements of newly synthesized poly(A)RNA content or synthesis rate do not yet exist for mammalian embryos. A crude estimate from Schultz' (1973) data suggests that about 40% of the 2-day morula and 4-day blastocyst ribosomes are polysomal. The amount of RNA in the 2-day embryo is about 34 ng, while that in the 4-day embryo is 123 ng (Manes, 1969), and most of this is necessarily ribosomal (see above). Hence the total amount of polysomal message might be about 0.5 ng in the 2-day embryo and 2 ng in the 4-day blastocyst. This change would help to explain the increase in total protein synthesis rate occurring at the blastocyst stage.

An issue regarding transcriptional activity in mouse embryos which has been the subject of much research and discussion and which should be mentioned here is the nature of the metabolic lesion in t^{12} mutants. This mutation is one of a series of lethal mutations discovered by Dunn and Gluecksohn-Waelsch (1953). t^{12} is among the earliest-acting known genetic defects. The morphogenesis of t^{12}/t^{12} homozygotes diverges from that of normal embryos in the morula stage (Smith, 1956; Mintz, 1964). Mintz (1964) and others suggested that t^{12} is a nucleolar defect which causes deficient ribosomal RNA synthesis. However, Hillman *et al.* (1970) and Hillman (1972) found that the nucleoli of t^{12}/t^{12} embryos are in fact normal in cytological appearance and as monitored by radioautography, i.e., in their RNA synthesizing activity, until the embryos begin to degenerate. Subsequent work showed that total RNA synthesis rate, including 28 S and 18 S ribosomal RNA synthesis, protein synthesis, and precursor uptake kinetics are all normal in t^{12}/t^{12} embryos (Hillman and Tasca, 1973; Erickson *et al.*, 1974). Thus, whatever the primary metabolic lesion in t^{12}/t^{12} homozygotes, its effects do not alter any of the quantitative parameters of transcription.

RNA Synthesis in the Embryos of Other Species

Only scattered observations on animals of other phylogenetic groups exist, and with one exception, no absolute rate measurements have been presented for these organisms. This exception is Drosophila. Fausto-Sterling et al. (1974) applied a high-pressure microchromatography system to analysis of nucleotide pool specific activity in Drosophila embryos and calculated absolute RNA synthesis rates from incorporation data. Their results, expressed as the amount of RNA synthesis during 1-hour labeling periods, showed that the absolute rate of total RNA synthesis per embryo drops by a factor of 3 during the period from preblastoderm to gastrula. After this it remains constant. On a per cell basis this decline in synthesis rate is of course greater. However, a large part of this early synthesis could be mitochondrial or could represent end-group addition, since radioautographs of dipteran eggs fail to display nuclear labeling until the blastoderm stage, even though the precursor is clearly available within the egg (Pietruschka and Bier, 1972; Zalokar, 1976). McKnight and Miller (1976) surveyed chromatin spreads from preblastoderm and blastoderm stage Drosophila embryos for transcription complexes visible in the electron microscope. Their observations are confined to those transcribed regions which are densely packed with polymerases and ribonucleoprotein fibrils. Such regions should account for a large amount of the instantaneous synthesis, however. They found relatively few such structures in chromatin from syncytial nuclei, and a fivefold increase was observed in blastoderm stage chromatin. Thus it seems likely that transcription does increase sharply after blastoderm formation. From the data of Fausto-Sterling et al. (1974) the rate of RNA synthesis in the whole embryo at the blastoderm stage is about 0.4 pg min^{-1}. The blastoderm forms when there are 256 nuclei, so the per nucleus rate from this point on would be $<1.6 \times 10^{-3}$ pg min^{-1}. We note that the Drosophila genome is almost 6 times smaller than the sea urchin genome, and the RNA synthesis rate observed is about 3 to 6 times lower than those listed in Table 5.2 for sea urchin embryos. The same general relation between absolute nuclear RNA synthesis rate and genome size appears to hold in the cells of Drosophila, sea urchin, and Xenopus embryos.

In Table 5.4 a variety of biochemical data relating to embryonic RNA synthesis in various organisms are listed. Radioautographic and other cytological observations are not included. Observations of this sort mainly concern the appearance of nucleoli, and as pointed out earlier, this may not provide a reliable index of the onset of ribosomal RNA synthesis.

While Table 5.4 cannot be regarded as inclusive in any sense, it provides a general scan of the descriptive information available. Nonetheless, many whole phyla are omitted for lack of data. The differences between sea urchin, amphibian, and mammalian embryos which we have reviewed in detail show, furthermore, that particular RNA synthesis patterns are likely to characterize each group of organism.

RIBOSOMAL RNA SYNTHESIS IN VARIOUS EMBRYOS

It should be clear by now that the point when ribosomal RNA synthesis is first observed (Table 5.4) cannot be naively regarded as the time when it first is activated, if it is activated at all. In many cases it has not been the authors' intent to discover the actual time of onset of ribosomal RNA synthesis, and the same is true of the other classes of RNA whose synthesis is considered in Table 5.4. The stages listed in Table 5.4 represent those periods by which some synthesis of the respective classes of RNA is *known* to occur. In certain examples, e.g., *Ilyanassa*, ribosomal RNA synthesis is detected biochemically only at the late gastrular stage, but fully formed nucleoli active in RNA synthesis are observed as early as the 16-cell stage of cleavage. The early onset of nucleolar activity is also reported in other gastropod species, including *Limnaea* and *Limax* (Raven, 1958), and *Cyclops* (Kiknadze, 1963). Among the most unequivocal ways to determine whether transcription in very early embryos is occurring is to observe the chromatin in the electron microscope under conditions in which transcription complexes can be visualized. Foe *et al.* (1976) used this approach to study early *Oncopeltus* embryos and observed both ribosomal and nonribosomal transcription complexes as early as the blastoderm stage. The ribosomal complexes have a predictable length and display tightly packed transcripts (see examples in Chapter 8). The nonribosomal transcription units can be identified by their greater length, which in this study averaged about 29,000 nucleotides, and by their less closely packed transcripts.

Table 5.4 shows that eggs vary by orders of magnitude in their ribosome content (i.e., total RNA). Several interesting cases of special mechanisms exist in which ribosomal RNA is provided to the embryo in unusual ways. In the teleost *Misgurnus fossilis*, as in most teleosts, the comparatively enormous yolky mass of the egg remains separated and noncellular during the period of germ layer formation. Aitkhozhin *et al.* (1964) showed that the ribosome content of the embryo itself increases during pregastrular development. However, during this time the total ribosomal RNA content of egg plus embryo remains constant at about 2200 ng, a relatively large value. The explanation for the change in embryo bulk RNA content is

TABLE 5.4. RNA Synthesis Patterns in Various Embryos

Species	RNA per egg at fertilization (ng)	Earliest stage at which synthesis of RNA class has been observed			
		Polysomal message	Heterogeneous RNA	Ribosomal RNA	Transfer RNA
Protostomes					
Mollusca					
Mulina lateralis (coot clam)		2-cell cleavage[a]	Cleavage–blastula[b]	Gastrula[b]	Mid–late cleavage[b]
Spisula solidissima[c] (surf clam)		2-cell cleavage	4-cell	Trochophore	
Nassaria (Ilyanassa) obsoleta (mud snail)	4	Gastrula[d]	Gastrula[d]	Gastrula[e]	Organogenesis[d]
Echiuroid					
Urechis caupo[f] (innkeeper worm)	14		Fertilization, early cleavage	Midgastrula	Blastula
Nematode					
Ascaris lumbricoides[g]	0.06	4-cell cleavage	After fertilization, precleavage	After fertilization, precleavage	
Insect					
Oncopeltus fasciatus[h] (milkweed bug)	300		Blastoderm	Blastoderm	Gastrula

Deuterostomes					
Echinoderm					
Strongylocentrotus purpuratus[i] (purple sea urchin)	3.3	Early cleavage	Early cleavage	Gastrula	Blastula
Teleost					
Misgurnus fossilis[j] (loach)	2200	Gastrula	Blastula	After gastrulation	Blastula
Amphibian					
Xenopus laevis[i] (clawed toad)	4000	Cleavage	Cleavage	Gastrula	Late blastula
Mammal					
Mus musculus[i] (mouse)	0.5		2-cell	4-cell	4-cell
Oryctolagus cuniculus[i] (rabbit)	20	16-cell	2-cell	16-cell	2-cell

[a] Kidder (1972b).

[b] Kidder (1972a). The author points out that ribosomal RNA synthesis may begin much earlier but be undetectable for the same reasons as in sea urchin embryos.

[c] Firtel and Monroy (1970).

[d] Collier (1960, 1965a,b); Collier and Yuyama (1969). It is assumed that stable poly(A)RNA is equivalent to messenger RNA. Labeled poly-(A)RNA was detected in 1-day (gastrula stage) embryos by Collier (1975a).

[e] Collier (1961), assuming that bulk RNA increase denotes ribosomal RNA synthesis; Koser and Collier (1972); Koser and Collier (1976).

[f] Gould (1969); Schwartz (1970); observation of tRNA synthesis in $5\frac{1}{2}$-hour embryos is based on _in vivo_ methylation from [methyl-^{14}C]-methionine of 4 S RNA.

[g] Kaulenas and Fairbairn (1968); Kaulenas et al. (1969). Ribosomal RNA synthesis occurs during a period of several days in which the eggs remain in the uterus. See text for further description.

[h] Harris and Forrest (1967); Foe et al. (1976). The latter observed in the electron microscope transcription of both heterogeneous RNA and ribosomal RNA in the chromatin of embryos between blastoderm and neurula stages.

[i] See text and Tables 4.2 and 5.3 for references.

[j] Aitkhozhin et al. (1964); Belitsina et al. (1964); Rachkus et al. (1971); Neyfakh et al. (1974); Solov'eva and Timofeeva (1974); Solov'eva et al. (1974).

claimed to be transport of ribosomes from yolk to embryo during cleavage and blastulation. In the electron microscope a gradient of ribosomes can be seen in the yolk, ending in a densely packed layer contiguous to the embryo. Another teleost, *Salmo gairdneri* (trout) has also been studied. Here ribosomal precursor synthesis can easily be measured as early as the blastula stage (Mel'nikova *et al.*, 1972). The ribosomal RNA precursor is actively methylated, aiding in its identification. Here, too, the nonembryonic yolk is important, in that conversion into mature ribosomal RNA is dependent on unknown yolk factors. In another enormously yolk-rich egg, that of the domestic fowl (*Gallus domesticus*), embryonic ribosomal RNA synthesis begins during cleavage (Wylie, 1972). In this organism the bulk RNA of the embryo rises about two orders of magnitude to over 100 μg as the cell number increases from a few thousand to over 60,000. The increase in bulk RNA is accounted for by new synthesis (Wylie, 1972). If the yolk, which is reported to contain 2000 μg of total RNA (Solomon, 1957), is ignored, the expanding chick embryo system resembles the mammalian embryos discussed above. From this comparison it can be seen that no simple correlation exists between the blastodisc form of development, in which the embryo is at first confined to a disc-like region on top of a relatively huge yolk-filled egg, and the mode of provision of ribosomal RNA.

A unique process of ribosomal RNA synthesis occurs in the *Ascaris* egg (Kaulenas and Fairbairn, 1968). This egg is fertilized in the uterus, and remains there for a period of many hours. After fertilization a shell forms around the egg, and the germinal vesicle undergoes its reduction divisions. A period of 12 to 24 hours is required for the egg to pass through the uterus, and pronuclear fusion and cleavage do not take place until an additional 40 hours have elapsed. The male pronucleus engages in active ribosomal (as well as other) RNA synthesis during the time that the oocyte reduction divisions take place. Newly synthesized ribosomal RNA was identified in these eggs by Kaulenas and Fairbairn (1968) on the basis of its nucleotide composition and sedimentation pattern, and dense accumulation of ribosomes can be seen in the egg cytoplasm surrounding the male pronucleus. The result of the pronuclear synthetic pattern is a 50% increase in the total RNA of the egg prior to pronuclear fusion. This egg, as noted in Table 5.4, begins with one of the smallest complements of ribosomal RNA known, only about 60 pg. Among the other RNA's synthesized by the male pronucleus are species which are template active in cell-free protein synthesis systems. Both these heterogeneous RNA's and the ribosomal RNA's are probably destined for use in cleavage or later. By cleavage the *Ascaris* egg can thus be said to contain a paternal as well as a maternal stockpile of pretranscribed RNA's.

GENERALIZATIONS AND A PARADOX REGARDING CLASSES OF NEWLY SYNTHESIZED EMBRYONIC RNA

Perhaps the most interesting general conclusions from Table 5.4 are that synthesis of heterogeneous RNA, undoubtedly largely nuclear RNA, is active soon after fertilization, and similarly that newly synthesized messenger RNA is already present on polysomes at the beginning of cleavage. Structural gene transcription evidently plays a role from the very beginning of development. This observation leads to a fundamental paradox which demands further exploration: Despite the early appearance of newly synthesized messenger RNA, it has so far proved impossible to demonstrate many early biological or molecular events which *require* new structural gene activity. Various interpretations of this paradox might be posed. Undoubtedly, the tests which have been applied are crude, and may well have missed the effects of hundreds of newly turned on embryonic structural genes. Another possibility is that the genes transcribed by the embryo belong in part to the same set as is transcribed in the preparation of the maternal messenger RNA during oogenesis. In this case embryo transcription would make little *qualitative* change in the messenger RNA population. We now turn to information gained from nucleic acid sequence homology experiments in the hope of further penetrating into this basic aspect of gene activity in early development.

6

RNA Sequence Complexity
and Structural Gene Transcription
in Early Embryos

The quantitative expressions needed to analyze renaturation and hybridization data are reviewed. Sequence complexity is defined for both single copy and repetitive sequence transcripts. Basic relations between sequence complexity and the rate of nucleic acid renaturation are developed for both single copy and repetitive sequence classes. The rate-limiting step in renaturation is the nucleation event, and the rate of this depends on sequence concentration. Data show that as predicted the observed second-order rate constant for renaturation varies inversely with sequence complexity. When renaturation is assayed by single strand nuclease resistance rather than by binding to hydroxyapatite, the disappearance of single strand nucleotides rather than totally single-stranded molecules is measured. For randomly sheared DNA the kinetics of renaturation measured in this way display non-second-order form. This result is important because it is needed to derive the kinetic form obeyed by DNA excess hybridization reactions with RNA. RNA excess hybridizations follow pseudo-first-order kinetics. From these reactions both the complexity and the prevalence of the hybridizing RNA's can be extracted. Data obtained in hybridization experiments with oocyte and embryo RNA's are reviewed. The single copy sequence complexities of all species of oocyte RNA's so far studied are similar (Table 6.1), suggesting that a certain set of transcripts is required for early development. Many (or all) of these stored oocyte transcripts are maternal messenger RNA's. Oocytes also contain a relatively complex set of repetitive sequence transcripts. Their function is unknown, but they persist beyond fertilization and can be recovered from the embryo. The presence of repetitive and nonrepetitive sequence transcripts in newly transcribed embryo RNA's is measured in DNA excess hybridization experiments. Messenger RNA is mainly single copy sequence transcript, though some messages are transcribed from repetitive structural genes. Messenger RNA contains no detectable interspersed repetitive and nonrepetitive sequences. In contrast, heterogeneous nuclear RNA's have an interspersed sequence organization similar to that of the DNA. The nuclear RNA of sea urchin gastrulae is very complex and represents over one-fourth of the single copy sequence in the genome. About one copy of each heterogeneous nuclear RNA species exists per nucleus at any one time. The single copy complexity of mammalian embryo RNA increases with developmental stage. Early competition hybridization experiments carried out with filter methods on a variety of embryos are reviewed in Table 6.2. These experiments show in general that the set of repetitive sequences transcribed during embryogenesis changes according to stage. Some oocyte RNA's compete well with the hybridization of early embryo transcripts, and this may be due to maternal histone messenger RNA. Polysomal messenger RNA in sea urchin embryos has a single copy sequence complexity large enough to code for about 10,000–15,000 diverse proteins. However this is only one-tenth the complexity of the heterogeneous nuclear RNA of the same stage. The complexities and the amounts of messenger and nuclear RNA in sea urchin embryos are compared in Table 6.3. The high complexity messenger RNA includes about 10% of the total mass of message, while most of the messenger RNA consists of a relatively small number of different messages, each present many times per cell. Each complex class messenger species is present on the average only about once per cell. At different stages of sea urchin embryogenesis, different sets of messenger RNA's are present in the polysomes, and these differ by several thousand diverse sequences. All sequences present in the gastrular polysomes are represented in the stored RNA of the oocyte. Subsets of the gastrular message set are present in pluteus and blastula, but no sequences are detected in pluteus other than those already present in gastrula. Some proteins needed in plutei thus must be synthesized earlier. Complexity of messenger RNA in several adult tissues is less than 35% of that in embryos. This suggests that the cost of development in terms of genomic information is high. A basic mystery is the functional role played by the very large number of diverse messenger RNA's being translated in early embryos.

In this chapter we review direct measurements regarding the DNA sequences transcribed during early development. The experimental approach considered here is based on hybridization of DNA with heterogeneous RNA fractions extracted from oocytes and early embryos. While the potential rewards of this type of experiment have long been appreciated, generally useful quantitative procedures for carrying them out became available only at the end of the 1960's. As detailed below, these advances were due mainly to new analyses of the process of nucleic acid renaturation provided by Britten and his associates and by Wetmur and Davidson. The following review is not intended to be encyclopedic. Reference to a good deal of earlier hybridization work is omitted, and we focus on several experimental studies which can be quantitatively interpreted in light of current knowledge. The majority of the experiments discussed here deal with sea urchin embryo RNA's and to a lesser extent, amphibian embryo RNA's. Unfortunately, few modern measurements exist for early embryo or oocyte RNA's from other animal groups.

Sequence Complexity and the Quantitative Analysis of Renaturation Experiments

DEFINITION OF SEQUENCE COMPLEXITY

A basic concept underlying this field is that of nucleic acid *sequence complexity*. The complexity of a population of nucleic acid molecules is the total length of diverse nonrepetitive sequence represented. Suppose, for example, that an RNA population consists of 100 molecules of sequence "a," 10 molecules of sequence "b," and 1 molecule each of sequences "c," "d," and "e." The complexity is the sum of the diverse sequences present, i.e., (a+b+c+d+e). Complexity is usually given in terms of nucleotides (for RNA) or nucleotide pairs (for DNA), but daltons or any other mass units can also be applied. If each of the species ("a" through "e") in our imaginary nucleic acid population were 10^3 nucleotides in length, the complexity would be 5×10^3 nucleotides. The term *representation* is often used for the frequency with which given transcripts occur in an RNA population. In this example the representation of sequence "a" is 100 times that of sequence "e."

Repetitive as well as nonrepetitive sequences in animal DNA are transcribed. A difficult issue may arise as to the meaning of complexity when the sequences in question are the typical moderately repetitive sequences found in all animal DNA's (Britten and Kohne, 1968a). It is now well known that these repetitive sequences are usually not perfect replicates (Britten and Kohne, 1968a; reviewed by Britten and Davidson, 1971; Davidson and Britten, 1973). Thus when such sequences are renatured, the duplexes formed include mismatched bases. The genome of the sea urchin *Strongylocentrotus purpuratus* provides a good example. Most of the moderately repetitive sequences in this DNA range in frequency of occurrence from about 100 to about 3000 times per haploid genome. About three-fourths of these repetitive sequences are only a few hundred nucleotides long, with a mean length around 300 nucleotides (Graham *et al.*, 1974; Britten *et al.*, 1976; see Chapter 1). Duplexes formed by renaturing these sequences melt 8°–10°C lower than do equally long native DNA duplexes (Davidson and Britten, 1973; Graham *et al.*, 1974). Since 1% base pair mismatch gives rise to approximately 1°C decrease in duplex thermal stability (reviewed by Britten *et al.*, 1974; Wetmur, 1976), it follows that an average of about 8–10% sequence divergence exists among these homologous repetitive sequences. The problem here is to find a useful definition of sequence complexity for a set of repetitive sequences homologous enough to react with each other, but divergent enough so that 1 out of every 10 or 15 nucleotides is different in individual copies of the sequence.

A useful concept is that of the *repetitive sequence family*. This is defined as the set of sequences sufficiently homologous to any individual repetitive sequence to form stable base-paired structures with this sequence when the DNA is renatured under given conditions. The complexity of a group of repetitive sequence families is simply the sum of the complexities of any one member from each of the individual repetitive sequence families. Thus, suppose a genome contains three repetitive DNA sequence families, each composed of 10 slightly divergent sequences, $a_1 \ldots a_{10}$, $b_1 \ldots b_{10}$, and $c_1 \ldots c_{10}$. The complexity of the repetitive DNA would be calculated as $(a + b + c)$, which under the renaturation conditions used is the same as $(a_3 + b_7 + c_9)$ or $(a_1 + b_1 + c_1)$, etc.

Later in this chapter we refer to several hybridization experiments which show that some fraction of the repetitive DNA sequence is represented in RNA. This, of course, means some fraction of the repetitive

DNA sequence *families*, since any RNA or DNA capable of reacting with one sequence of a repetitive sequence family can also react with all others. For example, an RNA transcribed only from sequence a_3 would hybridize with 33% of the total repetitive DNA in the example given above. The complexity of the RNA would be stated as 33% of the repetitive DNA complexity.

BASIC RELATIONS BETWEEN RENATURATION RATE CONSTANT AND SEQUENCE COMPLEXITY

We now consider the estimation of sequence complexity by measurement of renaturation kinetics. The main object of the following discussion is to review briefly the relations needed for analyses of relevant RNA–DNA hybridization and DNA–DNA renaturation experiments. For derivations of some of these relations and detailed physicochemical data on nucleic acid renaturation the reader is referred elsewhere: The technical foundations of this area are to be found in papers by Britten and Kohne (1967, 1968a,b) and by Wetmur and Davidson (1968). A useful review incorporating much recent information is that of Wetmur (1976), and special treatments of renaturation and hybridization kinetics for various particular circumstances are presented in papers by Britten *et al.* (1974), Smith *et al.* (1975), Britten and Davidson (1976), Davidson *et al.* (1975b), Galau *et al.* (1976b), and Rau and Klotz (1976) among others.

The rate-limiting step in renaturation is the bimolecular reaction of single-stranded regions bearing complementary nucleotide sequences. The process by which a fruitful collision of strand pairs occurs, the region of complementarity is recognized, and base pair formation begins is termed *nucleation*. Under most conditions, if not all (see, e.g., Rau and Klotz, 1976), the continuation of base pair formation to the end of the complementary region is very fast compared to the rate of nucleation. When assayed in certain ways, the kinetics of DNA renaturation appear to be approximately second order, and this provides convincing evidence that nucleation is rate limiting. Second-order renaturation kinetics were clearly demonstrated in a series of studies on prokaryote DNA's in which most sequences appear only once per genome. Britten and Kohne (1968a) observed second-order renaturation kinetics in experiments in which hydroxyapatite chromatography was used to follow the course of the reaction. Hydroxyapatite, a calcium phosphate complex, binds double-stranded DNA at certain phosphate buffer concentrations, while releasing single-stranded DNA fragments. In addition, Britten and Kohne (1968a) and Wetmur and Davidson (1968) demonstrated second-order kinetics for at least the initial portion of the reaction by measuring the decrease in the optical absorbance of DNA during renaturation. It should be noted here that earlier workers (see, e.g., Marmur et al., 1963) had also suggested second-order kinetics for the process of DNA renaturation.

Equation 6.1 describes by definition a second-order reaction. Here C is the concentration of nucleotides remaining single stranded (conveniently expressed in moles nucleotide liter^{-1}) at time t (in seconds), and k is defined as the observed second-order rate constant:

$$\frac{dC}{dt} = -kC^2 \tag{6.1}$$

The units of k are thus $M^{-1}sec^{-1}$. A very useful form of the solution to equation (6.1) is

$$\frac{C}{C_0} = \frac{1}{1 + kC_0 t} \tag{6.2}$$

where C_0 is the total DNA concentration or the concentration of single-stranded nucleotides at the initiation of the reaction (Britten and Kohne, 1967, 1968a). As mentioned above C/C_0 can be measured directly by optical hypochromicity or hydroxyapatite chromatography as well as by several other methods. It is the usual practice to extract the rate constant k from the data by least-squares methods, and this is illustrated in many examples later in this chapter.

For a given DNA the observed rate constant k is found to vary sharply with monovalent cation concentration. For example, the value of k is about 5 times higher at $0.6\ M$ Na$^+$ than at $0.18\ M$ Na$^+$. Tables for conversion of renaturation rate constants to their equivalent values under "standard conditions" (i.e., $0.18\ M$ Na$^+$, 60°C) are to be found in Britten et al. (1974). In this chapter all renaturation rates and related data are given after conversion to their values under standard conditions.

The value of the observed rate constant also depends significantly on the DNA fragment length. Wetmur and Davidson (1968) showed that the rate of the reaction varies directly with the square root of the fragment length. Arguments have been made (Wetmur and Davidson, 1968; Wetmur, 1971) that this length dependence means that the nucleation process is inhibited by limitations on the freedom of the incident nucleic acid strand to penetrate the region of solution within which the elements of another strand are likely to be found. This is known as the "excluded volume effect." The dimensions of the excluded volume are a function of the length and the flexibility of the nucleic acid chain under the particular conditions applied. Detailed interpretation of the excluded volume phenomenon remains a subject for future research, and quantitative explanations of the absolute values of the renaturation rates observed under given conditions have yet to be derived.

We now consider the renaturation of DNA's of differing complexity. Since the nucleation event is rate limiting, the rate of the reaction for each DNA depends directly on the concentration of each sequence in the mixture. The *sequence concentration* determines the frequency with which fragments bearing a given sequence encounter other fragments which include elements of complementary sequence. Therefore, for a given total DNA concentration, the greater the complexity the slower the reaction, since the concentration of each sequence is lower. The useful principle emerging from this logic is that the rate of renaturation is inversely proportional to the complexity of the renaturing nucleic acid. This provides an extremely powerful tool for measuring nucleic acid sequence complexity.

The relation between sequence complexity and sequence concentration can be seen in the following formalisms (Britten and Kohne, 1967; Britten, 1969): Consider a genome which contains only single copy sequence, and is G nucleotides in length. G is thus the complexity as well as the genome size. Any particular sequence, "i," occurs once per genome. Therefore, the concentration of any one nucleotide in this sequence is

$$C_i = \frac{C_0}{G} \tag{6.3}$$

This concentration determines the rate of duplex formation when the DNA is allowed to renature. Equation (6.2) states that for any given nucleotide in the sequence the fraction present on fragments remaining single stranded (SS) at time t is

$$\frac{C_{i(SS)}}{C_i} = \frac{1}{1 + k_i C_i t} = \frac{1}{1 + k_i (C_0/G)t} \tag{6.4}$$

It follows that the *observed rate constant* of renaturation for any sequence (or set of sequences) is inversely proportional to G, the genome size; i.e.,

$$k = \frac{k_i}{G} \tag{6.5}$$

Here k_i is the basic nucleation rate, which depends on salt concentration, temperature, microscopic viscosity, etc. (see the review of Wetmur, 1976 for a summary of these effects). The effect of fragment length is included in the observed rate constant k in two ways. As noted above it affects the nucleation rate, and the fragment length also determines the yield in base-paired nucleotides resulting from each fruitful collision (see Britten and Davidson, 1976). For any given set of conditions, k_i in equation (6.5) can be evaluated by measuring the rate of renaturation of a single copy DNA from a genome of known size. For example, under standard conditions (0.18 M Na$^+$, 60°C) 450 nucleotide fragments of E. coli DNA react with an observed second-order rate constant of 0.25 M^{-1} sec^{-1}, measured by hydroxyapatite assay (Britten and Kohne, 1967). Since the E. coli genome contains about 4.2 × 10^6 nucleotide pairs, the value of k_i for these conditions is 1.05 × 10^6 nucleotide pairs M^{-1} sec^{-1}. A frequently used relation by which an unknown genome size may be estimated is derived from equation (6.5). If G_1 and k_1 are the genome size and observed renaturation rate of a known DNA (e.g., E. coli DNA), and G_2 and k_2 are the equivalent parameters for an unkown DNA,

$$G_2 = \frac{k_1}{k_2} G_1 \qquad (6.6)$$

In the examples we have so far considered the complexity of the DNA equals the genome size, but this is of course a special case. The complexity of any DNA fraction in which all sequences are present in equal concentration can be calculated from equation (6.6).

Experimental verification of the relations shown in equations (6.5) and (6.6) is presented in Fig. 6.1. This graph is reproduced from a paper of Laird (1971) and demonstrates the inverse proportionality of genome size and the observed second-order rate constant. The data shown cover four orders of magnitude, and other observations concerning naturally occurring DNA's extend the proportionality down to complexities of a few nucleotides. Such low complexities are found in some satellite DNA's, the renaturation kinetics of which have been studied by many investigators including Waring and Britten (1966), Britten and Kohne (1967), Flamm et al. (1969), Hutton and Wetmur (1973), Brutlag and Peacock (1975), and Cordeiro-Stone and Lee (1976). We note that the proportionality shown in Fig. 6.1 between renaturation rate and genome size is not specifically dependent on the functional form of the reaction kinetics. As discussed briefly below the renaturation of randomly sheared DNA is actually not exactly a second-order process, though for most purposes the differences are slight. However, renaturation is dependent kinetically on the rate of occurrence of successful nucleations with which the pairing of complementary fragments begins. This basic fact is implicit in the relation symbolized in equation (6.5).

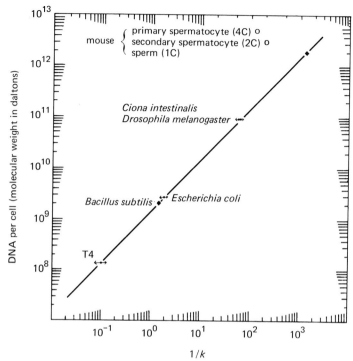

Fig. 6.1. Correlation between genome size and $1/k$ predicted by equation (6.5). k is the observed rate constant for renaturation, fit to the data by assuming second-order kinetics [see equation (6.2) and text]. Original sources of the genome size and renaturation rate data are given by Laird (1971). Observations were made by the hydroxyapatite procedure. Horizontal lines about each point represent the 95% confidence limits ($\pm 2\sigma$) for the rate constant determinations. The DNA had been sheared by passage through a needle valve at 12,000 lb in.$^{-2}$, and was thought to be about 400 nucleotides. However, the absolute values of the renaturation rates obtained suggest that the fragments may actually have been somewhat longer. Values of haploid DNA content per cell are plotted for *Ciona* and *Drosophila*, and for the mouse, where sperm was used for the genome size determination. From C. D. Laird (1971). *Chromosoma* **32**, 378.

RENATURATION OF REPETITIVE SEQUENCES

The concept of sequence concentration used in equations (6.3) and (6.4) also provides the basis for understanding the renaturation of repetitive sequences. Since the rate of the reaction depends on sequence concentration, a repetitive sequence present many times per genome will react at a rate which is proportionately higher than a sequence present once per genome. Suppose a particular class of repetitive sequences occurs at a frequency of F copies per genome. If all sequences of this repetition class are taken together, they occupy a fraction α of the genome, the total size of which we again term G. Following the definition for repetitive sequence complexity given above, the complexity of this set of repetitive sequences, N, is thus

$$N = \frac{\alpha G}{F} \qquad (6.7)$$

This relation of course reduces to $N = G$ for the case where the genome is entirely nonrepetitive. When the DNA is sheared, the sequence concentration, C_r, for a given family of repetitive sequences is [see equation (6.3)]

$$C_r = \frac{FC_0}{G} \qquad (6.8)$$

where C_0 is again the total concentration of nucleotides. Therefore, as in equation (6.4),

$$\frac{C_{r(SS)}}{C_r} = \frac{1}{1 + (k_i F / G) C_0 t} \qquad (6.9)$$

From this expression, the observed rate constant for reaction of the repetitive sequences can be seen to be proportional to F, the repetition frequency, and inversely proportional to G, the genome size. It follows from equations (6.9) and (6.4) that the value of F for a given class of repetitive sequences can be directly measured simply by renaturing whole, sheared DNA. That is,

$$F = \frac{k_r}{k_{sc}} \qquad (6.10)$$

where k_r is the observed rate at which the repetitive sequences in the whole DNA renature, and k_{sc} is the rate at which the single copy sequences renature in the same experiment. Equation (6.10) states that a repetitive DNA fraction renatures faster than a single copy sequence in the same genome in proportion to the number of copies comprising each repetitive sequence family. It will be noted that in a DNA which includes both repetitive and nonrepetitive sequence, k_{sc} is inversely proportional to the genome size. This is shown in Fig. 6.1 for mouse and *Drosophila* DNA.

Figure 6.2 illustrates the renaturation kinetics of a typical animal DNA, that of the sea urchin *Strongylocentrotus purpuratus*. The ordinate shows the fraction of the DNA fragments present in duplex-containing structures, as measured by binding to hydroxyapatite under temperature condi-

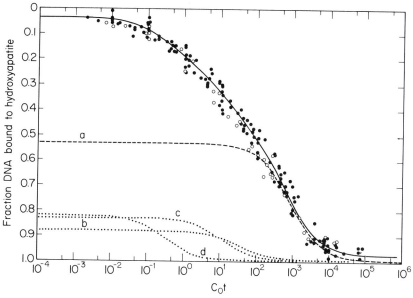

Fig. 6.2. Renaturation kinetics of 450 nucleotide long sea urchin DNA fragments. Ordinate shows fraction of fragments bound to hydroxyapatite after incubation to various C_0t's. C_0t (DNA concentration × time, in moles nucleotide liter^{-1}sec) is plotted on a log scale on the abscissa. Closed circles show data obtained at 60°C, and open circles show data obtained at 50°C. Reactions at 60°C were carried out in 0.12 M phosphate buffer (0.18 M Na$^+$) or are plotted at the C_0t values calculated to be equivalent to 0.18 M Na$^+$ on the basis of the amount of acceleration in rate expected for Na$^+$ concentrations above 0.18 M (Wetmur and Davidson, 1968; Britten *et al.*, 1974). Reactions at 50°C were run in 0.14 M phosphate buffer. A computer operating with a nonlinear least-squares program has been used to fit the renaturation data to equation (6.11). The dashed line (a) shows the renaturation in whole DNA of fragments bearing only single copy sequence. The rate constant for this reaction is 1.25 × 10^{-3} M^{-1}sec^{-1}, and this component includes 47% of the DNA. The dotted lines (b–d) show various repetitive sequence kinetic components. These components probably all represent averages of many individual sequence families differing in repetitiveness, and several alternative kinetic components can be resolved with little change in the root mean square error. Furthermore, each fragment renatures at the rate characteristic of the most highly repetitive sequence of recognizable length which it contains, and less repetitive sequences may also be present on it. For curve b in this particular solution the rate constant is 3 × 10^{-2} M^{-1}sec^{-1}; the average repetition of each sequence would be about 20–30, and the fraction of fragments bearing these sequences is 12%. For curve c the rate constant is 1 × 10^{-1} M^{-1}sec^{-1}; the average repetition of each sequence is about 80, and the fraction of fragments included is 17%. For curve d the rate is 3.3 M^{-1}sec^{-1}; the average repetition of each sequence is about 2600, and the fraction of fragments included is 16%. Data from D. E. Graham, B. R. Neufeld, E. H. Davidson, and R. J. Britten (1974). *Cell* **1**, 127; and F. C. Eden, D. E. Graham, M. J. Smith, G. A. Galau, W. H. Klein, E. H. Davidson, and R. J. Britten, unpublished data.

tions equivalent to those at which the renaturation was permitted to occur. The abscissa of Fig. 6.2 is calibrated on a log scale in units of concentration × time for the reacting DNA solutions; that is, the "C_0t" term of
equation (6.2). This mode of presenting renaturation data is known as a
"C_0t plot," and was introduced by Britten and Kohne (1968a). The units of
C_0t are customarily moles nucleotide liter^{-1} sec. Figure 6.2 shows that the
renaturation of sea urchin DNA occurs over a range of more than 6
decades in C_0t. Evaluation of equation (6.2) shows that a single kinetic
component can occupy only two decades between 10% and 90% of the
reaction. Thus, a number of individual kinetic components evidently contribute to the overall curve shown. Equation (6.10) shows that each of
these represents a particular repetition frequency class. The most slowly
renaturing fraction consists of fragments which contain recognizable
lengths only of nonrepetitive sequence (curve a). The dotted lines (curves
b, c, and d) show the faster kinetic components resolved from the overall
curve. These represent the reaction of sets of fragments bearing repetitive
sequences. Moving from right to left, each component of the reaction
represents the renaturation of increasingly repetitive sequences. It is important to realize that in an experiment such as that shown in Fig. 6.2, the
DNA fragments renature and become bindable by hydroxyapatite at rates
appropriate for the most highly repetitive sequence which they contain.
Thus, because repetitive and nonrepetitive sequences are interspersed
(Chapter 1), many fragments contain elements of both these sequence
classes. These will be bound to hydroxyapatite after incubation to C_0t's
permitting the reaction of the repetitive sequences only. Therefore the
quantity of interspersed repetitive sequence is overestimated and the
quantity of nonrepetitive sequence underestimated in hydroxyapatite as-
says of renaturation. The magnitude of this effect depends on the fragment length and on the spacing of the interspersed repetitive and nonrepetitive sequences in the genome.

The kinetic analysis shown in Fig. 6.2 is based on the principle that the
overall renaturation is the sum of a series of second-order components,
each behaving according to equation (6.2). The calculation is thus carried
out as shown in the following expression:

$$\frac{C}{C_0} = \beta + \frac{\alpha_1}{1 + k_1C_0t} + \frac{\alpha_2}{1 + k_2C_0t} + \frac{\alpha_3}{1 + k_3C_0t} + \cdots \quad (6.11)$$

where β is the fraction of DNA remaining unassociated at the termination of the reaction and each kinetic component, representing a fraction
α_j of the genome, renatures with the observed rate constant k_j. The various k's depend on the repetition frequency of the sequence components
they represent, as in equation (6.10). Customarily the parameters of equa-

tion (6.11) are derived from renaturation data by nonlinear least squares or equivalent procedures with the aid of a computer. The kinetic components calculated for the data shown in Fig. 6.2 are illustrated by the dotted lines, and the values of k_j and α_j in equation (6.11) are listed in the legend to the figure.

It is important to note that kinetic components such as those portrayed in Fig. 6.2 are only numerical averages. Without extensive additional analysis, for instance, it is impossible to assume that sequences present around 10^2 times per genome actually constitute a major class in sea urchin DNA. Rather, it is likely that there are repetitive sequences in this genome which are present over a range from 10^1 to 10^3 times, and for which the *average* representation is 10^2-fold repetition. In some DNA's more or less discrete repetitive sequence frequency classes have been demonstrated by means of physical isolation and detailed measurement of their reaction kinetics (reviewed by Britten and Davidson, 1971; Davidson and Britten, 1973). One example is the major repetitive sequence class of *Xenopus* DNA, consisting of sequences present 1×10^3 to 2×10^3 times per genome (Hough and Davidson, 1972).

An isolated kinetic fraction of DNA renatures more rapidly than it does in the presence of all the other sequences in the DNA, since its sequence concentration increases as it is purified. The observed (or calculated) rate constant for the renaturation of a purified frequency component is called "k_{pure}" where, as above, α is the fraction of the genome occupied by this component, and k is its observed renaturation rate constant in whole DNA.

$$k_{\text{pure}} = \frac{k}{\alpha} \tag{6.12}$$

The complexity of the repetitive component can either be calculated directly from k_{pure} by means of equations (6.5) or (6.6) or estimated from its repetition frequency and quantity by means of equation (6.7).

An important feature included in Fig. 6.2 is that data obtained at 50°C (open circles) are only slightly different from those obtained at 60°C (closed circles). This is a fairly typical, though not universal result. It cannot be assumed that the renaturation of a previously unstudied DNA will be insensitive to such criterion changes. When this is the case, as with sea urchin DNA, it means that modest change in the experimental conditions under which renaturation occurs has little or no effect on the apparent quantity of repetitive DNA or on its apparent degree of repetition. Though moderately repetitive sequences in animal DNA are typically divergent, the stability of repetitive duplexes formed by renaturation is high enough so as not to affect the results under usual conditions. We have already noted that the repetitive duplexes formed by renaturing sea urchin DNA melt about 8°–10°C below native DNA duplexes of the same length because of sequence mismatch. However, neither this nor the small additional decrease in stability because of the average duplex length in renaturation experiments carried out with fragments several hundred nucleotides long is sufficient to affect the renaturation rate. The reason for this is that the rate of renaturation remains close to the optimum over the whole range from 15°–30°C below the melting temperature of the renatured strand pairs (Bonner *et al.*, 1973; Wetmur, 1976). Only as the incubation temperature approaches the melting temperature does the rate drop sharply. Most of the duplexes scored in the experiments shown in Fig. 6.2 melt between 73°C and 83°C, and thus, as expected, renaturation at 50°C is found to be equivalent to renaturation at 60°C. Of course, if the temperature were increased above 60°C, some effect on the renaturation kinetics would be expected.

KINETICS OF DISAPPEARANCE OF SINGLE STRAND NUCLEOTIDES AND RNA HYBRIDIZATION IN DNA EXCESS

It was noted earlier in this discussion that the renaturation of randomly sheared DNA fragments is not an ideal second-order reaction, and this fact has some interesting and significant consequences. Random shearing produces fragments which begin at as many different sites as there are nucleotide pairs in the genome. Therefore, when any two fragments of DNA bearing complementary sequences react, single-stranded tails will almost always remain after the complementary regions have become paired. For the fragment length distributions produced by the usual random shearing procedures, the mean length of duplex resulting from a single nucleation event is about 55% of the mean single strand length (Smith *et al.*, 1975). The result is that, except at the beginning and end of a renaturation reaction, the fraction of *nucleotides* remaining single stranded is always significantly greater than the fraction of *fragments* remaining completely single stranded. It is the latter, of course, which is measured by the hydroxyapatite binding assay. Morrow (1974) and Smith *et al.* (1975) showed that the fraction of nucleotides remaining single stranded, S, is approximated by the expression

$$\frac{S}{C_0} = \left(\frac{1}{1 + kC_0 t} \right)^n \tag{6.13}$$

Here C_0 is the total DNA concentration and t the time, as above. The observed rate constant k also has exactly the same meaning as in equation (6.2). It is found empirically that the best value of n in equation (6.13) is about 0.45. The form of equation (6.13) is clearly non-second-order, except when the exponent $n = 1.0$, in which case equation (6.13) reduces to equation (6.2). If shearing occurs at specific rather than random locations, as, for example, when restriction enzymes are used to cut a single copy DNA, the value of n in equation (6.13) does in fact equal 1. Thus, Morrow (1974) showed that the disappearance of single-stranded nucleotides in SV40 DNA treated with restriction endonucleases follows perfect second-order kinetics.

In practice S is usually measured by the use of single-strand-specific nucleases or by optical methods (though the detailed analysis of complete kinetics as measured by optical hypochromicity has yet to be carried out). Most commonly used is the single-strand-specific Sl nuclease derived from *Aspergillus*. In Fig. 6.3 the renaturation kinetics of randomly sheared 700 nucleotide long *E. coli* DNA are shown. The two curves represent the kinetics of the reaction as assayed by hydroxyapatite binding and by Sl

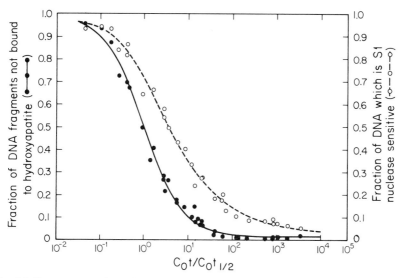

Fig. 6.3. Renaturation of *E. coli* DNA assayed by hydroxyapatite binding and Sl nuclease resistance. The DNA was randomly sheared to about 700 nucleotides. DNA duplex formation was measured by binding to hydroxyapatite at 60°C in $0.12\,M$ phosphate buffer (●), and by resistance to digestion with single-strand-specific Sl nuclease (○). Two sets of data were pooled by normalizing both to a hydroxyapatite $C_0t_{1/2}$ of 1.0 (i.e., $C_0t/C_0t_{1/2}$ for each data set). The solid line represents the least-squares solution of the renaturation kinetics as assayed by hydroxyapatite binding, according to equation (6.2) in text. The dashed line represents the least-squares solution for the Sl nuclease kinetics, according to equation (6.13) in text. The normalized rate constant k is 1.0 for both curves and the best value for n in equation (6.13) for the Sl nuclease curve is 0.453. From M. J. Smith, R. J. Britten, and E. H. Davidson (1975). *Proc. Natl. Acad. Sci. U.S.A.* **72**, 4805.

nuclease resistance. The hydroxyapatite data (solid circles) are fit according to equation (6.2), and the Sl nuclease data according to equation (6.13). Figure 6.3 demonstrates that at any given C_0t value less of the DNA is resistant to Sl nuclease than binds to hydroxyapatite. The explanation for the exact form of the Sl nuclease kinetics illustrated here is complex. This subject has been further developed by Smith *et al.* (1975) and Britten and Davidson (1976) but lies outside of the scope of the present discussion. It should be stressed that when renaturation of randomly sheared DNA is assayed by hydroxyapatite binding, the actual deviations from perfect second-order kinetics are so small as to be nearly undetectable, as can be seen in Fig. 6.3. Furthermore, when the Sl nuclease kinetics are fit with the form shown in equation (6.13), the basic relations between the observed rate constant, the complexity, and the repetition frequency remain unaffected. Thus, as mentioned previously, equations (6.5), (6.6), and (6.10) do not depend on the exact order of the reaction kinetics or the means of assay, so long as appropriate methods are used to evaluate the observed rate constant, k.

For our present purpose the main importance of the Sl nuclease kinetic determination illustrated in Fig. 6.3 is that it provides a basis for analysis of a very useful class of hybridization and renaturation reactions. These are reactions in which a labeled RNA or DNA present in trace quantities is reacted with excess DNA. Such reactions are required in order to determine whether an RNA is transcribed from repetitive or single copy sequences, or both. In an experiment of this type the excess DNA is termed the "driver DNA." The tracer is capable of reacting with all of the single-stranded DNA sequence present in the driver at any given time [i.e., the DNA whose concentration is denoted by S in equation (6.13)] and not solely with those fragments which remain completely single stranded at that time [i.e., the DNA whose concentration is denoted by C in equations (6.1) and (6.2)]. Therefore equation (6.13) must be used to determine the concentration of reactive driver DNA sequence. Furthermore it often happens that the rate of reaction of the tracer nucleic acid with single-stranded driver DNA sequence differs from that of the driver fragments with each other. For example, the tracer fragment length may be shorter, or the tracer may be RNA. For reasons which remain obscure, in DNA-driven hybridization reactions the rate of RNA–DNA duplex formation is significantly lower than that of DNA–DNA duplex formation even when fragment lengths are carefully taken into account (Melli *et al.*, 1971; Smith *et al.*, 1976). Several relevant examples of DNA excess hybridization experiments are discussed later in this chapter. The following expression has been derived for the purpose of analyzing such reactions (Davidson *et al.*, 1975b). Suppose the observed rate constant for tracer–

DNA duplex formation is called h, S is the concentration of single-stranded nucleotides in driver DNA as above, and U is the fraction of the tracer fragments remaining unreacted at time t. Then

$$\frac{dU}{dt} = -hUS \qquad (6.14)$$

Substituting equation (6.13) for S and solving, we have

$$\frac{U}{U_0} = \exp \frac{h\left[1 - (1 + kC_0t)^{1-n}\right]}{k(1 - n)} \qquad (6.15)$$

Note that k is the same observed rate constant as in equations (6.2) and (6.13), here applied to the driver DNA.

RNA EXCESS HYBRIDIZATION KINETICS

We turn now to the analysis of RNA excess hybridization reactions. RNA excess reactions are required for measurement of the sequence complexity of RNA populations and for comparisons of the sets of sequences present in various RNA populations. These reactions are usually carried out with labeled DNA, and the RNA excess is such that the concentration of RNA at the start of the experiment, R_0, remains essentially unchanged throughout. The reaction is thus pseudo-first-order in form. We let D represent the unreacted DNA tracer concentration at time t, R_0 the starting (and final) RNA concentration, and k_H the observed rate constant for RNA–DNA duplex formation. The rate of change in the concentration of unreacted DNA tracer is then

$$\frac{dD}{dt} = -k_H R_0 D \qquad (6.16)$$

Thus,

$$\frac{D}{D_0} = e^{-k_H R_0 t} \qquad (6.17)$$

Here D/D_0 is the fraction of DNA tracer remaining unreacted.

For each kinetic component of the form given by equation (6.17) the pseudo-first-order hybridization rate constant, k_H, is inversely proportional to the RNA complexity. The complexity of the class of RNA reacting with these kinetics can be measured independently, however, simply by determining the fraction of the total DNA tracer hybridized at the termination of the reaction. Thus if the RNA complexity is known, the concentration in the total RNA of those species driving the reaction can be calculated. For this calculation, we require the rate constant expected for a pure RNA of the measured complexity. This is obtained by use of a proportionality constant relating RNA complexity and the pseudo-first-order rate constant, k_H [see equation (6.5)].

Figure 6.4 displays a set of measurements which provides the necessary information (Galau et al., 1976b). Here two pseudo-first-order reactions are shown, which follow closely the kinetics described by equation (6.17). The open circles represent the reaction of ϕX174 ^{3}H-DNA tracer driven by excess ϕX RNA. The RNA had been transcribed enzymatically from the (−) strand of ϕX DNA. The complexity of the ϕX174 genome (and of the RNA transcript) is known to be about 5400 nucleotides, and k_H is evaluated from the data in Fig. 6.4 at about 170 $M^{-1}sec^{-1}$. The expected rate constant for the reaction of any other RNA of known complexity can

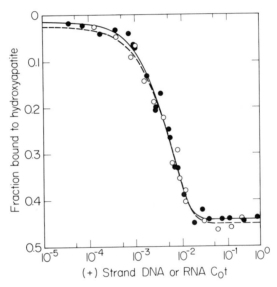

Fig. 6.4. Pseudo-first-order reactions driven by ϕX174 RNA and DNA. Trace quantities of 300 nucleotide long ϕX174 replicative form (RF) ^{3}H-DNA were reacted with excess ϕX (+) strand RNA or DNA in 0.12 M phosphate buffer at 60°C. The (+) strand DNA was obtained from mature phage and the RNA was transcribed in vitro using E. coli polymerase. The fraction of ^{3}H-DNA in duplex was assayed by binding to hydroxyapatite in the same phosphate buffer at 60°C. Data were fit to the pseudo-first-order function shown in equation (6.17) by least squares methods. The maximum possible reaction is 50% since both driver nucleic acids are complementary to only one strand of the ^{3}H-DNA tracer. (●—●) Reaction of ^{3}H-DNA with excess unlabeled 300 nucleotide (+) strand DNA. The best fit rate constant is 200 M^{-1} sec^{-1}. (O--O) Reaction of ^{3}H-DNA with excess unlabeled 360 nucleotide long (+) strand RNA. The best fit rate constant is 170 M^{-1} sec^{-1}. From G. A. Galau, R. J. Britten, and E. H. Davidson (1976b). Proc. Natl. Acad. Sci. U.S.A., in press.

be calculated by application of these data to equation (6.6), i.e., for these conditions and fragment lengths.

In the reaction displayed by the closed circles in Fig. 6.4 (+) strand ϕX174 DNA is used to drive the reaction of the same tracer. This reaction is directly analogous to the one just considered, in that like the (+) strand RNA, the driver DNA cannot react with itself. The reaction is therefore again pseudo-first-order, as demonstrated by the kinetics shown in Fig. 6.4. The significance of this reaction is that its rate is almost indistinguishable from that of the RNA-driven reaction. The best fit value of k_H for this reaction is 200 $M^{-1}sec^{-1}$, close to that expected on the basis of the second-order reaction kinetics. The experiment of Fig. 6.4 shows that in RNA excess reactions the basic rate of formation of RNA–DNA duplex is almost the same (probably within 20%) as the basic rate of formation of DNA–DNA duplex.

Once the complexity of an RNA is known and the expected hybridization rate is calculated, the fraction of the total RNA serving as driver can be easily obtained. The driver fraction which we shall call Df, is simply the ratio of the expected rate, $k_{H_{exp}}$, calculated for a pure RNA of the measured complexity, to the observed rate actually fit to the data points on the basis of the total RNA in the reaction mixture, $k_{H_{obs}}$

$$Df = \frac{k_{H_{exp}}}{k_{H_{obs}}} \qquad (6.18)$$

For example, suppose a hundredfold excess of ribosomal RNA were added to the ϕX RNA reaction shown in Fig. 6.4, and the RNA C_0t were calculated on the basis of total RNA concentration. The rate constant would then be 1.7 $M^{-1}sec^{-1}$ rather than 170 $M^{-1}sec^{-1}$, and equation (6.18) would show that the reacting RNA species constitutes 1% of the total RNA. Equation (6.18) provides information as to the quantity of the hybridizing RNA's in the preparation which often can be obtained in no other way. As will be illustrated below, valuable parameters such as the number of RNA's of a given sequence per cell can be calculated in this way.

Sequence Complexity of Oocyte RNA

SINGLE COPY SEQUENCE COMPLEXITY OF OOCYTE RNA'S

The RNA's stored in the mature oocyte can be considered the set of transcripts required as development begins. The complexity of this set of transcripts is therefore a question of fundamental interest. Studies have been carried out on the oocyte RNA's of two sea urchins (*Strongylocen-*

trotus purpuratus and *Arbacia punctulata*), the amphibian *Xenopus laevis*, and the echiuroid worm *Urechis caupo*. The complexities measured for these RNA's are listed in Table 6.1. The most extensive measurements shown are those of Galau *et al.* (1976c) and Hough-Evans *et al.* (1976) carried out on *Strongylocentrotus purpuratus* oocyte RNA's. The method used was RNA excess hybridization to a single copy DNA tracer as described above (Fig. 6.4). Thus the complexity is calculated from the amount of single copy DNA hybridized at the termination of the reaction (see Table 6.1 for details). The complexity of the RNA in the *Strongylocentrotus* oocyte is found to be about 37×10^6 nucleotides. In addition, Anderson *et al.* (1976), using similar procedures, measured a complexity which falls within about 20% of this value for the oocyte RNA of a different sea urchin species (Table 6.1). It seems, therefore, that the complexity of the sea urchin oocyte RNA lies in the range 30×10^6 to 40×10^6 nucleotides. These measurements show that the stored RNA of the sea urchin oocyte contains almost an order of magnitude more single copy sequence than does the whole *E. coli* genome.

Table 6.1 shows that the complexities of *Urechis* and *Xenopus* oocyte RNA's are similar to that of sea urchin oocyte RNA. This is a significant observation, in that echiuroid worms, sea urchins, and amphibians belong to the most diverse phylogenetic groups, shed eggs which vary greatly in total RNA content (see Table 5.4), and possess different sized genomes. With respect to the latter point, a pertinent comparison has been made by Rosbash *et al.* (1974). Their experiments showed that the complexity of *Triturus* oocyte RNA is about the same as that of *Xenopus* oocyte RNA, though *Triturus*, a urodele amphibian, contains about seven times more DNA in its genome. The oocyte RNA complexity data thus imply that in all these organisms a certain set of transcripts is required for the initial stage of embryogenesis. This is suggested by the fact that the complexities of all the measured oocyte RNA's fall in the range 25×10^6 to 50×10^6 nucleotides. We return to the developmental significance of oocyte RNA's later in this chapter, an issue which clearly revolves around their physiological function.

Table 6.1 also lists the percent of the total oocyte RNA accounted for by the sets of molecules whose complexities are given in column 2. We refer to this subfraction as the "complex class" of the total oocyte RNA. From the amount of the total RNA which is of the complex class the average number of copies of each complex class sequence present in the oocyte can be calculated. Table 6.1 shows that this parameter varies greatly according to species. For example, there appear to be three orders of magnitude more copies of each complex class RNA sequence in a *Xenopus* oocyte than in a sea urchin oocyte. However, on a per unit mass or per ribosome basis the number of copies of each RNA sequence is not different. The *Xenopus* oocyte has about a thousand times more ribosomes than does the sea urchin oocyte (Table 4.2), which compensates for the difference in the number of copies of each complex RNA sequence in the oocytes of these species. The *Urechis* egg closely resembles the sea urchin egg in the representation of the complex oocyte RNA sequences (Table 6.1). *Urechis* oocytes contain about four times the amount of total RNA as do sea urchin oocytes (Table 5.4). On this basis they include on the average slightly fewer complex maternal RNA sequences per ribosome. Nonetheless, the main result of these comparisons is that the quantity of each maternal sequence stored in the oocyte seems to be correlated with the ultimate capacity for protein synthesis represented by the stockpile of maternal ribosomes.

In Chapter 4 we estimated the quantity of maternal messenger RNA in the sea urchin oocyte to be about 50–100 pg (Table 4.2). From Table 6.1 it appears that the amount of complex oocyte RNA is 16.5–49.5 pg (i.e., 0.5–1.5% of the 3.3 ng of RNA per oocyte). In other words, the amount of complex class RNA is equivalent to between 15 and 100% of the probable quantity of maternal message in the oocyte. Therefore, if the complex oocyte RNA's are maternal message, they represent an appreciable fraction of the total store of maternal messenger RNA. In *Xenopus* oocytes the amount of stored poly(A)RNA is about 40–70 ng (Table 4.2), and this is equal to the quantity of complex RNA calculated from the data in Table 6.1 (see footnote k of Table 6.1). Furthermore, as might be expected from this equivalence, the oocyte poly(A)RNA is mainly nonrepetitive sequence transcript, a property which is characteristic of most *Xenopus* messenger RNA. This was shown directly by Rosbash *et al.* (1974), who transcribed the oocyte poly(A)RNA with reverse transcriptase. The resulting cDNA was then renatured with excess whole *Xenopus* DNA. The kinetics of the reaction showed that most of the cDNA had reacted with single copy sequence in the driver DNA. We conclude that it is likely that the poly(A)RNA of *Xenopus* oocytes studied by Rosbash *et al.* (1974) and the total oocyte RNA whose complexity was measured by Davidson and Hough (1971) are the same or largely overlapping RNA populations. If this is so, the complex RNA's probably constitute most of the maternal message in the *Xenopus* oocyte.

COMPLEXITY OF REPETITIVE SEQUENCE TRANSCRIPTS IN OOCYTE RNA

Hough and Davidson (1972) showed that at least 2% of *Xenopus* oocyte RNA consists of nonribosomal repetitive sequence transcripts. These

6. *RNA Complexity and Structural Gene Transcription*

TABLE 6.1. Complexity of Mature Oocyte RNA's

(1) Organism	(2) Complexity (nucleotides $\times 10^{-6}$)	(3) Percent of total RNA consisting of molecules of complexity in column 2	(4) Representation of complex RNA: Estimated number of molecules of each sequence per oocyte	References
Strongylocentrotus purpuratus	37 ± 4^a	$0.5–1.5^b$	1.6×10^{3} ^{c}	Galau *et al.* (1976c) Hough-Evans *et al.* (1976)
Arbacia punctulata	30^d	$1–2^e$		Anderson *et al.* (1976)
Urechis caupo	$31–47^f$	0.18^g	1×10^{3} ^{h}	Davis (1975)
Xenopus laevis	27^i 40^j	$\sim 1^k$	1.8×10^{6} ^{l}	Davidson and Hough (1971) Rosbash and Ford (1974), Rosbash *et al.* (1974)

^{a} The error estimate represents ± 1 standard deviation around the complexity value given. These measurements were made by the excess RNA–single copy DNA tracer procedure. As required, the kinetics of the reaction were fit well by equation (6.17). Termination occurs at about 3% of the single copy DNA hybridized. The single copy sequence length in this genome is 6.1×10^8 nucleotide pairs (Graham *et al.*, 1974). Thus the RNA complexity (on the basis of asymmetric transcription) is $(2)(0.03)(6.1 \times 10^8) = 37 \times 10^6$ nucleotides (Galau *et al.*, 1976c; Hough-Evans *et al.*, 1976).

^{b} This calculation (Galau *et al.*, 1976b) is based on application of equation (6.18), where $k_{H_{exp}}$ is calculated directly from the pseudo-first-order ϕX174 standard curve shown in Fig. 6.4 and from the complexity measured for the oocyte RNA (column 2 of this table). The uncertainty shown is due to ignorance of the length of the hybridizing RNA molecules, which would affect the $k_{H_{exp}}$.

^{c} Calculated using 1% as an average figure for the fraction of sea urchin oocyte RNA which is occupied by complex RNA species (column 3), and 3.3 ng as the content of total RNA per oocyte (Table 5.4). The representation is calculated as the number of nucleotides in the complex RNA class divided by the complexity (column 2).

^{d} This measurement was made by the excess RNA–single copy DNA tracer method (Anderson *et al.*, 1976), and pseudo-first-order kinetics were observed [equation (6.17)]. The complexity calculation was as in footnote a. It was assumed that 75% of the *Arbacia* genome is single copy, as is the case for *Strongylocentrotus*. Some uncertainty in this measurement exists with respect to the fraction of the single copy ^{3}H-DNA tracer able to react when the experiments were carried out. The listed value is based on the reactivity with excess DNA measured for the DNA fragments recovered from DNA–RNA hybrids.

^{e} Calculated from the kinetic data of Anderson *et al.*, (1976) by use of equation (6.18), as in footnote b.

^{f} Calculated from data of Davis (1975) who used the excess RNA–single copy DNA tracer method. Appropriate pseudo-first-order kinetics [equation (6.17)] were observed. The amount of single copy DNA in the *Urechis* genome is not known accurately. From renaturation data obtained in the writer's laboratory, as well as the data of Davis (1975), the single copy content most likely falls in the range of 40–60%. Davis showed that 4.3% of the single copy DNA hybridizes with oocyte RNA. The genome size of *Urechis* is 9.1×10^8 nucleotide pairs (Dawid and Brown, 1970). Thus the complexity of the oocyte RNA is $(2)(0.043)(0.4)(9.1 \times 10^8) = 3.1 \times 10^7$ nucleotides, to $(2)(0.043)(0.6)(9.1 \times 10^8) = 4.7 \times 10^7$ nucleotides.

[g] Calculated from data shown by Davis (1975) using equation (6.18) and the complexity of 47×10^6 nucleotides given in column 2.

[h] Calculated as in footnote c, using 14 ng total RNA per oocyte (Table 5.4) and the values given in columns 2 and 3.

[i] Calculated from data of Davidson and Hough (1971). These experiments were carried out by reacting increasing amounts of oocyte ^{3}H-RNA, labeled uniformly *in vitro*, with single copy DNA. At low RNA/DNA ratios the reactions are DNA driven. The reactions become RNA driven at higher RNA/DNA ratios, and the terminal values were obtained at these ratios. However, no kinetics for the RNA excess reactions were obtained. The authors concluded from the fraction of single copy DNA hybridized at termination (0.6%) that the complexity of *Xenopus* oocyte RNA is 20×10^6 nucleotides. However, it was shown by Davidson *et al.* (1973) that repetitive and nonrepetitive sequences are closely interspersed in *Xenopus* DNA (Chapter 1), and the fraction of the total DNA which is single copy sequence is hence underestimated when measured by hydroxyapatite binding. Rather than the 55% of the genome originally considered to be single copy sequence, we now apply more accurate value of 75% (Davidson *et al.*, 1973). Hence $(2)(0.006)(0.75)(3 \times 10^9$ nucleotides/genome) $= 27 \times 10^6$ nucleotides.

[j] Calculated from other data reported by Davidson and Hough (1971), and obtained by hybridizing RNA in large excess with single copy ^{14}C-DNA tracer. The fraction of the ^{14}C-DNA recovered as hybrid at terminal RNA C_0t's was 0.9%, representing 1.8% of the single copy sequence.

[k] Davidson and Hough (1971) estimated that ≤0.1% of the oocyte RNA is complementary to the single copy DNA sequence set reacting in their experiments. This estimate was based on the ratio of RNA to single copy DNA at which saturation of the hybridized single copy sequences occurred. However, it has since been shown (Smith *et al.*, 1976) that in DNA excess RNA–DNA hybridization is somewhat retarded, and thus at the lower RNA/DNA ratios (see footnote i) the amount of hybridization obtained was probably kinetically limited. This does *not* affect the terminal values upon which the complexity is based. However, it means that the representation values obtained by Davidson and Hough (1971) could be too low. Therefore, we use data of Rosbash and Ford (1974) and Rosbash *et al.* (1974) to obtain the value shown in column 3. These authors studied the complexity of that fraction of *Xenopus* oocyte RNA which contains poly(A) sequence (see Fig. 4.2, and Table 4.2). Their complexity measurements agree well with those shown in column 2. Their measurements were carried out by using reverse transcriptase to synthesize cDNA against the oocyte poly(A)RNA, and then determining the rate of reaction of the oocyte poly(A)RNA with the cDNA. The kinetics of the reaction show that all the poly(A)RNA drives the cDNA reaction. Since about 1% of the oocyte RNA is poly(A)RNA (i.e., 40 ng), this is taken as the content of oocyte RNA representing the sequences whose complexity is shown in column 2.

[l] Calculated as in footnote c, using 4 µg total RNA per oocyte (Table 5.4), taking 1% (column 3) as the fraction of total RNA which is of the complex class. For this calculation the value of 40×10^6 nucleotides (column 2) was used as the RNA complexity.

cannot be accounted for wholly by known RNA species such as histone messenger RNA's and transfer RNA's. In *Xenopus* oocytes the complexity of the total repetitive sequence transcripts is more than 3×10^4 nucleotides [calculated from data of Hough and Davidson (1972) taking into account the amount of hybridization observed and the 75% reactivity of the tracer DNA]. Similar observations were made on the oocyte RNA's of the amphibian *Engystomops pustulosus* (Hough *et al.*, 1973). The histone messages and transfer RNA's might account for about 30% of the observed repetitive sequence complexity. The nature of the remaining repetitive sequence transcripts remains hypothetical. It is known in other animal systems (e.g., Campo and Bishop, 1974; Klein *et al.*, 1974; Spradling *et al.*, 1974) that, aside from histone messages, up to about 20% of the total messenger RNA is transcribed from repetitive structural genes, and nonhistone maternal messages transcribed from such repetitive genes could well be included in the oocyte RNA. Alternatively these repetitive sequence transcripts could have some completely different function.

A number of older observations also indicate the presence of heterogeneous repetitive sequence transcripts in oocyte RNA's. These studies were carried out by filter hybridization procedures capable of detecting the reaction of repetitive sequences only. This limitation results from failure to achieve high enough C_0t to permit reaction of single copy sequences. DNA-driven filter reactions suffer from the problem that relatively low amounts of DNA can be loaded on the filters. Furthermore, it was long customary to permit the reactions to continue for only 16–24 hours. Nor have sufficient RNA concentrations and reaction times been applied in most RNA excess filter hybridization experiments to generate high RNA C_0t's. Reliable kinetics are generally not available from filter experiments in any case, since the reaction of filter-bound DNA can be shown to be retarded by large factors. To summarize, the filter hybridization experiments can tell us nothing about single copy transcripts, and information regarding repetitive sequence transcripts can be extracted from them only with great caution. Nonetheless, some of the experiments yield interesting qualitative conclusions relevant to the sequence content of oocyte RNA. These are now briefly reviewed.

RNA's extracted from *Xenopus* oocytes and uniformly labeled *in vitro* with dimethyl sulfate were hybridized with filter-bound DNA by Crippa *et al.* (1967). This experiment shows that about 3% of the total DNA sequence is represented by the repetitive sequence transcripts in oocyte RNA. In this case the result was within a factor of about two of that obtained by the modern RNA excess hybridization methods described above. That is, the data of Hough and Davidson (1972) show about 6% of the repetitive DNA sequence is represented in oocyte RNA, and repetitive

DNA is about 25% of the total, so that by this measurement about 1.5% of the total DNA would be represented by the repetitive sequence hybrids. A similar filter hybridization study was carried out on sea urchin oocyte RNA by Hynes and Gross (1972). They reported that somewhat more than 3% of the total DNA (i.e., 12% of the repetitive sequence) is homologous to dimethyl sulfate labeled oocyte RNA, but their reaction failed to terminate, and they reached no quantitative conclusion. Most of the other filter experiments dealing with repetitive sequence transcripts in oocyte RNA were carried out by the competition method. These experiments all show that the hybridization of labeled (repetitive sequence) transcripts from embryos can be competed more or less effectively by total oocyte RNA. A minimum conclusion is that the oocyte contains various heterogeneous repetitive sequence transcripts. Among the oocytes for which this conclusion probably can be drawn are those of sea urchins (Glišin *et al.*, 1966; Whiteley *et al.*, 1966; Whiteley *et al.*, 1970; Chetsanga *et al.*, 1970), the sand dollar *Dendraster excentricus* (Whiteley *et al.*, 1970; Mizuno *et al.*, 1974), the gastropod *Acmaea scutum* (Karp and Whiteley, 1973), the oyster *Crassostraea gigas* (McLean and Whiteley, 1974), and the tunicate *Ascidia callosa* (Lambert, 1971). An additional implication of these competition studies is that some of the repetitive sequence transcripts synthesized during early embryogenesis are homologous with those stored in the oocyte. This interesting result is examined later in the present chapter. We now consider a different aspect, viz., the persistence of maternal oocyte RNA sequences in later development.

OOCYTE REPETITIVE SEQUENCE TRANSCRIPTS IN EARLY EMBRYOS

Heterogeneous repetitive sequence transcripts originally stored in oocytes have been demonstrated in a series of low C_0t hybridization experiments performed on amphibian embryos. The first experiments of this kind were carried out on *Xenopus* RNA's by Crippa *et al.* (1967) and Davidson *et al.* (1968). It was shown that RNA's extracted from cleavage- and blastula-stage embryos compete almost completely with oocyte RNA's labeled during oogenesis. These results indicate that the concentration of the hybridizing RNA species in mature oocyte RNA is maintained throughout the cleavage and blastula stages. An example of this type of experiment is shown in Fig. 6.5a (Crippa *et al.*, 1967). It seems likely that the maternal transcripts are simply retained in the cytoplasm of very early embryo blastomeres. However, experiments such as those in Fig. 6.5a cannot distinguish between retention and replacement by new synthesis

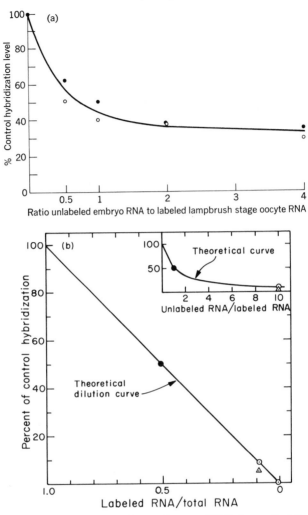

Fig. 6.5. Persistence in amphibian embryos of oocyte RNA's transcribed from repetitive sequences. (a) Total RNA from two cell (●) and blastula (○) stage *Xenopus* embryos was mixed in the ratios shown on the abscissa with oocyte RNA labeled *in vivo* during the lampbrush stage of oogenesis (see Chapter 8) and hybridized with DNA. The RNA was labeled by injection of ³²P-orthophosphate into the body cavity of the female. Three days later the lampbrush stage oocytes were removed, isolated from follicle cells, and the RNA extracted. Hybridization occurred in solution, and after ribonuclease treatment the hybrid duplexes were trapped on filters. In the experiments shown 6 μg of oocyte ³²P-RNA and 5 μg of DNA were present, plus the unlabeled RNA. According to measurements made under the same conditions this amount of oocyte ³²P-RNA is about twice that needed to saturate hybridizable sequences in the DNA. At saturation about 1.5% of the DNA is hybridized, suggesting that 3% of the total DNA sequence is complementary to the repetitive sequence transcripts

in somewhat later embryos. Hough *et al.* (1973) approached this issue by comparing RNA's extracted from oocytes and embryos of *Engystomops pustulosus*, a neotropic anuran in which oogenesis can be induced to occur synchronously in the laboratory (Davidson and Hough, 1969a). Under appropriate conditions the RNA's of a whole clutch of growing oocytes can be labeled by injection of precursor into the body cavity of the female. The labeled RNA's can later be recovered from the shed oocytes or embryos. Figure 6.5b shows a competition experiment in which oocyte RNA labeled during oogenesis is hybridized with a repetitive DNA fraction isolated from whole *Engystomops* DNA. The heterogeneous repetitive sequence transcripts in the oocyte include about 1% of the total oocyte RNA according to other experiments (Hough *et al.*, 1973). Figure 6.5b demonstrates that these maternal transcripts remain present throughout early development. Thus even as late as the tadpole stage *maternally labeled* embryo RNA's can be extracted from the embryos, and mature oocyte RNA competes with its hybridization with DNA. This would not be so if most of the labeled heterogeneous RNA species had been synthesized during embryogenesis, since as discussed in the following section amphibian oocyte RNA does not share detectable sequence homology with RNA's transcribed during early embryogenesis from repetitive sequences. In addition, other evidence presented by Davidson *et al.* (1968) and Davidson and Hough (1969a) precludes the possibility that the labeled RNA is newly transcribed. Therefore, at least for this

---

labeled in the oocyte (see text). At the saturation point RNA and DNA C_0t's (corrected for salt concentration) were about 10, and no further hybridization was obtained at RNA C_0t's of about 60. About 60% of the labeled RNA is competed by the embryo RNA's, and the same amount of competition is observed with unlabeled mature oocyte RNA (Davidson *et al.*, 1966). From M. Crippa, E. H. Davidson, and A. E. Mirsky (1967). *Proc. Natl. Acad. Sci. U.S.A.* 57, 885. (b) Maternal RNA was labeled by injection of ³H-uridine into the coelom of female *Engystomops pustulosus* during synchronous oogenesis (Davidson and Hough, 1969a). The labeled RNA's were extracted from oocytes, neurulae and tadpoles, and hybridized with an isolated repetitive sequence fraction of *Engystomops* DNA. Reactions were carried out in RNA excess and the hybrid content was assayed by hydroxyapatite binding. Excess unlabeled ribosomal RNA was present in all samples and the hybridization observed is of heterogeneous nonribosomal species. The relative amounts of unlabeled mature oocyte RNA indicated on the abscissa were added as competitor, and the reactions were run to kinetic termination. The solid line represents the expected fraction of control hybridization if the competing and the labeled RNA populations are identical. The insert shows the data plotted conventionally [i.e., as in (a)], with the ratio of unlabeled to labeled RNA on the abscissa. Both methods of presentation show that competition is close to that expected for homologous RNA populations. ³H-RNA's were extracted from mature oocytes (●); neurulae (○); and tadpoles (△). From B. R. Hough, P. H. Yancey, and E. H. Davidson (1973). *J. Exp. Zool.* 185, 357.

embryo, the maternal repetitive sequence transcripts themselves are retained, rather than being replaced by new synthesis.

To summarize this discussion, RNA's transcribed from both single copy and repetitive DNA sequences are known to be stored in the mature oocytes of several species. It is clear that the repetitive sequence transcript set includes a greater number of RNA species than can be accounted for as histone messenger RNA's and transfer RNA's. Other maternal messenger RNA's transcribed from repetitive structural genes could be included. However, the actual functional nature of these RNA's is unknown. They are inherited by the embryo and in amphibia persist far into early development. Maternal nonrepetitive sequence transcripts constitute an enormously diverse set of RNA's, the complexity of which is found to lie in the range 25×10^6 to 50×10^6 nucleotides in several phylogenetic groups. Their function also remains in part unknown. It is shown below that in the sea urchin a large fraction of the oocyte single copy sequence transcripts consists of RNA's complementary to structural gene DNA.

Repetitive and Nonrepetitive Sequence Transcripts in the Newly Synthesized RNA's of Early Embryos

ABSENCE OF DETECTABLE INTERSPERSED REPETITIVE SEQUENCE TRANSCRIPTS IN MESSENGER RNA

Recent findings have somewhat simplified the problem of understanding the cellular associations of repetitive and nonrepetitive sequence transcripts. In sea urchin embryos, as described below, and in several mammalian systems, it now seems clear that heterogeneous nuclear RNA includes covalently linked repetitive and single copy sequences. Thus the internal organization of nuclear RNA molecules resembles the interspersed sequence organization of the genomic DNA. Messenger RNA is clearly different from nuclear RNA in this respect, since it contains no detectable interspersed repetitive sequence elements. A minor fraction of messenger RNA's, including the histone messengers, is transcribed from structural genes which are repeated in the genome to various extents. This has been established for L cells by Greenberg and Perry (1971); for sea urchin embryos by Goldberg et al. (1973), McColl and Aronson (1974), and Nemer et al. (1975); and for HeLa cells, mosquito tissue culture cells, and rat myoblasts by Spradling et al. (1974), among others. Depending on the biological source, however, 70 to >95% of the messenger RNA is always found to be single copy sequence transcript. The purpose of the following paragraphs is to review the evidence supporting these conclusions, particularly in reference to the nuclear and messenger RNA's of sea urchin embryos.

The relevant experiments all involve the hybridization of RNA's with excess DNA. To determine the fraction of messenger or nuclear RNA transcribed from repetitive or nonrepetitive sequence, the labeled RNA is hybridized to DNA fragments at various DNA C_0t's, and the amount of hybrid is measured after ribonuclease treatment to destroy unhybridized RNA. Different methods are required to determine whether the repetitive sequence transcripts in the messenger RNA population are covalently linked to nonrepetitive sequence transcripts (interspersed arrangement), or, alternatively, are located on separate molecules. For this purpose it is desirable to omit any RNase treatment. In an interspersed arrangement, the whole of any message bearing a repetitive sequence would be associated with a hybridized region after incubation with excess DNA to low C_0t, but the covalently linked single copy RNA "tail" would be ribonuclease sensitive, since it would remain unhybridized. In this case, the amount of RNA associated with hybridized regions after low C_0t incubation should depend strongly on whether or not prior ribonuclease treatment has been applied. If the repetitive and nonrepetitive sequences are on separate messenger RNA molecules, ribonuclease treatment should make little or no difference in the amount of RNA associated with hybrid duplex. An experiment of this sort was carried out by Klein *et al.* (1974) on HeLa cell poly(A) messenger RNA. About 7% of the labeled messenger RNA hybridized with human DNA at DNA C_0t 40, and no distinction was found between samples which had been ribonuclease treated and those which had not. Therefore, unhybridized single copy RNA "tails" were absent from the hybrid structures formed at low DNA C_0t. Campo and Bishop (1974) also reached the conclusion that repetitive and nonrepetitive messenger RNA transcripts exist on separate molecules. These authors studied rat myoblast poly(A) messenger RNA's and relied on a completely different technical approach. Repetitive DNA of the rat was bound to Sepharose and reacted with poly(A) messenger RNA. Campo and Bishop showed that the repetitive and nonrepetitive sequence transcripts in the messenger RNA could be effectively separated by this means. Therefore, though the messenger RNA's may derive from either repetitive or nonrepetitive sequence, these classes of transcript are confined to separate molecules and are not interspersed in messenger RNA.

In Fig. 6.6 is displayed a DNA excess hybridization experiment carried out with messenger RNA's extracted from the polysomes of sea urchin gastrulae (Davidson *et al.*, 1975b). The samples were assayed for labeled messenger RNA–DNA hybrids by hydroxyapatite binding without prior ribonuclease treatment (see legend to figure). The renaturation kinetics of

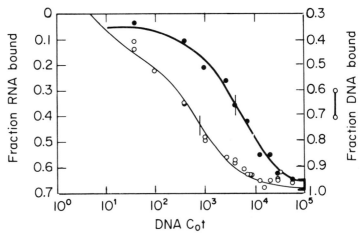

Fig. 6.6. Hybridization of polysomal messenger RNA from sea urchin gastrulae with excess DNA. Messenger RNA from *Strongylocentrotus purpuratus* gastrulae was labeled by incubating the embryos for 3 hours with ^{3}H-uridine. The ^{3}H-RNA was hybridized with 400 nucleotide sea urchin DNA fragments at 60°C in 0.5 M phosphate buffer to the indicated DNA C_0t's (abscissa). The samples were then brought to 0.2 M phosphate buffer, 8 M urea, 1% SDS, and analyzed by hydroxyapatite chromatography at 40°C in the urea phosphate buffer. By this means binding of nonhybridized RNA to the hydroxyapatite is abolished; only ^{3}H-RNA hybridized to DNA binds in this medium. The thin solid line (reproduced from data reported by Smith *et al.*, 1974; Galau *et al.*, 1974; and Davidson *et al.*, 1975b) shows the renaturation kinetics of whole sheared sea urchin DNA. Only DNA driver reactions occurring in the present experiments are shown (O). DNA reassociation is shown on the right ordinate, which extends from 0.3 to 1.0, as the fraction of DNA bound. Hybridization of messenger RNA with whole DNA is represented by closed circles. Hybridization is shown on the left ordinate, which extends from 0 to 0.7, as the fraction of ^{3}H-messenger RNA bound. Control experiments in which DNA excess hybridization is studied with prokaryote nucleic acids show that in these conditions the highest terminal value possible is 70–75% (Smith *et al.*, 1976). The heavy solid line through the closed circles represents the best least-squares fit to the hybridization data according to equation (6.15) in text. The rate constant derived from the least-squares fit is 2.8×10^{-4} $M^{-1}sec^{-1}$, the terminal value is 65–69%, and the RNA fragment length was 560 nucleotides. The rate constant for the single copy DNA sequence in the driver DNA is 1.2×10^{-3} $M^{-1}sec^{-1}$. Short vertical lines in each curve mark the half-reaction points for the single copy transition in whole DNA and for the messenger RNA hybridization. From E. H. Davidson, B. R. Hough, W. H. Klein, and R. J. Britten (1975b). *Cell* **4**, 217.

the driver DNA (thin solid line) are also shown, for comparison with the RNA–DNA hybridization kinetics. Messenger RNA hybridization data in Fig. 6.6 are fitted with equation (6.15). This form is required because, as can be seen, the basic rate of the messenger RNA–DNA hybridization is severalfold slower than that of the DNA–DNA renaturation. As noted in the first section of this chapter this behavior is typical in DNA excess hybridization reactions. Figure 6.6 shows that all but a few percent of the hybridized messenger RNA molecules react with only DNA fragments bearing single copy sequence. Therefore, most structural genes active in these embryos occur about once per haploid genome. In addition, since no RNase treatment has been included, covalently linked *repetitive* sequence transcripts long enough to hybridize are clearly absent from most messenger RNA molecules. In this respect the messenger RNA of sea urchin embryos resembles that of the mammalian tissue culture cells referred to above. Of course any of the messenger RNA populations studied in this manner could contain repetitive sequences 10 or 20 nucleotides long, since these could not have formed stable hybrid duplexes.

INTERSPERSED REPETITIVE SEQUENCE TRANSCRIPTS IN EMBRYO NUCLEAR RNA

At least 10% of the mass of the heterogeneous nuclear RNA in sea urchin gastrulae consists of repetitive sequence transcripts, though the major fraction of this RNA class is nonrepetitive sequence. This was shown by Smith *et al.* (1974), who carried out DNA excess hybridization experiments with 10 minute pulse-labeled nuclear RNA. Upward of 70% of the total labeled RNA could be hybridized at high DNA C_0t, a value close to the maximum possible hybridization expected in this type of reaction (Smith *et al.*, 1976). Smith *et al.* (1974) also studied the effect of prior ribonuclease treatment on the low C_0t hybridization of nuclear RNA molecules degraded to a length of 1000–3000 nucleotides. Over two-thirds of the nuclear RNA covalently associated with hybrid structures consists of unhybridized, ribonuclease sensitive single copy transcript. From this and other experiments Smith *et al.* (1974) concluded that $\geqslant 50\%$ of the nuclear RNA molecules have an interspersed sequence arrangement. Repetitive sequence interspersion is not a peculiarity of sea urchin embryo nuclear RNA. Evidence for similar kinds of sequence organization in mammalian heterogeneous nuclear RNA's was reported by Darnell and Balint (1970), Pagoulatos and Darnell (1970), and Molloy *et al.* (1974) for HeLa cell nuclear RNA and by Holmes and Bonner (1974b) for the nuclear RNA of rat ascites cells. The fact that animal cell nuclear RNA generally includes some repetitive sequence transcripts is well established by a number of earlier investigations which were not directly concerned with the sequence organization of the RNA. These include the studies on the rapidly labeled RNA of L cells (Shearer and McCarthy, 1967; Greenberg and Perry, 1971), of rat liver (Church and McCarthy, 1967;

Melli *et al.*, 1971), of frog embryos (Daniel and Flickinger, 1971), of mosquito cells (Spradling *et al.*, 1974), and of sea urchin embryos (Chetsanga *et al.*, 1970; McColl and Aronson, 1974).

The nuclear and messenger RNA's of embryonic systems thus differ in their basic sequence organization, as well as in length, synthesis rate, and turnover kinetics (Chapter 5). Messenger RNA molecules are either wholly single copy or wholly repetitive sequence transcripts, while nuclear RNA contains both classes of transcript in an interspersed sequence arrangement. These statements, of course, refer to the repetitive sequence elements detectable under the same experimental conditions as used to identify repetitive sequences in the genomic DNA.

The Complexity of Embryo Nuclear RNA and Developmental Changes in the Pattern of Nuclear RNA Transcription

Only a single detailed measurement of heterogeneous nuclear RNA complexity is currently available for any early embryo. Hough *et al.* (1975) carried out this measurement on the nuclear RNA of sea urchin gastrulae. The complexity was estimated from RNA excess hybridization with labeled single copy DNA tracer. Most of the hybridization reaction displays kinetics consistent with a single pseudo-first-order kinetic component [see equation (6.17)], and this is shown in Fig. 6.7. Here it is seen that the RNA's driving the main kinetic component of the reaction are complementary to about 10.7% of the single copy DNA tracer. Approximately 75% of the DNA tracer was present in fragments long enough to form stable duplexes in these experiments, and at least the single copy sequences in the nuclear RNA are known to be transcribed asymmetrically (Hough *et al.*, 1975). Therefore, $\geq 8.5\%$ of the single copy sequence in the genome is represented in the nuclear RNA, and the complexity is at least 1.74×10^8 nucleotides. Hough *et al.* (1975) showed that their results do not exclude the presence of additional very rare sequences representing the whole of the genome. However, if these are present each sequence could occur on the average only once per 30 cells (or less). It is also possible that among its 600 cells the gastrula contains some rare cell types producing specific sets of nuclear RNA's. Such transcripts would be present at lower concentrations than are most of the RNA's driving the hybridization reaction displayed in Fig. 6.7, and their complexity would not be included in the cited value.

The nuclear RNA at gastrula is at least five times more complex than is the total oocyte RNA of the same species of sea urchin (Table 6.1). This

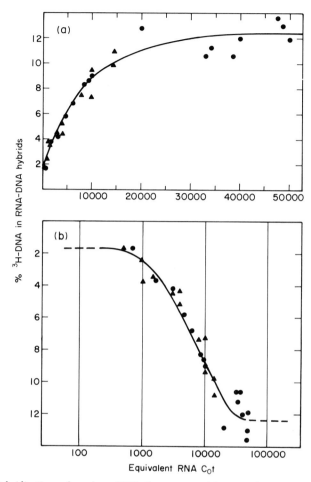

Fig. 6.7. Hybridization of nuclear RNA from sea urchin gastrulae with nonrepetitive
³H-DNA tracer. Two RNA preparations were used, as indicated by the different symbols. In
(a) the data are presented as a conventional linear saturation curve and in (b) the same data
are replotted on a log RNA C_0t scale as in Fig. 6.4. The solid lines show the pseudo-first-order
function [equation (6.17)] which best fits the data according to a least-squares analysis
assuming a single kinetic component. The rate constant for this fit is $1.11 \times 10^{-4} M^{-1}\text{sec}^{-1}$.
The reaction extends from about 1.7% of the ³H-DNA to about 12.4%, i.e., about 10.7% of
the ³H-DNA in the experiment is complementary to the RNA sequences whose rate constant
of hybridization is given above. The initial 1.7% hybridization may be due to reaction of a
small amount of repetitive sequence contaminating the nonrepetitive ³H-DNA preparation.
From B. R. Hough, M. J. Smith, R. J. Britten, and E. H. Davidson (1975). *Cell* **5**, 291.

result does not fit well with the possibility that the oocyte RNA is some form of nuclear RNA stored for use during early stages when the embryo has few nuclei of its own. Such an interpretation might be suggested by the fact that oocyte RNA, like nuclear RNA, contains a variety of repetitive sequence transcripts. The complexity of the repetitive transcripts of nuclear RNA is not satisfactorily known at present, and their diversity cannot be compared to that of the repetitive transcripts in the oocyte RNA.

It is interesting to compare the single copy complexity measured for nuclear RNA of sea urchin gastrulae to the nuclear RNA complexities measured in other tissues (see review in Davidson and Britten, 1973). The complexity of the nuclear RNA of a single mammalian cell type, the rat ascites cell, was measured by Holmes and Bonner (1974a) to be about 1.28×10^8 nucleotides. Values between 4×10^7 and 2×10^8 nucleotides have been obtained for complex multicellular organs, such as mammalian liver, kidney, and spleen, by other investigators. Much higher values were reported for nuclear RNA in rodent brain by Bantle and Hahn (1976) as well as by earlier workers (references in Davidson and Britten, 1973). The large complexity of sea urchin gastrular nuclear RNA is thus in no way unusual for animal cell nuclear RNA's, particularly if one considers that the gastrula includes a number of diverse anlagen or cell types.

By use of the relation shown in equation (6.18) it can be calculated that only about 2.5–3.0% of the total RNA is driving the hybridization reaction shown in Fig. 6.7. According to Aronson et al. (1972), a large fraction of the RNA associated with sea urchin embryo nuclei prepared by the procedures used by Hough et al. (1975) is ribosomal, even though this method involves removal of much of the outer nuclear membranes. High concentrations of a relatively small number of certain nonribosomal RNA sequences may also be present, and in addition the nuclear preparations probably contain some cytoplasmic fragments. Thus the low driver RNA concentration is not necessarily surprising. However, this result raises the issue of whether the rapidly turning over heterogeneous nuclear RNA's whose synthesis rates are shown in Table 5.2 are the same as the nuclear RNA's whose complexity is measured in the experiment of Fig. 6.7. To answer this question Hough et al. (1975) labeled the nuclear RNA for 10 minutes and compared the specific activity of the hybridized RNA with that of the total nuclear RNA. The ratio of the specific activity of the total nuclear RNA to the specific activity of the RNA hybridizing with the single copy tracer averaged about 0.03, almost exactly the same as the fraction of the total RNA driving the hybridization reaction according to the reaction kinetics. It follows that the rapidly turning over (i.e., rapidly labeling) RNA includes the class of nuclear RNA whose complexity is at least 1.74×10^8 nucleotides. If all the rapidly decaying nuclear RNA turns over at the same approximate rate, furthermore, the quantity of this RNA is probably no more than a factor of about 1.5 (considering possible errors) greater than the quantity of the complex RNA.

A portion of the rapidly turning over nuclear RNA could be made up of a few species of RNA, each present in a relatively large number of copies. Smith *et al.* (1974) suggested that about a third of the rapidly labeled nuclear RNA might belong to such a prevalent sequence class because this amount of additional hybridization was obtained at extremely high DNA to ^{3}H-RNA ratios. However, the physical chemistry of DNA excess hybridization reactions is not well known, and the hybridization could be affected by the extremely high DNA/RNA ratios used. From the complexity and the approximate quantity of the complex class of the nuclear RNA we can calculate the number of copies of each sequence existing at steady state, both for the case in which a prevalent nuclear RNA sequence class exists and for the case in which it does not. This calculation is limited by several uncertainties, but is nonetheless informative. We first assume that *all* of the rapidly turning over RNA belongs to the complex class. Its quantity is calculated in Chapter 5 from the nuclear RNA synthesis kinetics to be about 0.17 pg, or 3×10^8 nucleotides per nucleus. If all the nuclei in the embryo transcribe the same sequences, the complexity of the RNA in each nucleus is 1.74×10^8 nucleotides, and in this case there would be only 1 or 2 copies of each sequence per nucleus. Since in reality the nuclei may have diverse patterns of transcription, this is probably to be regarded as an underestimate by some small factor. On the average this factor cannot be greater than 3–4, since 28.5% of the total complexity is represented in the nuclear RNA of the gastrula. If we assume that a prevalent nuclear RNA sequence class exists, the average number of copies of each *complex class* sequence per nucleus would be slightly less. In either case, the conclusion seems clear: The complexity of the nuclear RNA is so large that only a very small number of copies of most kinds of sequence can exist at any one time per nucleus. This is an important result which will ultimately have to be taken into account in considering the function of heterogeneous nuclear RNA.

The same conclusion can be reached by an essentially independent route. From Table 5.2 we recall that an average rate of synthesis of rapidly turning over heterogeneous RNA per nucleus is about 3×10^{-3} pg min^{-1} at the gastrula stage. This represents about 9×10^4 nucleotides sec^{-1}. A. I. Aronson (personal communication) has shown that the rate of transcription of heterogeneous nuclear RNA in sea urchins is approximately 9 nucleotides sec^{-1} per polymerase. Thus there are about 10^4 polymerase molecules functioning at any one time in each gastrular cell nucleus. Since the length of DNA they transcribe is 1.74×10^8 nucleotide pairs, the polymerases are spaced more than 10^4 nucleotides apart in transcribed regions (or somewhat less, if diverse cells transcribe different regions of the genome). At 9 nucleotides sec^{-1} it would require almost 20 minutes to transcribe 10^4 nucleotides. However, 20 minutes is also the approximate half-life of nuclear RNA's in the sea urchin embryo (see Chapter 5), and these are on the order of 10,000 nucleotides long. Therefore, in the complex nuclear RNA sequence class the average number of RNA molecules of each species present at any one time is around one per nucleus.

COMPLEXITY OF MAMMALIAN EMBRYO RNA

Several observations on total RNA complexity exist for mammalian embryos at various stages. Since the most complex RNA in the cell is nuclear RNA, the total RNA complexity should equal the nuclear RNA complexity. Church and Brown (1972) found that total RNA of mouse embryos during preimplantation stages reacts with about 1% of single copy DNA tracer. According to this measurement the complexity of this RNA is about 4×10^7 nucleotides, assuming asymmetric transcription. This value is similar to the complexity of the oocyte RNA's shown in Table 6.1. After implantation total RNA complexity rises rapidly. By term Church and Brown found some 10% of the single copy DNA hybridizable by total embryo RNA. This is almost certainly a considerable underestimate, since there are many cell types which are rare in the whole embryo at the time of birth, and any RNA species peculiar to these cell types might never achieve high enough RNA sequence concentration to be able to react. Similarly, Gelderman et al. (1971) reported that at least 6% of the single copy DNA is complementary to neonatal mouse embryo RNA, and they pointed out that the actual value might easily be twice this. Assuming asymmetric transcription this would represent up to 25% of the single copy sequence, or over 5×10^8 nucleotides. Most of the complexity in the total RNA of the newborn or term mouse embryo can be accounted for as brain RNA (Hahn and Laird, 1971; Brown and Church, 1971, 1972; Church and Brown, 1972; Bantle and Hahn, 1976). Schultz et al. (1973a) have carried out similar experiments on rabbit embryos. RNA from late preimplantation blastocysts (6 day) hybridizes with 1.8% of the single copy DNA tracer, and this value rises to about 2.5% in 12-day postimplantation embryos. It seems clear that as development progresses and the biological complexity of the mammalian embryo increases, the complexity of the nuclear RNA also increases.

COMPLEXITY OF REPETITIVE SEQUENCES TRANSCRIBED
DURING EMBRYOGENESIS

We now briefly consider a rather obscure area much in need of modern measurements. This is the complexity of the repetitive transcripts in embryo nuclear RNA. Most of the studies which are available were carried out with filter hybridization methods. Some of the severe problems encountered in interpreting this type of experiment are noted above. However, in the absence of more current forms of data, it is worth reviewing some of these experiments, since at least their qualitative conclusions remain informative.

Chetsanga *et al.* (1970) carried out filter hybridizations with excess RNA transcribed *in vitro* with *E. coli* polymerase from sea urchin embryo chromatin. They observed that at apparent termination (RNA $C_0t \leqslant 80$) about 4–6% of the DNA is hybridized by the blastula chromatin transcript, and 8–9% of the DNA is hybridized by the pluteus chromatin transcript. The *in vitro* transcripts largely resemble the natural transcripts according to competition data (Chetsanga *et al.*, 1970). At least the *in vitro* transcripts do not appear to include a great excess of sequences beyond those normally transcribed. However, it is clearly unsafe to assume that the *in vitro* transcripts are completely asymmetric (Reeder, 1973), and RNA–RNA duplex formation could have decreased the amount of hybridizable RNA. Since about 25% of sea urchin DNA is repetitive sequence (Graham *et al.*, 1974), these data suggest that a minimum of 20% of the genomic repetitive sequence families, but probably not all, are represented in the chromatin-primed RNA.

Some evidence also exists with respect to the complexity of the repetitive transcripts in amphibian embryo RNA's. In a study by Greene and Flickinger (1970) total RNA was extracted from several relatively late stages of *Rana pipiens* embryo, and then labeled *in vitro* with ^{3}H-dimethyl sulfate. This RNA was hybridized in excess with long filter-bound DNA at relatively low RNA C_0t's (<500 M sec; compare Fig. 6.7). The fraction of the total DNA hybridized at the highest RNA C_0t's at the RNA/DNA ratios used was 1.5–3% for various embryonic stages between gastrula and tailbud. The earlier data of Denis (1966) are approximately in agreement with those of Greene and Flickinger (1970). Denis labeled *Xenopus* embryos *in vivo* and attempted to quantitate the mass of the labeled hybridized RNA by measuring the precursor pool specific activity. At gastrula stage up to 2% of the DNA was hybridized, and by tailbud about 5% of the DNA was hybridized with even higher values obtained when swimming tadpole RNA was tested. These experiments were carried out in a three-dimensional DNA– agar matrix, and the amount of hybridization observed is less likely to have been kinetically limited than in a two-dimensional mat of filter-bound DNA. The RNA's studied by Denis (1966) and Greene and Flickinger (1970) were extracted from whole embryos rather than isolated nuclei and of course could have represented prevalent messenger RNA's transcribed from repetitive structural genes. However, Daniel and Flickinger (1971) obtained more or less similar results in experiments on isolated nuclear RNA of *Rana pipiens* embryos. About 4% of the DNA was hybridized by this RNA at the tailbud stage. These experiments suggest that, as for the sea urchin, an appreciable fraction of the total repetitive sequence is represented in amphibian embryo RNA, probably nuclear RNA. The fraction of total DNA which is repetitive is 25% in *Xenopus* (Davidson *et al.*, 1973) and may be slightly larger in *Rana* species (Straus, 1971). Thus 10– 20% or more of the repetitive sequence families may be included in embryo transcripts.

COMPETITION HYBRIDIZATION EXPERIMENTS WITH EMBRYO RNA'S

The main contribution of the low C_0t competition hybridization studies has been to suggest that as development progresses changes occur in the sets of repetitive sequence transcribed in the embryo. We will argue that the most likely general interpretation is that the sets of sequences represented in nuclear RNA change during embryogenesis. Many of the competition experiments upon which this general conclusion rests are summarized in Table 6.2. Here the labeled RNA preparations are referred to as the "reference RNA's." Table 6.2 classifies experiments according to whether they were performed in RNA or DNA excess with respect to the reference RNA (i.e., whether sufficient reference RNA was present to saturate the homologous sites in the DNA). If so, RNA sequence excess was achieved, and the RNA C_0t's attained were probably sufficient to more or less terminate the repetitive sequence reaction. Thus, in reference RNA excess experiments, no additional repetitive sequence is hybridized when the total RNA concentration or RNA/DNA ratio is increased. The amount of hybridization observed reflects both the repetitiveness and the complexity of the repetitive sequence families represented. Where the DNA is in excess over the reference RNA, the hybridization observed mainly involves RNA species transcribed from relatively highly repeated sequences. A problem which arises in such experiments is that addition of homologous or partially homologous competitor RNA eventually increases the RNA C_0t, so that transcripts from other repetitive sequences may start to react. This alters the reference RNA hybridization. Thus addition of homologous unlabeled RNA results in the expected specific activity dilution curves only in RNA excess competition experiments. This can be seen clearly in Table 6.2, where the first listing in each set of experiments shows the competition by unlabeled RNA extracted from the same stage as the labeled reference RNA.

The main result of every experiment shown in Table 6.2 is that reference RNA hybridization is competed less effectively by RNA's extracted from other developmental stages. The differences in RNA populations are in some cases impressively large. The most convincing examples are the cases of reference RNA excess in which homologous RNA competes quantitatively with the reference RNA just as expected in a dilution experiment. An example concerning *Xenopus* embryo RNA's is shown in Fig. 6.8. Here the reference RNA is labeled between ovulation and blastula stages (cf. Chapter 5). Figure 6.8a shows that the reference RNA is present in excess, since addition of further radioactive blastula RNA results in no additional hybridization. Increasing ratios of unlabeled blastula RNA yield a nearly ideal dilution of the radioactivity in the hybridized RNA (Fig. 6.8a and b). In Fig. 6.8c it is shown that oocyte RNA contains a completely distinct set of transcripts which do not compete at all with the embryo transcripts (though they hybridize with a significant fraction of the repetitive DNA, as noted above). Several other cases are shown in Table 6.2 in which no competition is observed with RNA from an earlier stage. In most of these examples the concentration of the noncompeted reference RNA sequences must have been at least an order of magnitude greater in the reference RNA than in the RNA from earlier stages. Similarly some labeled RNA species are found to be absent at a later stage of development (Table 6.2). These species are likely to have existed in a small number of copies and to have turned over at moderate or high rates. In most of the experiments shown in Table 6.2 the reference RNA's were labeled for 1 hour. An additional property of some of the labeled reference RNA's discussed earlier is that they represent appreciable fractions of the repetitive DNA sequence. Combining all these characteristics, a likely interpretation is that the hybridized reference RNA consists largely of interspersed repetitive sequence elements transcribed in heterogeneous nuclear RNA. Its complexity may be too high to be accounted for solely as messenger RNA's deriving from repetitive structural genes. If this assumption is correct the experiments summarized in Table 6.2 could indicate that partially distinct regions of the genome are represented in nuclear RNA at each stage.

In many of the organisms listed in Table 6.2 there seems to be a high degree of homology between oocyte RNA and labeled RNA's extracted from very early embryos. Several such experiments have been carried out

TABLE 6.2. Competition Hybridization Experiments concerning RNA's Transcribed from Repetitive Sequences during Embryogenesis

(1) Organism	(2) Labeled reference RNA	(3) Competing RNA[a]	(4) Reference RNA or DNA in excess[b]	(5) Competing RNA reference RNA (mass ratio[c])	(6) Fraction of reference RNA hybridization observed[d]	Reference
Echinoderms *Strongylocentrotus purpuratus*	Blastula[e]	Blastula Gastrula Prism Oocyte	RNA	6 6 6 6	0.12 0.50 0.46 0.12	Glišin et al., 1966
	Prism (1-hour label)	Prism Blastula Gastrula Oocyte	DNA	70 70 70 70	0.59 0.76 0.59 0.82	Whiteley et al., 1966
	Prism (1-hour label)	Prism Oocyte Prism Oocyte	RNA? DNA	4.5 4.5 70 70	0.29 0.85 0.20 0.50	Whiteley et al., 1970
	Blastula[f] (transcribed in vitro)	Blastula Pluteus Oocyte	RNA	6 6 6	0.34 0.62 0.39	Chetsanga et al., 1970
	Pluteus[f] (transcribed in vitro)	Pluteus Blastula Oocyte	RNA	6 6 6	0.13 0.40 0.42	
Dendraster exentricus	Prism (1-hour label)	Prism Oocyte	DNA	70 70	0.15 0.48	Whiteley et al., 1970
Amphibia *Xenopus laevis*	Blastula (cumulative label from ovulation...)	Blastula Cleavage Gastrula O...	RNA	3 13 13 7.5	0.30 0.99 0.75 1.0	Davidson et al., 1968 (Fig. 6.8)

					Reference
Gastrula (1-hour label)	Gastrula	DNA	10.6	0.47	Denis, 1966
	Cleavage		10.6	1.0	
	Tailbud		10.6	0.61	
Tailbud (1-hour label)	Tailbud	DNA	10.6	0.44	
	Cleavage		10.6	0.89	
	Gastrula		10.6	0.67	
	Neurula		10.6	0.56	
	Tadpole		20.2	0.33	
Rana pipiens Larva (labeled *in vitro* with ^{3}H-dimethyl sulfate)	Larva	RNA	2.4	0.45	Greene and Flickinger, 1970
	Larva		6.9	0.14	
	Neurula		2.4	0.70	
Tunicate *Ascidia callosa* Tadpole (1-hour label)	Tadpole	DNA	60	0.55	Lambert, 1971
	Unhatched tadpole		60	0.72	
	Oocyte		60	0.75	
Mollusc *Crassostraea gigas* Veliger (1-hour label)	Veliger	DNA	32	0.50	McLean and Whiteley, 1974
	Swimming larvae		32	0.70	
	Oocyte		32	0.78	
Teleost *Misgurnus fossilis* Gastrula (2-hour label)	Gastrula	DNA	100	0.44	Rachkus *et al.*, 1969
	Blastula		100	0.67	
	Oocyte		100	0.56	

[a] Unlabeled RNA homologous to the reference RNA is listed first.

[b] Reference RNA excess judged by experiments showing apparent saturation of the DNA at the reference RNA/DNA ratios used.

[c] Data listed are for the highest ratio of competing RNA to reference RNA reported.

[d] The expected fraction, for the case of perfect homology, is calculated as [reference RNA/(reference RNA + competing RNA)] or [competing RNA/reference RNA (value in column 5) $+1]^{-1}$.

[e] The labeling period is not stated by the authors except that it was a "pulse label."

[f] Reference RNA transcribed *in vitro* from chromatin preparations, using *E. coli* polymerase.

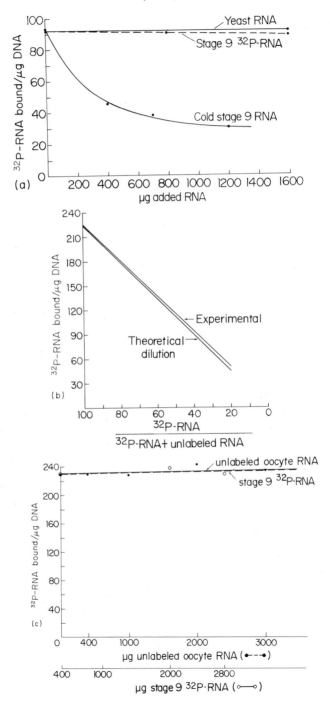

on sea urchin embryo RNA's. Glišin *et al.* (1966) and Chetsanga *et al.* (1970) found that oocyte RNA competes with the hybridization of labeled blastula RNA as well as does unlabeled blastula RNA, and significant competition was observed by Whiteley *et al.* (1970). Oocyte RNA is a poor competitor for the hybridization of labeled RNA from later stages (see Table 6.2). The shared transcript population seems to be a preferentially labeled set of rather prevalent sequences, since the amount of reference RNA hybridization blocked by oocyte RNA is far greater in DNA excess than in RNA excess experiments. This was shown by Whiteley *et al.* (1970) and Hynes and Gross (1972) [though an unresolved contradiction persists between these results and the RNA excess results of Glišin *et al.* (1966)]. A class of RNA which can be labeled extensively in 1 hour and is dominant at the blastula stage though nearly absent later on, is the histone message set. We know that histone messenger RNA is present in relatively high concentration in mature sea urchin oocytes (Chapter 4). It will also be recalled that oocyte RNA has been shown to compete effectively with the hybridization of histone messenger RNA (Farquhar and McCarthy, 1973; Skoultchi and Gross, 1973). Much of the homology between blastula transcripts and oocyte RNA could thus be due simply to histone message. An additional item of evidence is derived from the experiments on amphibian embryo RNA's shown in Table 6.2 and Fig. 6.8. As will be recalled from the discussion in Chapter 4, the quantitatively important histone message synthesis seen in sea urchin embryos is not observed in amphibian embryos. In contrast to the sea urchin results, *Xenopus* oocyte RNA and newly synthesized blastula RNA do not compete, as shown in Fig.

Fig. 6.8. Lack of homology between stored oocyte RNA's and repetitive sequence transcripts of the *Xenopus* blastula. (a) Increasing amounts of unlabeled stage 9 RNA were added to hybridization mixtures containing 1 μg ^{3}H-DNA and 400 μg total ^{32}P-RNA extracted from stage 9 blastulae. Labeling was by injection into the body cavity of the mother just prior to ovulation (see Chapter 5), and therefore the total high molecular weight ^{32}P-RNA accumulated between ovulation and blastulation is extracted. Reactions were carried out in liquid medium (0.05 M Tris, 0.9 M NaCl, 0.09 M Na citrate) at 60°C for 16 hours. After incubation the DNA and hybridized RNA were trapped on nitrocellulose filters, and these were treated with RNase to remove unhybridized labeled RNA. The amount of ^{32}P-RNA present is sufficient to saturate homologous DNA sites as shown by absence of further hybridization when additional ^{32}P-RNA is added (dashed line). Presence of equal quantities of heterologous yeast RNA has no effect. The competition by unlabeled homologous RNA is almost exactly as predicted on the basis of a simple dilution of labeled with unlabeled sequences. This is shown directly in (b). (c) Experiments carried out exactly as in (a) but with unlabeled oocyte RNA added as competitor and 400 μg stage 9 ^{32}P-RNA as reference RNA. No competition is observed. The same oocyte RNA (labeled *in vitro*) was shown to hybridize normally, i.e., to about 1.5% of the total DNA in other experiments (see text). From E. H. Davidson, M. Crippa, and A. E. Mirsky (1968). *Proc. Natl. Acad. Sci. U.S.A.* **60**, 152.

6.8. Results consistent with these were also reported by Denis (1966), who found that total RNA from early *Xenopus* cleavage stages fails to compete with the hybridization of gastrular reference RNA.

The era of the low C_0t filter competition experiments is now past, and it is worth summing up their overall results, as these are pertinent to our present subject. The major conclusion is that repetitive sequence transcripts change with developmental state. Since at least some of these are probably interspersed nuclear RNA sequences, basic alterations in transcriptional activity appear to occur. RNA's transcribed in early embryos are also present in some oocytes. We concluded that stockpiled histone messenger RNA's may be a prominent component of this group of sequences. Other heterogeneous repetitive sequence transcripts in oocyte RNA persist throughout early development but are not transcribed extensively or at all in early embryos. This RNA class could include structural gene transcripts. If so, structural genes of several unknown species must be included, since more than 10–20% of the total repetitive sequence families are represented in this RNA class. The actual functional nature of these repetitive sequence transcripts remains a subject for speculation, and none of the above conclusions can be considered as anything but highly tentative pending quantitative hybridization studies by more modern methods.

The Complexity of Polysomal Messenger RNA in Early Embryos

In principle the problem of measuring the number of messenger RNA's could be approached by determining the number of diverse proteins synthesized in a given system. So far, however, even the most advanced two-dimensional electrophoretic analyses lack sufficient resolution to accomplish this task by at least an order of magnitude. The only direct method available is measurement of messenger RNA complexity. A structural gene is by definition a DNA sequence on which a messenger RNA is transcribed. Since most structural genes are single copy sequence elements, their number can always be measured by messenger RNA hybridization with single copy DNA. In this way the size of the structural gene set required under given developmental conditions can be estimated.

COMPLEXITY OF POLYSOMAL MESSENGER RNA IN SEA URCHIN GASTRULAE

A measurement of the complexity of the polysomal messages present in the 600 cell sea urchin gastrula is shown in Fig. 6.9. These data were obtained by Galau *et al.* (1974), using the same RNA excess hybridization procedures as applied to the measurements of nuclear RNA complexity in these gastrulae (Fig. 6.7). Figure 6.9a shows an overall C_0t curve of sea urchin DNA, similar to that illustrated in Fig. 6.2. The renaturation kinetics of the single copy component in the total DNA are again indicated by the dashed line in Fig. 6.9a. The reaction with total sea urchin DNA of the single copy DNA tracer used for the hybridization experiments is also shown (open symbols). It can be seen that very little repetitive sequence is present in this tracer preparation. As indicated in the legend, the rate of this reaction is almost exactly the same as that of the single copy component in whole DNA. The most important data in Fig. 6.9a are indicated by the closed triangles. These represent the renaturation with total sea urchin DNA of ^{3}H-DNA tracer recovered from hybrids with polysomal messenger RNA. The kinetics of their reaction show clearly that the formerly hybridized tracer DNA fragments belong to the single copy sequence class. It follows that the fraction of the single copy DNA tracer occupied by the messenger RNA at termination of the hybridization reaction provides a direct measure of the sequence complexity of the messenger RNA. This type of control experiment is known as a "playback" and has routinely been carried out in all complexity measurements made by the RNA excess method.

In Fig. 6.9b and c the hybridization kinetics of the messenger RNA-driven reaction are shown. These reactions follow the pseudo-first-order kinetics described by equation (6.17) assuming a single kinetic component in the RNA. The polysomal message preparations were obtained by a procedure designed to minimize the possibility of contamination with nuclear RNA's. This method involves separation of polysomes from monosomes in velocity sedimentation gradients, followed by dissociation of the polysomes with puromycin (Blobel, 1971; Goldberg *et al.*, 1973). Its success in providing polysomal messenger RNA preparations free of nuclear RNA is demonstrated most succinctly by comparison of the hybridization of the messenger RNA with that of the nuclear RNA shown in Fig. 6.7. More than ten times more tracer DNA hybridizes with the nuclear RNA than with the messenger RNA, a result which could not have been obtained had the message preparation included significant nuclear RNA contamination. Calculations show that RNA's with the complexity of nuclear RNA could be present at the level of no more than one copy in over 10^3 cells in these messenger RNA preparations (Hough *et al.*, 1975).

At termination about 1% of the single copy tracer has hybridized with the gastrular messenger RNA. Since 75% of the ^{3}H-labeled single copy tracer can be reacted (Fig. 6.9a), and messenger RNA transcription is asymmetric; 2.7% of the single copy sequence is represented in the gastru-

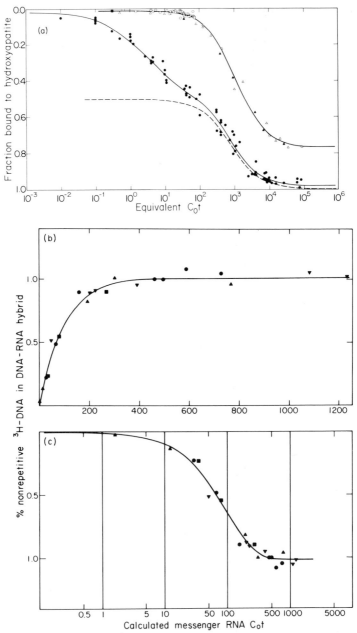

Fig. 6.9. Complexity of polysomal messenger RNA in sea urchin gastrulae. (a) Reassociation of various *Strongylocentrotus purpuratus* DNA fractions. Reassociation of whole 450-nucleotide long DNA (●) at 60°C. The curve is a least-squares fit to the data utilizing three

lar messenger RNA. Graham *et al.* (1974) showed that the single copy sequence content of the *Strongylocentrotus* genome is about 6.1×10^8 nucleotide pairs. The complexity of the messenger RNA set present in the gastrular polysomes is thus 17×10^6 nucleotides. This is sufficient to represent 1×10^4 to 1.5×10^4 distinct messages, assuming 1200–1800 nucleotides as the length of a typical messenger RNA. Data reviewed in Chapter 5 showed the mean length of sea urchin embryo messenger RNA to be about 2000 nucleotides, or about 1800 nucleotides not counting poly(A) tracts. The median length, also minus poly(A) tracts, is about 1200 nucleotides. We conclude that there are 10,000–15,000 structural genes whose transcripts are required to operate a sea urchin gastrula. In other words 10,000–15,000 diverse proteins are probably being synthesized in the gastrula. It is now apparent, furthermore, that the messenger RNA and the nuclear RNA of the gastrula are very different RNA populations. Table 6.3 summarizes the distinctions between these classes of RNA. Except for the complexity data considered here, the properties of nuclear and messenger RNA listed in Table 6.3 have been discussed earlier.

PREVALENT AND COMPLEX CLASSES OF EMBRYO MESSENGER RNA

From the kinetic data shown in Fig. 6.9 Galau *et al.* (1974) calculated the average frequency of occurrence of each species of message [equation (6.18)]. It is clear that some structural genes are far more extensively represented by polysomal messenger RNA than are most. About 90% of

---

kinetic components (two of these are not shown, cf. Fig. 6.2). Reassociation of the non-repetitive sequence component in the whole DNA is indicated by the dashed line. Its second order rate constant is $1.3 \times 10^{-3}\ M^{-1}sec^{-1}$. Reassociation of nonrepetitive ^{3}H-DNA with a 10,000-fold excess of 450-nucleotide long DNA ($\triangle$) and with itself ($\bigcirc$) is also shown. The curve through these data is a least-squares second-order fit with a rate constant of $0.91 \times 10^{-3}\ M^{-1}sec^{-1}$. Also shown are data for the reassociation of ^{3}H-DNA extracted from RNA–DNA hybrids with whole 450 nucleotide long DNA ($\blacktriangle$) and with itself ($\blacksquare$). A fit to these data was identical to that for the nonrepetitive ^{3}H-DNA reaction, except that the second-order rate constant was calculated to be slightly larger at $1.01 \times 10^{-3}\ M^{-1}sec^{-1}$. (b) and (c) Hybridization of nonrepetitive ^{3}H-DNA with messenger RNA. Four messenger RNA preparations are symbolized. Two methods of presenting the same data are shown in (b) and (c). The curves are calculated according to equation (6.17). The pseudo-first-order rate constant of this fit is $0.010\ M^{-1}sec^{-1}$. The hybridization reaction extends from 0.006 to 1.016% of the nonrepetitive ^{3}H-DNA. That is, the extent of the observed reaction is 1.01% of the ^{3}H-DNA in the experiment. Polysomal ribosomal RNA is included in the messenger RNA preparations, and the messenger RNA C_0t shown is calculated on the basis that messenger RNA is 2% of the total RNA. However, as noted in Chapter 5, a more correct estimate is probably 4%. The fraction of the messenger RNA driving the reaction is about 10%. From G. A. Galau, R. J. Britten, and E. H. Davidson (1974). *Cell* **2**, 9.

TABLE 6.3. Comparison of Heterogeneous Nuclear and Messenger RNA of the 600-Cell *Strongylocentrotus purpuratus* Gastrula[a]

	Nuclear RNA	Messenger RNA
1. Steady state content:		
Per cell (nucleotides)	3×10^8	2.1×10^8
Per embryo (nucleotides)	1.8×10^{11}	1.3×10^{11}
2. Apparent mean length (nucleotides)	8×10^3	1.8×10^3
3. Number of molecules:		
Per cell	3.7×10^4	1.2×10^5
Per embryo	2.7×10^7	7.2×10^7
4. Synthesis rate (pg min^{-1} per embryo)	~2.5	1.5×10^{-1}
5. Decay rate (min^{-1})	2.3×10^{-2}	2.1×10^{-3}
6. Sequence complexity (nucleotides)	1.74×10^8	1.7×10^7
7. Complex class	⩾60%	~10%
8. Number of molecules of each complex class sequence per cell[b]	~1	~1

[a] Data from text and tables in Chapters 4, 5, and 6.
[b] Assuming all cells in the embryo are similar.

the polysomal messenger RNA in the gastrula, the "prevalent message class," may consist of about 1×10^3 to 2×10^3 species, while the other 10% is made up of 1×10^4 to 1.5×10^4 species referred to above. Not much is yet known about the complexity of the prevalent message set. The study of Brandhorst (1976) in which newly synthesized proteins of sea urchin embryos are analyzed by two-dimensional gel electrophoresis resolves about 400 proteins. Most of these are probably translated from prevalent messenger RNA's. Other relevant data were reported by Nemer et al.(1974), who hybridized poly(A)-containing messenger RNA with cDNA transcribed from this messenger RNA fraction (see Chapter 5 for discussion of poly(A) messenger RNA in sea urchin embryos). From the kinetics of the hybridization reaction they concluded that the poly(A)RNA is mainly prevalent class message, and that it consists of about 1400 diverse species. Additional prevalent species could of course be included in the large fraction of messenger RNA lacking poly(A) sequences. Assuming that 4% of the polysomal RNA is messenger RNA, as above, the experiments of Nemer et al. (1974) show that about 60 copies of each prevalent poly(A) messenger RNA are present per cell. In contrast each complex class messenger RNA is present only 1–10 times per gastrula cell. This calculation is based on the assumption that all 600 cells of the gastrula contain the same species of message. If this is not the case, then of course some messages would be more concentrated in certain cells and absent in others. Prevalent and complex message classes are not peculiar to sea urchin embryos, but have been identified in most of the other tissues and cell types in which messenger RNA complexities have so far been investigated. These include Friend cells (Birnie *et al.*, 1974), HeLa cells (Bishop *et al.*, 1974), *Drosophila* larvae (Levy and McCarthy, 1975), and several mouse organs and cell types (Ryffel and McCarthy, 1975). It should also be noted that the messenger RNA complexities determined for these cell types and tissues mostly fall within the range of 5×10^6 to 20×10^6 nucleotides. Thus, neither the complexity nor the population structure of the gastrular messenger RNA is atypical.

An issue which deserves some comment is whether the complex class messages are physiologically meaningful. An alternative proposition might be that complex class messenger RNA molecules are the result of "leakage" from repressed genes, similar to that known to occur in inducible prokaryote systems in the repressed state. However, several kinds of evidence suggest that these messages are functionally important. As shown below the sets of complex class messages present at different stages of development and in diverse tissues are clearly distinct. They differ from one another in specific ways, and by millions of nucleotides of sequence complexity. Thus, the presence of complex class messenger RNA's must be regulated. Most single copy sequences in the genome are completely excluded from the polysomal message set (e.g., >97% in the case of the gastrula). Furthermore, calculations show that a number of physiologically important enzymes in mammalian liver cells are probably maintained by complex class messenger RNA's present in only about 1–10 copies per cell (Galau *et al.*, 1976c). Among these enzymes are acetyl-CoA carboxylase, alanine aminotransferase, xanthine oxidase, and NAD glycohydrolase. Thus complex class messenger RNA's play a functional role in liver. Though the "leakage" hypothesis cannot be rigorously excluded, the evidence summarized here appears to render it unlikely. Complex class messages are an ubiquitous feature of animal messenger RNA populations, and of course the large majority of active structural genes produce messenger RNA's belonging to the complex messenger RNA class. Regulation of their expression must be considered a basic aspect of cell differentiation.

Unfortunately, no information on the complexity of polysomal messenger RNA in early embryos other than sea urchin yet exists. Some related measurements carried out on the total cytoplasmic poly(A)RNA of *Drosophila* larvae were reported by Levy and McCarthy (1975). These, however, concern relatively advanced (third instar) larvae and thus fall outside the main focus of this discussion. Levy and McCarthy synthesized cDNA against the total cytoplasmic poly(A)RNA of the larvae. The complexity of the larval poly(A)RNA was estimated from the kinetics of the reaction between it and the cDNA. It was concluded that the poly(A)RNA complexity is on the order of 8×10^6 nucleotides. Furthermore, the larval cDNA did not cross-react completely with cytoplasmic poly(A)RNA from whole adult flies or from Schneider tissue culture cells. About 10–15% of the cDNA sequences were either not represented or were poorly represented in these other RNA's, suggesting the existence of some larval-specific messages. The data also show clearly that these putative larval-specific messages belong to the complex sequence class. In a general sense the structure of the *Drosophila* larval messenger RNA population appears to be similar to that of the sea urchin embryo.

CHANGES IN STRUCTURAL GENE SETS ACTIVE AT VARIOUS STAGES OF SEA URCHIN DEVELOPMENT

We now consider to what extent the sets of structural genes represented in the polysomes at various stages of embryogenesis differ. One method of obtaining such information is to prepare cDNA from the polyadenylated messenger RNA of a given stage and to react it with messenger RNA's of other stages, as in the study of Levy and McCarthy (1975) just mentioned. This procedure has been applied to several adult animal tissues by Bishop and his associates (e.g., Hastie and Bishop, 1976), but so far not to early developing systems. Galau *et al.* (1976c) followed an alternative route. Here the set of single copy DNA sequences complementary to the total messenger RNA of the sea urchin gastrula was isolated and partially purified, and was then reacted with other RNA preparations. We refer in the following paragraphs to the single copy DNA fraction representing the gastrular messages as *mDNA*. A single copy DNA fraction totally stripped of the gastrular message sequences was also prepared by Galau *et al.* (1976c) and is referred to as *null mDNA*. Reaction of a nongastrular messenger RNA preparation with mDNA measures the fraction of the particular set of 1×10^4 to 1.5×10^4 structural genes expressed in gastrula which is also represented in the test messenger RNA population. Reaction with null mDNA, on the other hand, measures the number of structural genes expressed which are different from those whose transcripts comprise the gastrular messenger RNA. Galau *et al.* (1976c) showed that when carried to termination the sum of the mDNA and null mDNA reactions equals the overall complexity of the messenger RNA preparation as required. This could be measured independently with total single copy DNA tracer, and provided an external control on the hybridization behavior of the mDNA and null mDNA tracers.

The behavior of the gastrular mDNA and null mDNA tracers is illustrated in Fig. 6.10a and b. It can be seen in Fig. 6.10a that 57% of the

[3]H-mDNA reacts with the gastrular message, or 63% of the reactable ^{3}H-mDNA compared to the 1.36% of the reactable ^{3}H-DNA calculated from the experiments shown in Fig. 6.9. Thus more than an order of magnitude increase in sensitivity to differences in sequence content is obtained when excess RNA reactions are carried out with different messenger RNA preparations. As before, the reactions appear to follow single component, pseudo-first-order kinetics. Figure 6.10b demonstrates the total absence of reaction between the null mDNA and the gastrular messenger RNA. The mDNA tracer reacts primarily with messages belonging to the complex class, since the prevalent class of messages are of too low complexity to account for much of the hybridization.

The reaction of mDNA and null mDNA tracers with polysomal RNA from a number of stages and tissues, and with total oocyte RNA, were studied by Galau *et al.* (1976c). An example is shown in Fig. 6.10c and d. Here the reactions of the mDNA and null mDNA with oocyte RNA and ovary polysomal RNA are illustrated. The ovary had previously been denuded of mature oocytes, which in any case possess few polysomes (see Chapter 4). The dashed line in Fig. 6.10c shows the reaction of mDNA with gastrular messenger RNA, from Fig. 6.10a. The ovary polysomes include about 80% of the message sequences found in the gastrular polysomes (Fig. 6.10c). However, essentially all the messenger RNA sequences in the polysomes of the gastrula are represented in the stored RNA of the oocyte. This striking result provides direct evidence that the complex RNA of the oocyte (Table 6.1) consists in large part of maternal message. That is, since the mDNA consists of structural gene sequences, the oocyte RNA transcripts hybridizing with it are also structural gene sequences. Figure 6.10d shows that the oocyte RNA contains an even larger set of sequences which do not react with gastrular messenger RNA. Whether these are also transcribed from structural gene sequences is yet unknown. The oocyte RNA complexities measured in the experiments shown in Fig. 6.10c and d add up to the total oocyte RNA complexity shown in Table 6.1.

The experiments of Galau *et al.* (1976c) are summarized in Fig. 6.11. Here the closed bars show the size of those structural gene sets represented in the messenger RNA's of various tissues and cell types which are also subsets of the gene set represented in gastrular messenger RNA. Similarly, the open bars give the size of the structural gene sets represented in each tissue which are excluded from the gastrular gene set. The total height of each bar indicates the overall complexity of the messenger RNA's in each tissue. Complexity is expressed in three ways in Fig. 6.11. The left-hand ordinate is calibrated as the total length of single copy nucleotide sequence, and the two right-hand ordinates as percent of gas-

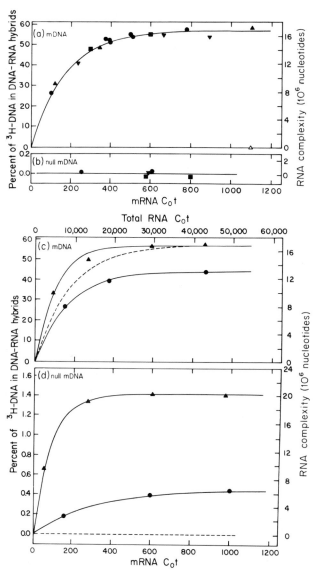

Fig. 6.10. Hybridization of sea urchin embryo RNA's with mDNA and null mDNA. Hybridization of gastrula mDNA (a) and null mDNA (b) with four different gastrula messenger RNA preparations. Messenger RNA C_0t's shown are calculated as indicated in the legend to Fig. 6.9. The line drawn through the data of (a) is a least-squares fit of a pseudo-first-order hybridization reaction with the assumption that the line intercepts the origin.. Different gastrula messenger RNA preparations are indicated (●), (▲), (▼), and (■). The last preparation was prepared by EDTA (rather than puromycin) release. Also shown (△) is a reaction

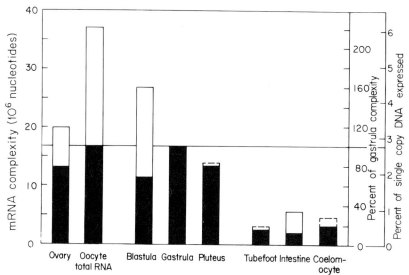

Fig. 6.11. Sets of structural genes active in sea urchin embryos and adult tissues. The closed portion of each bar indicates the amount of single copy sequence shared between gastrula messenger RNA and the RNA preparations listed along the abscissa. These data are obtained from the mDNA reactions described in the text. The open bars show the amount of single copy sequence present in the various RNA's studied but absent from gastrula messenger RNA. These data derive from the null mDNA reactions described in text. Dashed lines indicate the maximum amount of null mDNA reaction which could have been present and escaped detection, in terms of complexity, for cases where no apparent null mDNA reaction was observed. The overall complexity for each RNA is indicated by the total height of each bar. Complexity is calibrated in three ways along the three ordinates shown. From left to right these are as nucleotides of single copy sequence, as percent of gastrula messenger RNA complexity, as percent of total single copy sequence. From G. A. Galau, W. H. Klein, M. M. Davis, B. J. Wold, R. J. Britten, and E. H. Davidson (1976c). *Cell* 7, 487.

---

mixture which was treated with RNase to destroy DNA–RNA hybrids before assay. The pseudo-first-order rate constant in (a) was $5.7 \times 10^{-3} M^{-1}\text{sec}^{-1}$, and the terminal value is $56.9 \pm 1.6\%$ (standard deviation). No reaction is evident in (b), as expected (see text). (c) and (d) Hybridization of gastrula mDNA and null mDNA with total oocyte RNA (▲) and ovary messenger RNA (●). The dashed lines of (c) and (d) are the hybridization reaction of gastrula messenger RNA with gastrula mDNA, and null mDNA, respectively, from (a) and (b). The solid lines of (c) and (d) are least-squares fits of pseudo-first-order hybridization reactions, assuming an intercept of zero for each case. The pseudo-first-order rate constants for the oocyte RNA reactions in (c) and (d), respectively, were $2.1 \times 10^{-4} M^{-1}\text{sec}^{-1}$ and $2.2 \times 10^{-4} M^{-1}\text{sec}^{-1}$. The terminal values were $57.5 \pm 2.0\%$ and $1.42 \pm 0.01\%$. The pseudo-first-order rate constant for the ovary messenger RNA reactions in (c) and (d), respectively, were $6.3 \times 10^{-3} M^{-1}\text{sec}^{-1}$ and $3.1 \times 10^{-3} M^{-1}\text{sec}^{-1}$. The terminal values were $45.0 \pm 2.9\%$ and $0.47 \pm 0.01\%$. From G. A. Galau, W. H. Klein, M. M. Davis, B. J. Wold, R. J. Britten, and E. H. Davidson (1976c). *Cell* 7, 487.

trular messenger RNA complexity and as percent of total single copy sequence. Figure 6.11 shows that each of the RNA preparations investigated contains a unique set of structural gene transcripts and leads to several interesting and relevant conclusions.

The polysomal messenger RNA of the pluteus appears to contain few if any sequences not already represented in the gastrular polysomes. Since the gastrular sequences are all included in oocyte RNA, this must be true of the pluteus messenger RNA set as well. The blastular message set is larger than the gastrular one. It is not yet known whether the nongastrular message sequences present in the blastula polysomes are also represented in oocyte RNA. In any case the overall complexity of the structural gene sets represented in the embryo polysomes seems to *decline*, surprisingly, as development proceeds. This is reminiscent of the falling rate of heterogeneous nuclear RNA synthesis over these same stages, as shown in Table 5.2.

The gastrular and blastular messenger RNA's which are homologous to oocyte RNA's could be maternal transcripts or could be newly synthesized by the embryo. The messenger RNA synthesis kinetics discussed in Chapter 5 show that essentially all the message in the sea urchin gastrula can be accounted for by new synthesis. However, these data are not accurate enough to preclude the possibility that only prevalent class messenger RNA's are newly synthesized, while approximately 10% of the message which constitutes the complex messenger RNA class persists as maternal transcripts throughout early development. To decide this issue Galau *et al*. (1976a) labeled the messenger RNA by exposing the embryos to ^{3}H-guanosine and ^{3}H-uridine for various periods of time during the blastula–gastrula stage. The complex class messages were then isolated as hybrids formed in messenger RNA excess with single copy ^{32}P-DNA. The hybridized messenger RNA's were found to be labeled, showing that they are newly synthesized, and since their specific activity was about the same as that of the total RNA, it was concluded that complex and prevalent class messages turn over at about the same rate (see Chapter 5). Thus by the blastula–gastrula stage both complex and prevalent class messages are mainly the products of embryo genome transcription. The difference in their representation within the cell must be due primarily to differences in the rates of messenger RNA synthesis (including processing). This is an important result, since it means that the maternal (i.e., oogenetic) program for structural gene transcription is also operative in the embryo. In this case the major role of the complex maternal RNA's would be mainly to tide the embryo over the early period and would end when the embryo develops the capacity to replace the maternal transcripts with its own. This in turn requires that the embryo inherits some form of sequence-specific regulatory instructions. These would specify subsets of the same group of structural genes as are transcribed during the growth of the oocyte for continued transcription in the embryo. A model for this in the sea urchin is of course the histone messages. As discussed in Chapter 4, these are inherited as maternal messages, but are also transcribed actively in the embryo genomes early in development.

The findings summarized in Fig. 6.11 shed new light on the data reviewed in Chapters 2 and 3. We now see that the gastrula and pluteus complex message sets are subsets of the maternal message set. This of course may not be true of the blastula message set. The nongastrula messages indicated in Fig. 6.11 for the blastula stage (open bar) could be partly or completely distinct from the oocyte RNA sequence set. However, the fact that all of the gastrula, all of the pluteus, and at least part of the blastula message sets are included in the maternal RNA sequence set leads us to suspect that it may be difficult to detect early embryo genome function by investigating the morphogenetic effects of enucleation, actinomycin treatment, and species hybridization. That is, qualitative developmental changes may not result from such operations if the same messages are being transcribed in the embryo nuclei as are inherited in the maternal RNA. This may partly explain the paradox which was raised by early investigations and has remained unresolved to the present time: Little evidence exists for qualitative control by the embryo genome over the events of early development, and yet it is clear that the embryo genomes are required. Much evidence shows that messenger RNA synthesis in embryos is both active and necessary. Interpretation of the morphogenetic and biochemical results of some species hybrid experiments now poses an interesting problem. Where early differences between the maternal and paternal forms exist, these experiments always reveal the dominance of maternal patterns far into development. This is due in part simply to the activity of maternal proteins and messenger RNA's, and it could also arise from the imposition of slightly different (maternal) patterns of transcriptional control on the embryo structural genes. However, a major cause must be that much of the activity of the paternal genome (i.e., embryo genome) is devoted to expression of structural genes for the same proteins as coded for by the maternal message. Thus the early developmental characteristics of the *paternal species* normally depend in part on *maternal message*, and cannot be expected to appear solely as the result of the activity of the hybrid *embryo* genomes.

Figure 6.11 also shows that certain modulations or alterations in the sets of messenger RNA's loaded on polysomes take place during early development. Thus somewhat different subsets of the gastrular message set (and thus of the maternal message set) are represented in pluteus and in blastula. If partly post-transcriptional at early stages but increasingly controlled at the transcriptional level later in development, such modulations could explain some of the experiments reviewed in Chapter 3. Among these are the studies in which both actinomycin-sensitive and actinomycin-insensitive alterations in the overall patterns of protein synthesis are detected. However, it is apparent that the patterns of protein synthesis observed in all these experiments must refer primarily to messenger RNA's of the prevalent class. Whether changes in the composition of the complex messenger RNA set would affect the overall profiles of newly synthesized proteins is doubtful. The disposition of prevalent class messages during early development may or may not be similar to that of the complex class messages. We recall that Brandhorst (1976) showed that the newly synthesized protein species resolvable on two-dimensional gels are generally the same whether the proteins are extracted from unfertilized eggs, from zygotes, or from blastulae. However, a new set of proteins is clearly being synthesized by the gastrula stage. A large amount of other gel electrophoresis data reviewed in Chapter 3 leads to the same conclusion, and suggests that the new proteins are translated from novel embryo messages. It remains to be seen whether significant numbers of *new* embryo structural genes coding for prevalent messages are transcribed during the blastula–gastrula stage, and whether the complex class nongastrula message set present in blastulae (Fig. 6.11) is likewise the result of new embryonic transcription.

According to Fig. 6.11 little or no additional messenger RNA sequence is found in the pluteus compared to gastrula. Therefore many of the proteins needed to construct the much more complex structures of the pluteus must be translated during the gastrula stage and even earlier, and similarly, proteins needed by gastrulae must be synthesized in all the pregastrular stages. Figure 6.10 shows that most of the gastrular messages are present, and are being translated even in the polysomes of the ovary. Thus the proteins needed at given stages of embryogenesis may be synthesized over long periods. These data suggest that there are important post-translational assembly mechanisms operating in embryonic morphogenesis.

Figure 6.11 also compares the messenger RNA's of three adult nonreproductive tissues of the sea urchin, namely, tubefoot, coelomocyte, and intestine, to the set of messenger RNA's present in the gastrular polysomes. These adult tissues consist of several different cell types, each present in significant concentration. In addition there may be rare cell types whose specific messenger RNA populations do not contribute significantly to the total complexities measured. Nonetheless, the structural gene sets active in the adult nonreproductive tissues are clearly severalfold smaller than those represented in reproductive cells and embryos. The complexity of the messenger RNA populations in these three adult non-reproductive tissues ranges from about 15 to 35% of that of gastrular messenger RNA. We conclude that there are differentiated adult tissue cell types which require less than about 5000 structural genes to operate. This statement refers to the usual circumstances of tissue maintenance and to messengers present at a frequency within 1–2 orders of magnitude of the typical complex sequence class messenger RNA's. In comparison, the process of embryonic differentiation can be seen to be relatively very expensive in terms of the number of structural genes and of the number of complex class messenger RNA species required.

A shared subset of gastrular structural gene transcripts seems to be present in all the adult nonreproductive tissues (Fig. 6.11). Galau *et al.* (1976c) carried out further experiments on this structural gene subset and showed that the sequences present in gastrular polysomes and also in each of the adult nonreproductive tissues are the same sequences. Thus when the adult messenger RNA preparations were mixed and hybridized with gastrular mDNA, about the same amount of hybridization was observed as with the individual adult tissue messenger RNA preparations. According to these experiments the structural genes in this ubiquitous subset represent only about 0.3% of the single copy DNA sequence. The number of such genes in the sea urchin genome is no more than 1000–2000, and they represent less than one-sixth of the set of structural genes required by the gastrula.

A final point to be drawn from the experiments shown in Fig. 6.11 is that they provide an estimate of the total cost, in structural genes, in structural genes, of early development in the sea urchin. As a crude maximum measure, the complexity of oocyte RNA approximates the structural gene information needed to program and carry out development from oogenesis to the feeding pluteus stage. This amounts to about 6% of the total single copy sequence, or, as Table 6.1 shows, to some 30×10^6 to 40×10^6 nucleotides. The oocytes from all species studied contain RNA's of approximately equal complexity. It is therefore reasonable to generalize this estimate to other creatures and other modes of development. Having begun the measurement of the complexity of gene activity in early development, our problem is to understand its functional nature. The basic mechanisms by which the active genes are regulated and the ways in which their products build the embryo remain unknown.

7

Localization of Morphogenetic Determinants in Egg Cytoplasm

Localization as used here refers to specification of embryonic cell fate according to the sector of egg cytoplasm inherited by embryonic blastomeres. This chapter is confined to discussion of the localization phenomenon in early development. The localization phenomenon was defined by classical cell biologists, who showed in cell lineage studies that each blastomere lineage gives rise only to specific differentiated cell types. This was confirmed for certain species in isolated blastomere experiments. Three cases of localization of morphogenetic potential are considered in detail. The first of these concerns ctenophores, in which the potentialities to differentiate prominent cilia ("comb plates") and light-producing photocytes are localized to different embryonic macromeres. These determinants are segregated independently of each other and are localized progressively to their final positions during early cleavage. The second case considered concerns the polar lobes of certain molluscan eggs, particularly *Ilyanassa*. A first cleavage polar lobe contains morphogenetic determinants for coelomic mesoderm, and its ablation prevents differentiation of many mesodermal derivatives. Some of the lobe-dependent structures are formed by direct descendants of cells inheriting polar lobe cytoplasm, while the morphogenesis of others results from inductive interactions between these descendants and other cell lineages. The polar lobes of several species contain special membrane-bound particles visible in the electron microscope, but the relation of these organelles to the morphogenetic determinants in the lobe has not been demonstrated. The lobe determinants appear cortical in nature and could be associated with the egg plasma membrane. Polar lobe removal strongly affects protein synthesis patterns in the pregastrular embryo, but does not affect cell division rate or differentiation of many tissues and cell types. At least some of the effects of lobe removal on protein synthesis are insensitive to actinomycin, and it is possible that the lobe contains a qualitatively special set of maternal messenger RNA's. The third case considered in detail is localization of germ cell determinants. In amphibian eggs these are located at the vegetal pole and can be destroyed by UV irradiation. Their existence is demonstrated by restoration of germ cell production in embryos developing from UV-irradiated eggs, by means of injection of vegetal pole cytoplasm from normal eggs. Injection of polar cytoplasm containing germ cell determinants can also induce germ cells in *Drosophila*, and these can be shown to give rise to viable gametes by use of genetic markers. In both amphibian and *Drosophila* eggs the polar plasm contains characteristic "polar granules," but the relation of these to germ cell determinants remains obscure. The question of the universality of cytoplasmic localization is discussed. Except for acoel turbellarians and mammals, where no evidence for localization exists, the localization phenomenon is found to occur in all areas of the phylogenetic map. Examples presented include (besides the cases discussed above) cephalopod molluscs, dipteran insects, sea urchins, ascidians, and amphibians. Localization is a developmental mechanism which is as ancient evolutionarily as the divergence of the protostomes and deuterostomes (i.e., during early Cambrian or before). Evidence suggests that the localized morphogenetic determinants are associated with the egg cortex in many phylogenetic groups. The history of interpretation of the localization phenomenon is discussed, beginning with the perceptive insights of nineteenth century investigators. Their considerations led to the correct view that localization results from genomic readout of developmental information during oogenesis. Special problems in interpreting the localization phenomenon are discussed. The regulative potential of embryonic systems is a manifestation of intercellular interaction. Many organisms displaying striking localization phenomena also have great regulative capacities, while others do not. All embryos utilize intercellular interaction as a developmental mechanism, and most rely on localization as well. The classical dichotomy between "mosaic" and "regulative" eggs is erroneous. In some organisms localization patterns are established progressively, during early cleavage, while in others they are pre-formed to a larger extent in the uncleaved egg. Localization patterns are sometimes extremely easily disturbed by manipulation (as could be the case in mammalian eggs), but this cannot be taken as evidence against their existence. Localization could function by qualitative regional sequestration of special sets of proteins, maternal messenger RNA's, or gene regulatory agents. In order to account for the diverse forms of differentiation arising through the action of cytoplasmic localization, however, the patterns of gene activity in embryo cell lineages must ultimately be affected.

Localization as used here is the specification of cell fate according to the sector of egg cytoplasm inherited by an embryonic cell lineage. Localization is a widespread, if not universal, mechanism of organizing the earliest embryonic structures. The localization phenomenon is particularly interesting to contemporary students of development because it suggests that specific programs of development are sequestered in the egg cytoplasm. As the early blastomeres divide up the egg cytoplasm, they appear to inherit "instructions" for various kinds of cell differentiation, including specific patterns of macromolecular synthesis. The main problems posed by the localization phenomenon are the molecular nature of these "instructions," the level(s) of control at which they operate, their subcellular location, and the means by which they are distributed to given regions of the egg or early embryo. In one form or another these problems have been studied ever since the foundation of the field of cellular developmental biology. The classic review of this area remains E. B. Wilson's monumental treatise, "The Cell in Development and Heredity" (1925). Wilson's insights into this problem are still relevant. In the following review the discussion of classical literature relies to a large extent on Wilson's synthesis of earlier results. Rather than attempt to deal encyclopedically with the subject of localization, the approach chosen here is to focus on those systems which have received experimental attention in recent times. A number of closely related processes, including inductive interactions between differentiated tissue layers, integumentary pattern formation, and the morphogenesis of specific larval and adult structures, are not considered here. Novel analyses of some of these subjects have recently been presented by Wolpert (1969), Lawrence (1973), Morata and Lawrence (1975), and Crick and Lawrence (1975), and the reader is referred to these sources for reviews and discussion. It is important to point out that many of these complex processes of later development originate as cytoplasmic localizations of developmental potential. Early localizations often serve to create distinct populations of embryonic cells, and these subsequently interact with each other inductively. Thus, pattern formation and embryonic morphogenesis frequently occur by means of interactions between cell lineages which were initially determined by means of localization patterns in the cytoplasm of eggs and early embryos.

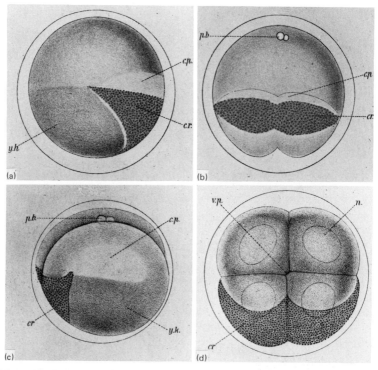

Fig. 7.1 (a)–(d). Living eggs of *Cynthia (Styela) partita*: (a) Right side view of fertilized egg showing the formation of the crescent (*cr.*) from the yellow hemisphere (*y.h.*); in (a)–(c) the future dorsal pole is below. The yellow crescent marks the posterior end. Above the yellow crescent is an area of clear protoplasm (*c.p.*). (b) First cleavage of an egg, viewed from the posterior region and showing the form taken by the yellow crescent during the division, and also, the enlargement of the area of clear protoplasm and its extension toward the polar bodies (*p.b.*). (c) Left side view of egg of same stage as (b) showing the lateral limits of the yellow crescent, the clear protoplasm in the upper (future ventral) hemisphere, and the yolk (*yk.*) in the lower. The anterior portion of the lower hemisphere is composed of light gray material; this is the gray crescent and gives rise to chorda and neural plate. (d) Four-cell stage seen from the vegetal pole (*v.p.*); the yellow crescent covers about half of the posterior blastomeres. (*n*), nucleus. From E. G. Conklin (1905). *J. Acad. Nat. Sci. Philadelphia* **13**, 1.

Classical Definition of the Localization Phenomenon

The cases of localization which most impressed classical experimentalists were those in which areas of future cell fate could be mapped out on the uncleaved egg cytoplasm. A spectacular example is Conklin's (1905) study of development in the ascidian *Cynthia (Styela)*. In this egg pigmented areas of cytoplasm corresponding to a morphogenetic fate map for the cells of the early embryo can be distinguished. Some of Conklin's elegant hand-drawn figures are reproduced in Fig. 7.1. These display the

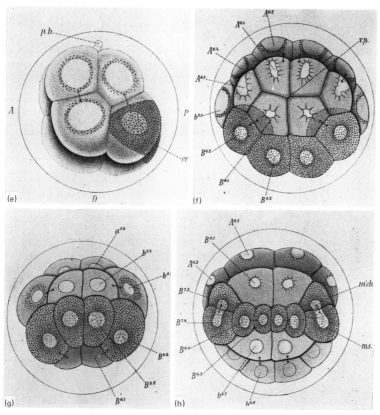

Fig. 7.1 (e)–(h). (e) 8-cell stage viewed from the right side showing a small amount of yellow protoplasm around all the nuclei. Note the crescent. *A*, interior; *P*, posterior; *D*, dorsal. (f) 22-cell stage from the vegetal pole; the embryo now contains 4 mesoderm cells (yellow), 10 endoderm, chorda, and neural plate cells (gray), and 8 ectoderm cells (clear). (g) Same stage viewed from the posterior region. (h) 44-cell stage, posterior view, showing separation of another mesenchyme cell (*m'ch.*) from a muscle cell (*ms.*). From E. G. Conklin (1905). *J. Acad. Nat. Sci. Philadelphia* **13**, 1.

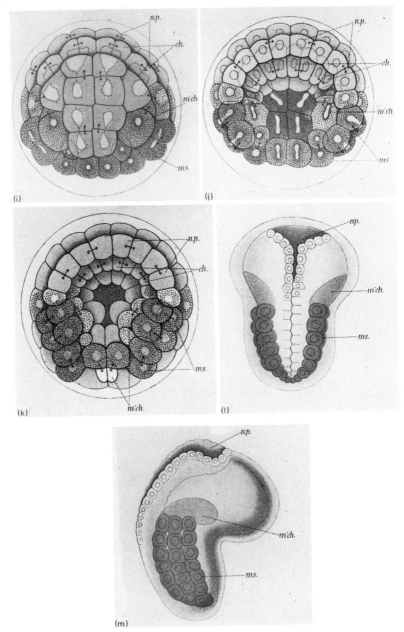

Fig. 7.1 (i)–(m). (i) 74-cell stage, dorsal view, showing division of 4 chorda (*ch.*) and 4 neural-plate (*n.p.*) cells; there are 10 mesenchyme and 6 muscle cells, besides 10 endoderm cells. (j) 116-cell stage showing the beginning of gastrulation, and also, the neural plate, chorda,

relationship between the various pigmented regions of the egg cytoplasm and the tissues ultimately formed from these regions. Five kinds of cytoplasm can be observed: a dark yellow cytoplasm eventually included in the tail muscles of the larva, a light yellow material later segmented into the coelomic mesoderm of the larva, a light gray substance inherited by notochord and neural plate cells, an opaque gray material segregated into the endoderm cell lineage, and a transparent cytoplasm later present only in ectodermal cells. The total cytoplasmic mass of the embryo remains constant through gastrulation, which is well under way by the 180-cell stage. Delineation of the presumptive tissue areas appears to be accomplished simply through the partitioning of the cytoplasmic materials present in the egg. As early as the 64-cell stage, in fact, the separation of the five recognizable kinds of egg cytoplasm into their respective cell lineages has been completed. The definitive distribution of these cytoplasmic substances in the uncleaved egg is set up within a few minutes after fertilization. It is not present when the egg is first shed. The process by which the various regions of cytoplasm appear is illustrated in Fig. 7.2. The unfertilized egg already possesses polarity in one axis so that the sperm is able to enter only at the bottom, but the demarcation of the anterior–posterior axis depends on the acentric movement of the two pronuclei. The fusion nucleus comes to lie near the future posterior end and there the yellow crescent cytoplasm later incorporated in the embryo's first mesodermal stem cells is localized.

EMBRYO CELL LINEAGE

The concept of cytoplasmic localization led to the growth of interest in the embryology of animals in which, unlike the case of the chick or frog, the fate of each cell lends itself to study. Complete cell lineages were worked out for the embryos of several protostomial invertebrates, including the annelids *Clepsine* (Whitman, 1878), *Arenicola* (Child, 1900), and *Tubifex* (Penners, 1922) and the mollusc *Crepidula* (Conklin, 1897). References to many early cell lineage studies are given by Wilson (1925; see Costello, 1956; Reverberi, 1971b). In the embryos first studied in this manner the cells are relatively few and are visually easy to distinguish. In Fig. 7.3, a diagram of the complete cell lineage of *Tubifex rivulorum* is reproduced, after the work of Penners (1922). This diagram shows that

muscle, and mesenchyme cells. (k) Late gastrula; the yellow cells in the midline are mesenchyme cells, the others, muscle cells. (l) Young tadpole seen from dorsal side, neural groove open in front and closed behind, small-celled mesenchyme in front of large muscle cells. (m) Same stage as preceding seen from the right side, showing neural groove, mesenchyme, and three rows of muscle cells. From E. G. Conklin (1905). *J. Acad. Nat. Sci. Philadelphia* **13**, 1.

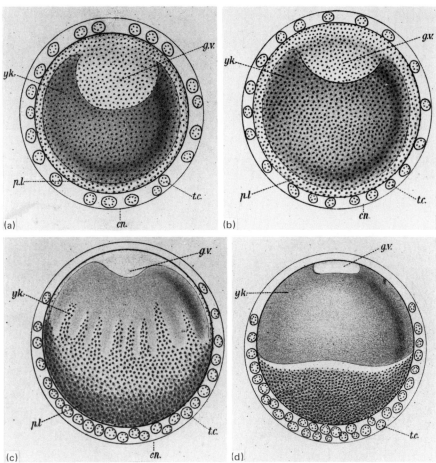

Fig. 7.2. Figures of the living eggs of *Cynthia* (*Styela*) *partita*; maturation and fertilization: (a) Unfertilized egg before the breakdown of the germinal vesicle (*g.v.*), showing central mass of gray yolk (*yk.*), peripheral layer (*p.l.*) of yellow cytoplasm, test cells (*t.c.*), and chorion (*cn.*). (b) Similar egg during the disappearance of the nuclear membrane, showing the spreading of the clear cytoplasm of the germinal vesicle at the animal pole. (c) Another egg about 5 minutes after fertilization, showing the streaming of the peripheral protoplasm to the lower pole where the spermatozoon enters, thus exposing the gray yolk (*yk.*) of the upper hemisphere; the test cells are also carried by this streaming to the lower hemisphere. (d) Later stage in the collection of the yellow cytoplasm. Clear cytoplasm lies beneath and extends a short distance beyond the edge of the yellow cap. (e)–(g) Successive stages of the same egg drawn at intervals of about 5 minutes; viewed from the vegetal pole. In (e) the area of yellow cytoplasm is smallest, and the sperm nucleus, (♂*n.*) is a small clear area. (f) and (g) show stages in the spreading of this yellow cytoplasm until it covers nearly the whole of the lower hemisphere [yellow hemisphere (*y.h.*)]; at the same time the sperm nucleus and aster move toward one side and the crescent (*cr.*) begins to form at this side. From E. G. Conklin (1905). *J. Acad. Nat. Sci. Philadelphia* **13**, 1.

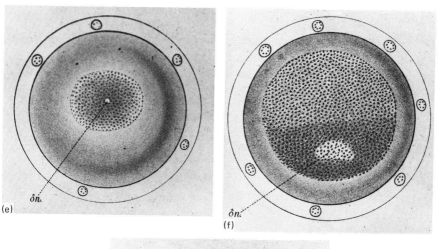

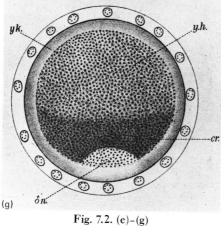

Fig. 7.2. (e)–(g)

ectodermal, endodermal, mesodermal "germ band," and ectodermal "germ band" develop from early segregating cell lineages. Thus the lineage of a particular specialized cell, e.g., a cell of the coelomic mesodermal column, can be traced back to one specific blastomere, in this case the 4d blastomere. That is, the embryonic mesodermal germ band cells constitute two clones descended from the myoblast stem cells (Myr and Myl in Fig. 7.3).

The orientation of the *Tubifex* egg is established before first cleavage. In this, *Tubifex* resembles a variety of other organisms, as the axis of the egg coincides with the axis of the embryo, owing to some process by which at least one axis of polarity is established before fertilization or at the latest by

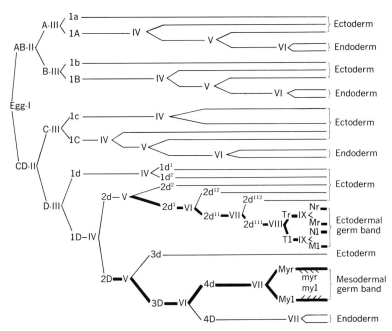

Fig. 7.3. Diagram of the cell lineage of *Tubifex*. The heavy lines in the lower part of the diagram give the history of the ectodermal and mesodermal germ bands. The designations represent names of individual cells. Myr stands for right myoblast; Myl for left myoblast; Tr and Tl, right and left teloblasts; Nr and Nl, right and left neural germ band; and Mr and Ml, right and left primary myoblasts. Roman numerals denote the number of cleavages which have occurred at each fork in the diagram. From A. Penners (1922). *Zool. Jahrb. Anat.* **43**, 323.

the onset of cleavage. Penners reported that it is possible to discern exactly which part of the cytoplasm of the uncleaved egg is going to be distributed into the D quadrant. In this organism the uncleaved egg possesses two areas of "pole plasm" clearly distinguishable by eye, and these are normally inherited only by the CD blastomere at the first cleavage (see Fig. 7.3). The pole plasms fuse around the blastomere nucleus at this stage and are shunted into the D blastomere at the second cleavage. Penners guessed that the morphogenetically important 2d and 4d cells owe their specific character to the presence of these pole plasms. However, it was difficult to show that the cytoplasmic determinants are really part of the special pole plasms visible to the microscopist. Commenting on Penners' studies, Morgan (1927) pointed out that the microscopically visible cytoplasmic inclusions in the egg of *Cumingia*, a bivalve mollusc, can be moved about in various orientations by centrifuging the egg, without in the least affecting subsequent morphogenesis. Early localization in *Cumingia* is as sharp and as determinate as in *Tubifex*.

MORPHOGENETIC POTENTIALITIES OF ISOLATED BLASTOMERES AND SETS OF BLASTOMERES

The simplest implication of the early localization and cell lineage studies was that the embryo is from the beginning a mosaic of determined cell lineages. To test this concept blastomeres were isolated from early embryos, and attempts were made to culture them in order to determine whether they would continue to differentiate in the expected way. One of the most successful examples of this kind of experiment is illustrated in Fig. 7.4, which is compiled from Wilson's (1904) report on experiments with embryos of the mollusc *Patella coerulea*. In this study Wilson isolated a number of different presumptive cell types in low calcium seawater and compared their subsequent development to that expected if they had remained in the context of the whole embryo. The isolated blastomeres follow their normal developmental fate as shown in Fig. 7.4. Here it can be seen that primary trochoblasts isolated from *Patella* embryos carry out

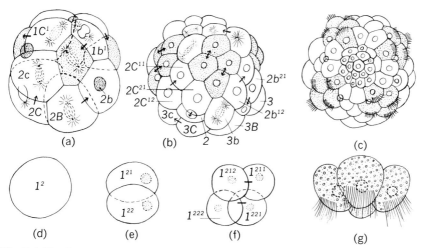

Fig. 7.4. Trochoblast differentiation in the normal embryo of *Patella* after isolation. The lineage of each cell is defined by its number, e.g. 1c¹ (see Wilson, 1904). (a)–(c). Normal development of *Patella*. (a) 16-cell stage, from the side (primary trochoblasts shaded); (b) 48-cell stage; (c) ctenophore stage, about 10 hours, from upper pole, primary trochoblasts ciliated. (d)–(g). Isolated primary trochoblasts cultured *in vitro*. (d) primary trochoblast; (e) result of first division; (f) after second division; (g) product of (f). After E. B. Wilson (1904). *J. Exp. Zool.* **1**, 197.

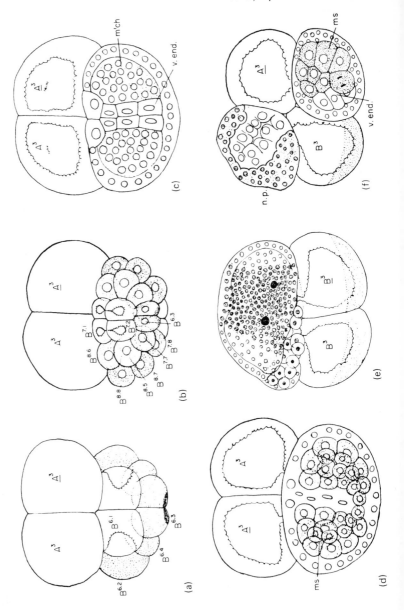

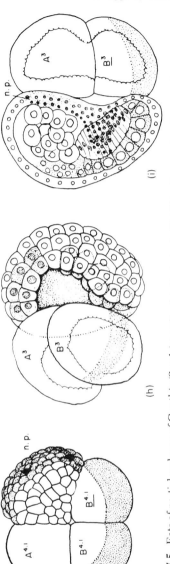

Fig. 7.5. Fate of partial embryos of *Cynthia (Styela) partita*. (a) Posterior half of 32-cell stage, dorsal view (cf. Fig. 7.1f–h). The cleavage of this half is altogether normal. Anterior blastomeres (A³) killed by spurting from a pipette in the 4-cell stage, fixed 1 hour later. (b) Posterior half of 76-cell stage (cf. Fig. 7.1i). Spurted in the 4-cell stage, fixed 2 hours later. Two rows of yellow crescent cells are present, the inner being mesenchyme, the outer muscle cells; the anterior pair of mesenchyme cells ($B^{8.6}$) are larger than normal. There are two pairs of caudal endoderm cells ($B^{7.1}$ and $B^{7.2}$). A pair of clear ventral ectoderm cells is visible in the midline behind B. (c) An embryo, spurted in the 4-cell stage, fixed at 4 hours, deep focus, showing the double row of ventral endoderm cells (v. end.) in the midline, and on each side of this a mass of mesenchyme cells (m'ch). (d) Ventral view of posterior half-embryo of the same stage as the preceding, showing the muscle (ms.) and mesenchyme cells beneath the ectoderm and on each side of the strand of ventral endoderm. (e) Anterior half-embryo, dorsal view. Spurted in the 4-cell stage, fixed 22 hours later. The yellow crescent is plainly visible in the injured cells. Sense spots are present, but the neural plate never forms a tube. The chorda cells lie in a heap at the left side. There is no trace of muscle substance or of a tail in this anterior half embryo. Normal larvae of this stage are undergoing metamorphosis. (f) Left anterior and right posterior (diagonal) quarter embryo, dorsal view; spurted in the 4-cell stage, fixed 5 hours later. The anterior quarter shows thickened ectoderm cells, probably neural plate (n.p.) around the endoderm cells; in the posterior quarter are 8 muscle and 3 caudal endoderm cells. (g) Right anterior dorsal eighth embryo, 14 hours after injury, showing endoderm, chorda, and neural plate cells with sense spots. (h) Right half-gastrula of about 200-cell stage; spurted in the 4-cell stage and fixed 3 hours later. The neural plate, chorda, and mesoderm cells are present only on the right side *and in their normal positions and numbers.* (i) Living left half-embryo, dorsal view, showing the endoderm cells forming exogastrulae and the yellow crescent cells at the surface. From E. G. Conklin (1905). *J. Exp. Zool.* **2,** 145.

the correct number of cell divisions and later become ciliated on the same schedule they normally would have followed. Wilson (1904) concluded:

> The history of these cells gives indubitable evidence that they possess within themselves all the factors that determine the form and rhythm of cleavage, and the characteristic and complex differentiation that they undergo, wholly independently of their relation to the remainder of the embryo.

Consistent results were obtained with other isolated cell types, and with partial embryos. Examples include isolated one-sixteenth embryo macromeres which produced endodermal gut rudiments and isolated apical progenitors which differentiated *in vitro* into apical sensory and ectodermal cells.

Another interesting case was provided in the further studies of Penners (1926) on the annelid *Tubifex rivulorum* (reviewed by Morgan, 1927). As shown in Fig. 7.3 both the neural ectodermal germ band and the mesodermal germ band derive from the D quadrant of the embryo. At the 4-cell stage the D macromere is the largest, and with respect to both size and rate of division, its products remain distinct from those of the A, B, and C quadrants. Penners found that each of these blastomeres would continue its normal course of development even if all the others were killed *in situ* by UV microbeam irradiation. Thus if A, B, and C are killed, the D macromere nonetheless adheres to its unique cleavage pattern. It gives rise to the primary neural ectoderm germ band stem cell 2d and to the primary mesodermal stem cell 4d (Fig. 7.3), and after this to the ectodermal and mesodermal germ bands. Similarly, if 4d is individually killed, the ectodermal germ bands form, but the embryo lacks coelomic mesoderm. The converse experiment, however, shows that the mesodermal stem cell (4d) derivatives possess the capability of producing the ectodermal germ band even if 2d has been eliminated. Essentially similar results have been reported from isolated blastomere experiments carried out on other annelid eggs (see review of Reverberi, 1971b). These include the marine forms *Sabellaria* (Hatt, 1932) and *Nereis* (Costello, 1945).

In Fig. 7.5 are illustrated some of the blastomere deletion experiments performed by Conklin on *Styela* embryos. These studies concern the fate of individual blastomeres and sets of blastomeres in eggs in which the other blastomeres have been killed. Conklin found that the surviving blastomeres appear to establish cell lineages which differentiate in their respective normal directions, though the overall organization of the embryo is of course affected. The drawings reproduced in Fig. 7.5 show that each embryo fraction contains the potentiality of forming certain presumptive tissue types, such as notochord, mesoderm, neural plate, gut, or ectoderm, even though the isolates were often not cultured long enough to establish completely their potentialities for differentiation (see, however, Fig. 7.5e). Except for some minor details of cell lineage assignment, the 1905 experiments of Conklin have recently been confirmed and extended by Reverberi, Ortolani, and others of their colleagues [extensive references on cell lineage data and an excellent recent review are to be found in an article by Reverberi (1971c)].

The classic experiments which defined the localization problem suggested that at least in certain embryos the cleavage planes separate cells whose descendants manifest certain morphogenetic potentialities from cells whose descendants manifest other morphogenetic potentialities. In these cases the embryo, by early cleavage, displays a "determinate" character. It is probable that embryo genome function is required for the realization of some of these potentialities. The classical localization experiments therefore suggest that something is partitioned into the blastomeres which ultimately affects the nature of genome function in their descendant cell lineages.

Localization in the Eggs of Ctenophores

The simplest animals in which cytoplasmic localization has been clearly identified are those belonging to the radiate phylum Ctenophora. Isolated ctenophore blastomeres were studied as early as 1880 by Chun and were the subject of investigation by Driesch and Morgan (1896), Fischel (1898, 1903), and Yatsu (1912). These early experiments as well as more recent blastomere isolation studies are reviewed by Reverberi (1971a). The cell lineage of ctenophore embryos is now well known (Ortolani, 1964). At the 8-cell stage the embryo consists of a pair of external cells denoted "E" cells, and located on either side of the four inner cells, which are denoted "M" cells (see Fig. 7.6). The derivatives of E cells and M cells have different morphogenetic fates. This is true both in normal development and in isolated blastomere experiments. The e micromeres given off by the E macromeres give rise to various structures, the most notable of which are the rows of large swimming cilia or "combs" which soon appear in the embryo. The m micromeres derived from M macromeres form other structures, including mouth, apical organ, and most spectacularly, photocytes. These cells are specialized to produce light. The appearance of cleavage-stage embryos of the ctenophore *Mnemiopsis leidyi* is shown in Fig. 7.6a. A diagram illustrating the location of the e micromeres from which the comb plate cilia derive, and the m micromeres from which the photocytes derive, can be seen in Fig. 7.6b (Freeman, 1976).

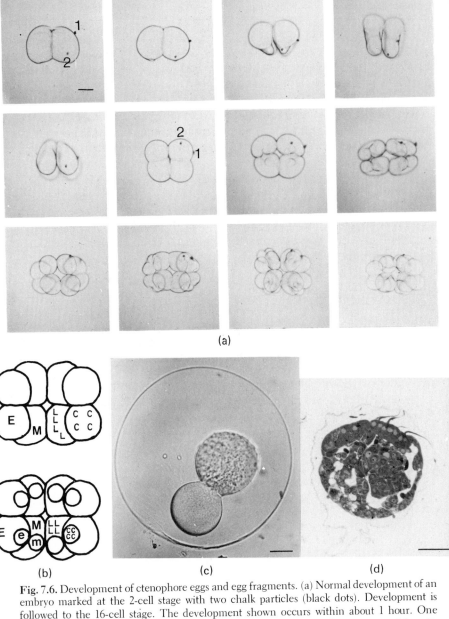

(a)

(b)   (c)   (d)

Fig. 7.6. Development of ctenophore eggs and egg fragments. (a) Normal development of an embryo marked at the 2-cell stage with two chalk particles (black dots). Development is followed to the 16-cell stage. The development shown occurs within about 1 hour. One chalk particle is segrated to an E cell, the other to an M cell. The embryo is pictured from its side at the beginning, but by the 4-cell stage it has rotated so that it is now viewed from the oral pole. Scale bar, 100 μm. (b) The localizations of developmental potential for photocyte

LOCALIZATION OF DETERMINANTS FOR COMB PLATE CILIA

Following the somewhat inconclusive suggestions of classical workers, Farfaglio (1963a,b) proved that the ciliated combs derive from e micromeres. The number of rows of comb plate cilia formed by blastomere isolates depended strictly on the number of E cells present, and Farfaglio concluded that only the clone of cells descending from the first micromere given off by an E macromere (e_1) can become ciliated. By means of a long series of isolated blastomere experiments Freeman and Reynolds (1973) similarly found that the derivatives of isolated E macromeres but not M macromeres can produce comb plate cilia. When isolated and cultured *in vitro*, the comb plate cilia appear on the same time schedule as normally observed in the whole embryo. This result is similar to that reported by Wilson (1904) for the isolated trochoblasts of *Patella*, as described above.

Freeman and Reynolds (1973) also studied egg fragments which had been cut into yolk-containing and yolk-free portions after centrifugation in sucrose gradients. As shown in Fig. 7.6c these eggs assume a "dumbbell" shape, and they can be dissected by hand. The nonyolky half-eggs, termed "cortical fragments" usually cleaved, and E and M macromeres formed. A cross section of a partial embryo developing from a "cortical" fragment and bearing comb plate cilia is shown in Fig. 7.6d. E and M macromeres produced by cleavage of "cortical fragments" give rise to cells bearing comb plate cilia. Thus the cytoplasmic substances responsible for this differentiation remain associated with the "cortical" egg cytoplasm. In other experiments Freeman (1976) showed that as early as the 2-cell stage, the comb plate cilia determinants in *Mnemiopsis* eggs are localized at the aboral pole. Removal at the 2- or 4-cell stage of the aboral cytoplasmic region normally segregated into E macromeres decreases the ability of the remaining fragment to form comb plate cilia. However, if

---

and comb plate cilia cell differentiation at the 8- and 16-cell stages. L indicates the localization of developmental potential which specifies photocyte differentiation. C indicates the localization of developmental potential which specifies comb plates. These cleavage stages are viewed from the aboral pole. M and E identify the macromeres, while m and e identify the micromeres. Note that at the 16-cell stage the potential to form photocytes is still associated with the M macromeres. From G. Freeman (1976). *Dev. Biol.*, **49**, 143. (c) Photograph of a dumbbell-shaped egg formed by centrifugation in sucrose–seawater. The cortical and yolky portions of the egg can be clearly distinguished. Dumbbell-shaped eggs maintain their configuration for only about 10 minutes after centrifugation, then there is a gradual return to a spherical shape and a normal cytoplasmic distribution. The bar indicates 50 μm. (d) Cross section through a 28-hour cortical fragment. Note the comb plate cilia. The bar indicates 50 μm. From G. Freeman and G. T. Reynolds (1973). *Dev. Biol.* **31**, 61.

the aboral region is removed just as the first cleavage is beginning, comb plate cilia are still formed. This result suggests that the localization of the potential for comb plate cilia production is progressive. It becomes final by the 2- to 4-cell stages, though the determinants for comb plate cilia are not actually segregated into the E cells until the 8-cell stage.

LOCALIZATION OF DETERMINANTS FOR PHOTOCYTE DIFFERENTIATION

The ability to produce light is a characteristic of all known ctenophores. This phenomenon begins early in embryogenesis. To study light production in embryos quantitatively, Freeman and Reynolds developed a microscope-mounted image intensifier. Light production is the differentiated function of photocytes. These cells appear and become active only after about 8 hours into development. This is about the same time as comb plate cilia first appear. Prior to this the molecular and cytological apparatus responsible for bioluminescence must be incomplete, since light production cannot be induced. In Fig. 7.7a a cross section of the 8-hour embryo is shown in which both photocytes and comb plate cilia are visible. Bursts of light emanating from the aboral regions of the embryo can be seen in Fig. 7.7b (Freeman and Reynolds, 1973). Extensive blastomere isolation experiments carried out by these authors show that by the 8-cell stage the ability to give rise to photocytes is confined to the M macromeres. Isolated M macromeres develop normally *in vitro* and eventually give rise to functional photocytes. By deleting M macromeres at successive stages, Freeman and Reynolds (1973) also found that the potential for photocyte differentiation is shunted into only one of the two M macromeres formed at the sixth cleavage from each preexisting M macromere. In addition they centrifuged *Mnemiopsis* eggs into "dumbbell" forms as shown in Fig. 7.6c and then bisected them in order to study the initial localization of the factors responsible for photocyte production. Only those cortical fragments also containing some yolky egg cytoplasm can give rise to M macromeres which ultimately produce photocytes. This result contrasts to the requirement for comb plate ciliary differentiation in egg fragments. As noted above comb plate production by these fragments requires solely cortical cytoplasm. Therefore the factors promoting photocyte differentiation are initially localized in different regions of the egg cytoplasm from those promoting comb plate differentiation. Furthermore, the factors responsible for comb plate and photocyte differentiation assume their final localization at different times. In 2- and 4-cell embryos the photocyte factors are not yet localized in those regions of the cytoplasm which are to be segregated to the M cells. This was shown by

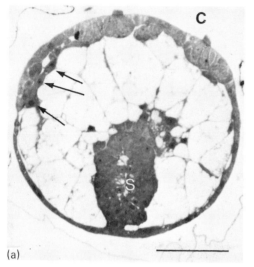
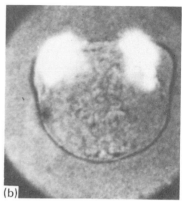

(a)   (b)

Fig. 7.7. Photocytes and light production in *Mnemiopsis* embryos. (a) Longitudinal section through 8-hour embryo at the developmental stage where light production first begins. Most of the internal portion of the embryo is occupied by large yolky cells. S, stomadeum. C, comb plate cilia. The arrows point to the photocytes. The bar indicates 50 μm. (b) Side view of embryo at developmental stage when light production is first detected. The embryo was viewed with black illumination and stimulated by a weak electric pulse. The embryo is 170 μm in diameter. The light-producing regions are in the aboral portion of the embryo. Photographed after image intensification using Polaroid film. From G. Freeman and G. T. Reynolds (1973). *Dev. Biol.* **31**, 61.

removing these regions and later measuring light production (Freeman, 1976). Evidently the photocyte and comb plate factors become localized by processes which are independent of each other.

The mechanisms by which the progressive localization of the comb plate and photocyte determinants occurs is associated in some way with the cleavage process. Thus Freeman (1976) found that 2,4-dinitrophenol and cytochalasin B, agents which block cleavage reversibly, also affect localization. When the eggs resume development following removal of either drug, a variety of cleavage forms can be found. M and E blastomeres were isolated from these, and their ability to form comb plates and to produce photocytes was measured. The extent to which either type of cytoplasmic factor is localized in the appropriate region of the egg was found to depend on the way the treated eggs had cleaved.

These experiments concern two highly specific types of cell differentiation. As shown in Figs. 7.6 and 7.7, production of comb plate cilia and photocyte bioluminescence are clearly measurable activities of highly specialized cells. The cells which carry out these activities become functional only many hours and many cell divisions after the 8-cell stage, when the responsible cytoplasmic determinants are finally segregated into the M and E macromeres. These determinants were originally present in the cytoplasm of the egg and are progressively localized to their final positions during the period between fertilization and third cleavage. There is as yet no evidence on the nature of the morphogenetic factors involved. However, it is significant that there are at least two distinct such factors or sets of factors, those responsible for photocyte differentiation and those responsible for comb plate differentiation. Presumably these could include enzymes and structural proteins, maternal messenger RNA's, transcriptional regulation signals, or any combination of these.

Cytoplasmic Localization in the Eggs of *Ilyanassa* and Certain Other Molluscs

CLASSIC EXPERIMENTS ON THE MORPHOGENETIC SIGNIFICANCE OF POLAR LOBE CYTOPLASM

In one of his most remarkable papers Wilson (1904) described a series of experiments with early embryos of the scaphopod mollusc *Dentalium*. The object was to investigate the qualitative control over morphogenesis apparently exercised by the cytoplasm of the egg. These experiments made use of a peculiarly convenient phenomenon which occurs early in cleavage in *Dentalium*, and in several other spiralians, both molluscs and annelids. This phenomenon is the transient extrusion during first cleavage of a "polar lobe," containing vegetal pole egg cytoplasm. In favorable organisms the lobe is attached by only a thin strand of protoplasm to one of blastomeres, the CD blastomere, into which it flows as the cleavage is completed. This is shown in Fig. 7.8, a photograph of living "trefoil" first cleavage embryos of the gastropod mollusc *Nassaria* (*Ilyanassa*) *obsoleta*. As in *Tubifex* (see Fig. 7.3), the CD blastomere alone carries the capacity to give rise to the mesodermal columns on which the development of various organs and ultimately of body form depends. At the second cleavage a polar lobe is once more transiently extruded, and as the cleavage terminates, it flows back into the D cell. Of the four blastomeres now present only the D quadrant lineage retains the ability to give rise to the mesoderm stem cells. The lobe is again briefly extruded at several subsequent cleavages, always returning to the D macromere. Crampton (1896), then a student of Wilson, had found that in first cleavage *Ilyanassa* eggs the polar lobe could easily be separated from the remainder of the embryo without interfering with the ability of the embryo to continue cleavage, but that the resulting embryos appeared to lack the mesodermal

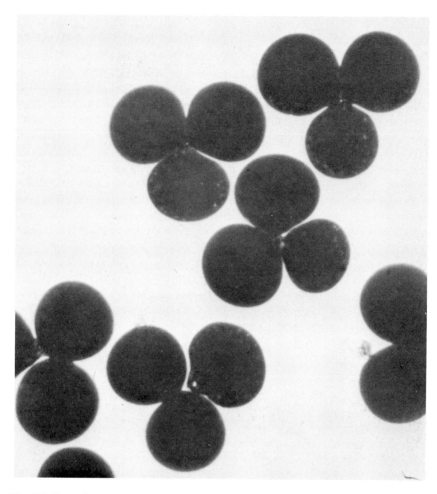

Fig. 7.8. First cleavage "trefoils" of *Ilyanassa obsoleta*. At this stage of first cleavage, the cytoplasmic polar lobe can be removed, leaving the AB and CD cells. As seen in the phase microscope, the polar lobe is the more refractile of the three bodies in each embryo, and in this picture can be identified by the light spots around its circumference. From E. H. Davidson, G. W. Haslett, R. J. Finney, V. G. Allfrey, and A. E. Mirsky (1965). *Proc. Natl. Acad. Sci. U.S.A.* **54**, 696.

stem cells. Since removal of the polar lobe involves the removal of no nuclear components, this important though preliminary result seemed to warrant further investigation. The most advanced of the lobeless embryos described by Crampton in his brief account attained the age of 48 hours. In this species differentiated organ primordia do not appear until much later, between 4 and 7 days of development. A clear result of Crampton's study was that the extirpation of the polar lobe at first cleavage causes the permanent loss of the special division schedule and the asymmetric appearance of the D blastomere lineage. After removal of the polar lobe, all four blastomere lineages behave exactly alike.

Wilson's study of these phenomena in *Dentalium* resulted in a significant demonstration of cytoplasmic localization. Isolated blastomere experiments showed that the morphogenetic character of each individual blastomere lineage is set from the beginning, at least from the point of the appearance of the cleavage planes which separate the early blastomeres. If the egg cytoplasm sequestered in the first polar lobe at the trefoil stage is removed, the morphogenetic value of the D blastomere and its descendants is altered. Wilson found that lobeless *Dentalium* embryos fail to develop the main coelomic mesoderm bands. Examination of postgastrular lobeless embryos revealed the absence of all major organ primordia which are directly or indirectly derivatives from this mesoderm. He reported that the lobeless embryos lack mouth, shell gland, and foot as well as the mesodermal primordium itself. Wilson pointed out that the lobeless embryos develop in exactly the same way (or fail to develop in exactly the same way) as do embryos deriving from isolated AB blastomeres or from single A, B, and C blastomeres. The latter single blastomere embryos cannot produce mesoderm either, since they lack the cytoplasm contained in the D blastomere. It follows that the cytoplasm extruded in the polar lobe contains the mesodermal determinants and that these cytoplasmic elements endow the D quadrant lineage with its particular morphogenetic potentialities. Since the nuclei in the embryo presumably are equal in their genomic content, the nucleus of the AB cell must contain information for the creation of mesodermal cells and the further differentiation of the mesodermal clones, just as does the nucleus of the CD cell. Therefore, the cytoplasm special to the CD cell must in some way determine the eventual utilization of this nuclear information, either directly or indirectly. Wilson (1904) also showed that the polar localization of the morphogenetically significant cytoplasm is established by means of a series of cytoplasmic movements a few minutes before the onset of first cleavage. The significant cytoplasm is already present in the egg when it is shed. However, just as in *Styela* (Fig. 7.2), its ultimate pattern of localization depends on cytoplasmic redistribution after fertilization.

Following these classic studies on *Dentalium* the effect of polar lobe removal was investigated in the eggs of a number of species (reviewed by Cather, 1971). Characteristic defects are observed following polar lobe removal in the annelid *Sabellaria* (Hatt, 1932; Novikoff, 1938), the lamellibranch mollusc *Mytilus edulis* (Rattenbury and Berg, 1954), and the gastropod mollusc *Bithynia tentaculata* (Cather and Verdonk, 1974). Many of Wilson's key results on *Dentalium* were recently confirmed by

Verdonk (1968). However, in this respect the most intensely studied spiralian species is now *Ilyanassa*, the organism originally the subject of
Crampton's (1896) experiments. Observations on the role of the polar lobe
in the early development of *Ilyanassa* are relatively detailed, and as reviewed below have stimulated several attempts to approach the localization phenomenon at a molecular level.

MORPHOGENETIC DETERMINANTS IN THE POLAR LOBE
CYTOPLASM OF *ILYANASSA* EGGS

Current knowledge of the morphogenetic significance of *Ilyanassa* polar
lobe cytoplasm is due mainly to the researches of Clement (1952, 1956,
1962, 1963, 1967, 1968). Wilson's early conclusions with respect to *Dentalium* were reproduced and then greatly extended by Clement. In Fig.
7.9a–g, the later cleavage and cell lineage of *Ilyanassa* eggs are diagrammed, with special emphasis on the morphogenetically significant D
quadrant (Clement, 1952). From the D quadrant cells arise the primary
mesentoblasts (labeled "ME" in Fig. 7.9e) and ultimately the organs forming with the participation of coelomic mesoderm. The primary mesentoblasts arise from 4d. Figure 7.9h–j also shows the symmetrical cleavage of
lobeless embryos, in which the D quadrant cells cannot be morphogenetically distinguished, just as originally reported by Crampton (1896).
Clement (1952) carried out histological studies on the veligers developing
from eggs from which the polar lobe had been removed at the trefoil
stage, and found that many organs and tissues are missing from the lobeless larvae. Additional observations on the veligers developing from lobeless embryos have been reported by Atkinson (1971). Lobeless and normal
veligers can be compared in the drawings shown in Fig. 7.10 (Atkinson,
1971). Lobeless larvae fail to organize heart, intestine, statocyst, operculum, velum, external shell, eyes, and foot. On the other hand, lobeless
larvae possess active muscle, nerve ganglia and nerve endings, stomach,
some velar tissue with cilia, digestive gland, mantle gland tissue, and
pigment cells. The muscles in the lobeless larvae derive from "ec-
tomesoderm," which in contrast to coelemic mesoderm arises from the
second and third micromere quartets and is regarded as a vestige of the
remote evolutionary origins of mesoderm in precoelomate radial animals
(Hyman, 1951). Removal of the first cleavage polar lobe cytoplasm thus
does not simply block all differentiation, only *certain* differentiation.

By removing the D macromere at successively later stages of development, Clement (1962) was able to locate the stages at which the morphogenetic determinants originally present in this macromere are shunted
into its descendants. Of particular interest are those cells which are the
direct ancestors of tissues for which the lobe carries determinants. As this

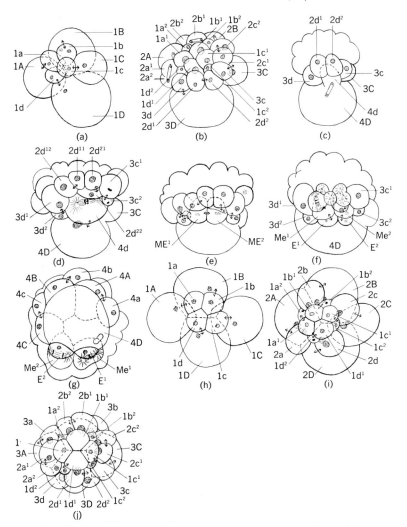

Fig. 7.9. D quadrant cell lineage in *Ilyanassa* embryos. Upper case letters refer to macromeres and lower case to micromeres. Preceding numerals indicate the macromere division stage at which the micromeres are given off, and superscripts indicate subsequent micromere division products. E^1 and E^2, primary entoblasts; Me^1 and Me^2, primary mesentoblasts. Me and E cells derive from the ME^1 and Me^2 stem cells. (a)–(c) Normal cleavage of *Ilyanassa*. From camera lucida drawings of stained whole mount preparations. ×196. (d)–(f) Normal cleavage showing the early derivatives of the first mesentoblast progenitor cell, 4d. (g) The egg has been oriented so that the vegetal pole is toward the observer. The division of Me^1 and Me^2 will produce the primordial mesoderm stem cells. (h)–(j) Cleavage after removal of the polar lobe at the trefoil stage. From A. C. Clement (1952). *J. Exp. Zool.* **121**, 593.

operation is carried out at progressively later cleavages, the degree of differentiation displayed by the embryos improves. By the time the fourth derivative of the D macromere, 4d, is given off (see Fig. 7.9) removal of the whole 4D macromere has no *qualitative* effect on later differentiation, and the resulting embryo is normal except for its small size. This experiment also serves to eliminate the possibility that the effects of polar lobe removal are due to some general injurious effect on the embryo, e.g., starvation for substrates carried in the polar lobe. Further evidence on this point comes from centrifugation experiments (Clement, 1968). Nucleated vegetal pole egg fragments produced by centrifugal force and containing only a small fraction of the original yolky cytoplasm were found to give rise to lobe-dependent, differentiated structures. Clement (1962) found that removal of the D macromere before the 2d cell is given off results in as severe an inhibition of morphogenesis as removal of the first cleavage polar lobe or of the whole D quadrant. The morphogenetically significant polar lobe contents therefore appear to be shunted into 3d and particularly into 4d. It is the latter cell which is the direct ancestor of the primary mesentoblasts (Fig. 7.9). Only embryos in which 4d is normally formed produce heart and intestine at the veliger larval stage.

Embryos from which the D macromere is deleted after the formation of 3d but before the formation of 4d display velum, eyes, foot, and some shell development. However, most of these tissues are not formed directly from the 3d cell lineage. An indirect or inductive effect involving the 3d descendants is the most likely cause of this result. Induction was shown explicitly to be involved in eye differentiation in *Ilyanassa* by Clement (1967). It is clear from the detailed experiments of Cather (1967) that this is also true of shell gland formation in *Ilyanassa*. The shell gland is of ectodermal derivation, and Cather demonstrated that any combination of ectoderm and endoderm can form shell, though neither can carry out this function alone. The formation of shell by A, B, and C quadrant ectoderm is repressed by the presence of the polar lobe cytoplasm in the D macromere, however. Shell formation is normally confined to the D quadrant ectoderm derivatives, but *induction* of this activity must be initiated during the third quartet stage by the 3D macromere. In other molluscs shell gland is known to develop by inductive interaction of archenteron and certain regions of the external larval wall [reviewed by Raven (1958) and Hess (1971)]. Another type of polar lobe effect, described by Cather (1973), concerns the regulation of apical cilia. Normally a tuft of these cilia appears at the top of the *Ilyanassa* larva, confined to a special plate of cells. However, if the polar lobe or the D macromere is removed, cilia develop over the whole upper end of the embryo. Thus the polar lobe normally inhibits ciliation on cells other than the apical plate cells, just as it is responsible for confining shell gland differentiation to certain cells.

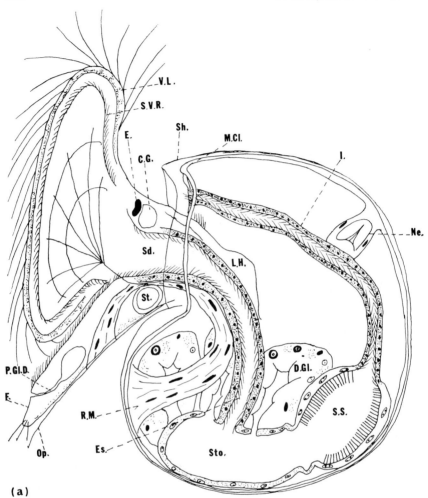

Fig. 7.10. Normal and lobeless *Ilyanassa* larvae. (a) Composite reconstruction of normal 8-to 9-day veliger larva of *Ilyanassa*, with left velar lobe omitted. (b) Reconstruction of a lobeless larva based on whole mount and sectional material. Ap.A., apical area; Bi. M., birefringent mass; C.G., cerebral ganglion; D. Gl., digestive gland; E., eye; Es., esophagus; F., foot; Gl. Cell, ectodermal gland cell; I., intestine; L. H., larval heart; M.B., muscle block; M. Cl., mantle collar; Ne., nephridium; Op., operculum; P.P., posterior protusion; P. Gl.D., pedal gland duct; R.M., retractor muscle; S.S., style sac; Sd., stomodeum; Sd. I., stomodeal-like invagination; Sh., shell; St., statocyst; Sto., stomach; S.V.R., secondary velar row; V.L., velar lobe. From J. W. Atkinson (1971). *J. Morph.* **133**, 339.

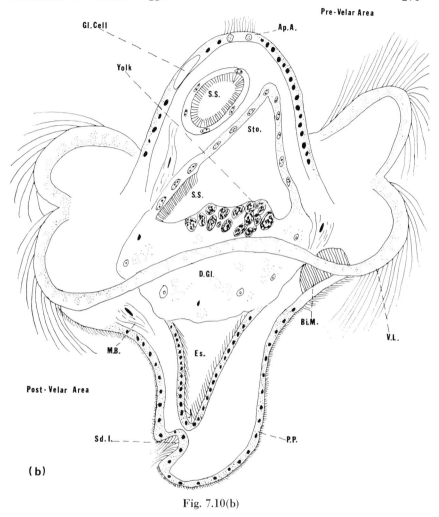

Fig. 7.10(b)

This brief review shows that the polar lobe cytoplasm influences development in several ways: (a) It affects cleavage patterns and thus the initial orientation of the embryo. (b) Via the 4d cell and the mesentoblasts it gives rise to the coelomic mesoderm, and thereby to the tissues derived directly from this mesoderm. (c) It is indirectly responsible for several structures arising by inductive interaction between mesoderm and derivatives of the A, B, and C quadrants. (d) It results in inhibition of certain differentiations in cells other than the appropriate ones. A distinction is made here between self-differentiation and differentiation occurring as a result of inductive interactions between cells which are already dissimilar. Self-differentiation is classically illustrated by the isolated blastomere experiments. In the present context the differentiation of the mesoderm anlage from D quadrant cells is an example of self-differentiation. It is clear that some very complex secondary interactions of an inductive nature are affected by polar lobe removal. The simplest and most precisely defined aspect of localization in *Ilyanassa*, however, is the presence of polar lobe determinants which control the self-differentiation of the D quadrant cell lineages.

CYTOLOGICAL OBSERVATIONS ON POLAR LOBES

The molluscan polar lobe has been the subject of a number of cytological and other studies designed to yield information on the subcellular location of the morphogenetic determinants. By centrifuging *Ilyanassa* eggs Clement (1968) obtained vegetal fragments in which the polar lobe was filled with clear cytoplasm rather than the normal yolk-filled polar lobe endoplasm. As noted above these partial embryos are able to develop lobe-dependent structures. This experiment suggests that the morphogenetic factors may be cortical or bound to the cortex; at least they are not moved out of the polar lobe region along with the yolky endoplasm by low centrifugal forces. These experiments do not distinguish whether the morphogenetic determinants are actually bound to the plasma membrane of the egg or are associated with the subcortical cytoplasm. As used in this chapter the term "cortex" should be taken to include either possibility.

Evidence suggesting cortical localization also exists for *Dentalium* eggs. One of the lobe-dependent structures in embryos of this organism is the apical tuft, a group of prominent cilia growing out of a small plate of cells at the top of the trochophore larvae. The apical tuft forms from descendents of both the 1c and 1d micromeres (Wilson, 1904; Geilenkirchen *et al.*, 1970; Van Dongen and Geilenkirchen, 1974). Nonetheless, the apical tuft determinants are localized to the polar lobe and the D quadrant, suggesting that some sort of intercellular interaction is ultimately involved in apical tuft formation (Van Dongen and Geilenkirchen, 1974). Verdonk (1968) showed that centrifugation sufficient to displace the polar lobe endoplasm fails to move the apical tuft determinants from the *Dentalium* polar lobe. Successive ablation experiments demonstrated that these determinants are located in the upper (animal) half of the polar lobe at first cleavage. Thus the bottom 60–70% of the lobe can be cut off and apical tufts will still develop, but removal of more than this blocks the appearance of apical tufts (Geilenkirchen *et al.*, 1970; Verdonk *et al.*, 1971). After second cleavage, the apical tuft determinants are found toward the animal side of the D macromere, i.e., the region where the 1d micromere will be given off. The specificity of its topographical location is consistent with a cortical or cortex-associated position for the lobe determinants. However, no direct evidence supports such an hypothesis.

In *Bithynia tentaculata*, another gastropod mollusc, the polar lobe is extremely small compared to the *Ilyanassa* polar lobes shown in Fig. 7.8. The *Bithynia* polar lobe occupies <1% of the egg volume. Its morphogenetic determinants nonetheless appear similar to those of the *Ilyanassa* polar lobe. Thus, when the lobe is removed, the embryos fail to form mesentoblasts or coelomic mesoderm bands, and all derivatives thereof (Cather and Verdonk, 1974). They lack eyes, foot, intestine, organized shell, operculum, etc., while successfully developing digestive gland, ganglia, and some muscle. In *Bithynia*, however, the developmental capacity of the C macromere is the same as that of the D macromere. Lobe-dependent structures form if either the C or D macromeres are present (Verdonk and Cather, 1973; Cather and Verdonk, 1974). What is significant here is the fact that the polar lobe cytoplasm is distributed to *both* C and D macromeres in *Bithynia*, according to cytological observations, while in *Ilyanassa* and *Dentalium* it is distributed only to the D macromere.

The small polar lobe of *Bithynia* contains cytologically unique structures. These have been studied by Dohmen and Verdonk (1974) and several of their preparations are reproduced in Figure 7.11. The lobe contains a structure staining densely for RNA, as shown in Fig. 7.11a, which was termed the "vegetal body." An electron micrograph of the polar lobe and the vegetal body is shown in Fig. 7.11b. Figure 7.11c shows the vegetal body at higher magnification. Here it can be seen that the vegetal body consists of numerous small membrane-bound vesicles as well as some dense, non-membrane-bound particles of about the same size. The vegetal body is shunted into the CD cell at first cleavage. It completely disappears or dissolves prior to second cleavage and the separation of the C from the D macromere. This behavior correlates with the equal distribution of morphogenetic potential between the C and D blastomeres in *Bithynia*. Similar membranous vesicles and electron-dense particles have been observed in polar lobes of *Dentalium* (Reverberi, 1970) and *Ilyanassa* (Pucci-Minafra *et al.*, 1969). Dohmen and Lok (1975) also described membranous vesicles similar to those of *Bithynia* polar lobes in the polar lobes of *Crepidula* eggs. They found that the *Crepidula* polar lobes contain in addition complex aggregates, i.e., membrane-free granular particles similar to the polar granules associated with the localized germ cell determinants in other eggs (see section on Localization of Germ Cell Determinants in this chapter). The various particulate structures

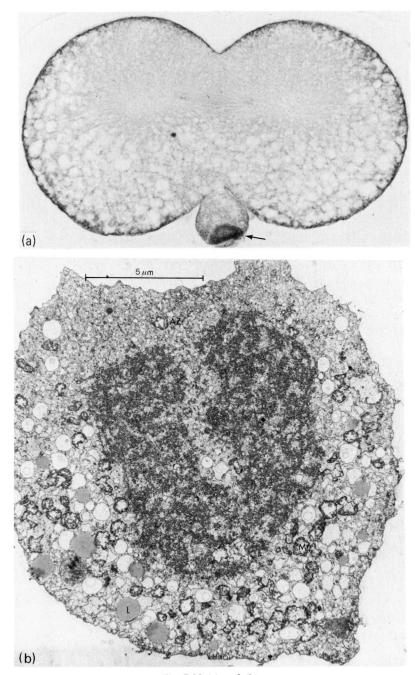

Fig. 7.11 (a) and (b)

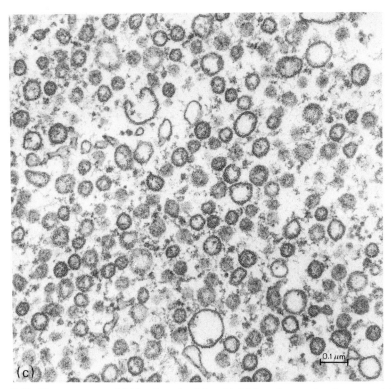

Fig. 7.11. Fine structure of *Bithynia* polar lobe. (a) Light micrograph of an egg at first cleavage, stained for RNA with methyl green–pyronin. The vegetal body (arrow) is densely stained. × 350. (b) Electron micrograph of the first polar lobe with vegetal body. Fixed in glutaraldehyde and osmium tetroxide, and stained by the uranylacetate-lead method. AZ, attachment zone; L, lipid; M, mitochondrion. ×6300. (c) Detail of a vegetal body, showing small vesicles. Most vesicles are completely or partially filled with a dark-staining substance, possibly RNA. Fixed and stained as in (b). ×68,250. From M. R. Dohmen and N. H. Verdonk (1974). *J. Embryol. Exp. Morph.* **31**, 423.

identified in the investigations reviewed here, plus a high concentration of mitochondria, thus distinguish the polar lobe cytoplasm. Unfortunately, any functional relation between the cytoplasmic inclusions visible in the electron microscope and morphogenetic determinants of the polar lobe remains completely hypothetical.

EFFECTS OF POLAR LOBE REMOVAL ON
MACROMOLECULAR BIOSYNTHESIS

The *Ilyanassa* embryo is one of the few cytoplasmic localization systems which have begun to be investigated at the molecular level. The evidence so far available concerns distinctions between lobeless and normal embryos. A basic fact of importance is that lobeless and normal embryos develop for several days at the same rate insofar as any gross measurements are concerned. Davidson et al. (1965) and Cather (1971) showed that lobeless and normal embryos of the same ages contain about the same number of cells during early development. Thus the rate of cleavage and DNA synthesis is not appreciably modified by removal of the polar lobe. Measurable incorporation of labeled uridine into ribosomal RNA also begins at the same time in lobeless and normal embryos. Ribosomal RNA synthesis in *Ilyanassa* embryos was studied by Koser and Collier (1972), Collier (1975a), and Newrock and Raff (1975). Under the conditions used by the latter authors, detectable synthesis of ribosomal RNA could be observed by about 27–30 hours, but not before 24 hours in both lobeless and normal embryos. This of course does not necessarily mean that the ribosomal genes are activated only at 27–30 hours for the reasons given in the discussions of ribosomal RNA synthesis in Chapter 5. However, the factors responsible for the visibility of ribosomal RNA synthesis by this point operate similarly in lobeless and normal embryos. These factors include the behavior of the precursor pools, the rate of increase in cell number, any changes in permeability, and the absolute ribosomal RNA synthesis rates. About the same amount of bulk (ribosomal) RNA is present in lobeless and normal embryos at 3 days of development (Collier, 1975b). After this the lobeless embryo lags in its accumulation of total RNA, according to Collier (1975b). Similarly the DNA content is the same up to 4 days of development in lobeless and normal embryos and after this begins to increase faster in normal embryos (Collier, 1975b). Morphogenetically the overall rate of development of lobeless and normal embryos is also similar. Thus in both types of embryos apical cilia appear at 48 hours and velar structures at 6 days (Newrock and Raff, 1975). It is clear that removal of the polar lobe causes specific, qualitative changes in the differentiation of certain of the embryonic cells. Gross changes in the rate of growth occur only after organogenesis has begun, no doubt as a secondary effect of the absence of several organs and tissues normally present.

As might be expected, the lack of certain differentiated tissues in lobeless embryos is reflected in an altered protein composition. Freeman (1971) studied the forms of alkaline phosphatases and esterases in *Ilyanassa* embryos, as defined by visualization of enzymatic activity on electrophoretic gels. In normal embryos alkaline phosphatase activity occurs in only one form, by the criterion of mobility, up to day 7. Organogenesis is well advanced by this stage. However, in normal day 9 veliger larvae at least three different forms of this enzymic activity have appeared. A single form of esterase is similarly present up through day 5 in normal embryos, while by day 7 at least seven different forms have appeared. A striking difference was seen in lobeless embryos. Only the original form of alkaline phosphatase is evident in 10 day lobeless veligers, and 2 of the 7 normally present forms of esterase are also completely missing in these veligers.

General patterns of protein synthesis in lobeless and normal *Ilyanassa* embryos have been examined by Teitelman (1973) and by Newrock and Raff (1975). As reviewed in Chapter 3 Teitelman (1973) found that the spectrum of proteins synthesized in these embryos greatly alters as organogenesis begins, at around 4 days of development. This conclusion was based on double label gel electrophoresis experiments (see Chapter 3 for discussion of this type of experiment). The most interesting result was that large differences exist in the pattern of proteins synthesized in 5 day lobeless embryos as compared to normal embryos (Teitelman, 1973). By day 5, of course, normal embryos already display some of the lobe-dependent structures and appear morphologically different from lobeless embryos.

It is now clear that the polar lobe cytoplasm qualitatively affects protein synthesis from very early in development. Donohoo and Kafatos (1973) compared the spectrum of proteins synthesized in embryos grown for 4 hours from isolated AB and CD blastomeres. Intact embryos would contain 12 cells at this stage. The spectrum of proteins synthesized by these two classes of partial embryo was clearly distinct, as illustrated in Fig. 7.12a and b. In addition, Newrock and Raff (1975) showed that pregastrular lobeless embryos differ from normal embryos in their patterns of protein synthesis. The earliest embryos studied were labeled for 5 hours about 24 hours after the start of cleavage. Some of these experiments are reproduced in Fig. 7.12c and d, and similar results were obtained by Newrock and Raff (1975) with embryos labeled at 48 hours of development (gastrulae). These data demonstrate that the patterns of protein synthesis are affected by polar lobe removal long before its morphological effects on embryonic organogenesis are evident. This result is consistent with that of Donohoo and Kafatos, since their experiments show that even in early cleavage, the embryo quadrant receiving the polar lobe cytoplasm synthesizes a distinct set of proteins.

DOES THE POLAR LOBE CONTAIN A SPECIFIC SET OF MATERNAL MESSENGER RNA'S?

One possible explanation for the effects of polar lobe removal is that a qualitatively special set of maternal messenger RNA's is carried in the polar lobe. *Ilyanassa* embryos almost certainly contain maternal messenger RNA. As discussed in Chapter 4 the isolated polar lobe itself carries

out protein synthesis (Clement and Tyler, 1967) and contains active polysomes (Geuskens, 1969). Additional evidence for maternal messenger
RNA in *Ilyanassa* eggs is the fact that the same amount of labeled amino
acid is incorporated into the proteins of normal and actinomycin-treated
cleavage stage embryos (Newrock and Raff, 1975). The actinomycin
treatment applied in these experiments was sufficient to eliminate all high
molecular weight RNA synthesis. Actinomycin-treated 48-hour embryos
synthesize different kinds of proteins than do untreated embryos, however. This is shown in Fig. 7.12e (Newrock and Raff, 1975). Because of the
various effects of actinomycin on polysomes, and other possible side effects (see Chapters 2 and 3 for discussion of the interpretation of actinomycin experiments), this result is not conclusive, but it suggests that
newly transcribed embryo messenger RNA's are important in *Ilyanassa*
embryos. Nonetheless, when both lobeless and normal embryos are
treated with actinomycin and their newly synthesized proteins are compared, highly significant differences are observed. An experiment of this
kind is reproduced in Fig. 7.12f (Newrock and Raff, 1975). According to
this experiment the effect of the polar lobe on protein synthesis during
cleavage may be at least partly independent of embryo genome transcription. Newrock and Raff (1975) concluded that specific sets of maternal
messenger RNA's, or factors which select a specific set of maternal messages for translation, are sequestered in the polar lobe. Either explanation
is consistent with the failure of actinomycin to obliterate the differences in
protein synthesis patterns between normal and lobeless embryos. On the
other hand, these results leave open the possibility that transcriptional
control processes are also involved in the polar lobe effect. The experiment shown in Fig. 7.12f indicates that some differences in protein synthesis still exist between lobeless and normal embryos in the presence of
actinomycin. However, it is not demonstrated that all or even most of the
qualitative differences between lobeless and normal embryos persist in the
presence of actinomycin. The polar lobe could contain specific maternal
messenger RNA's, some of which are translated to produce gene regulatory agents. It is also possible that lobeless and normal embryos differ in
their sensitivity to actinomycin, so that the differences observed depend
on response to this drug. Despite these caveats, however, the experiments
of Newrock and Raff (1975) define a number of critical objects for future
research. Among these is clearly a comparison of the set of structural
genes being transcribed in lobeless and normal embryos. Unfortunately,
very little is so far known about transcription in lobeless as opposed to
normal *Ilyanassa* embryos. Early experiments of Davidson *et al.* (1965)
showed that the rate of incorporation of label into total RNA fails to increase as rapidly in lobeless embryos as in normal embryos, beyond 24

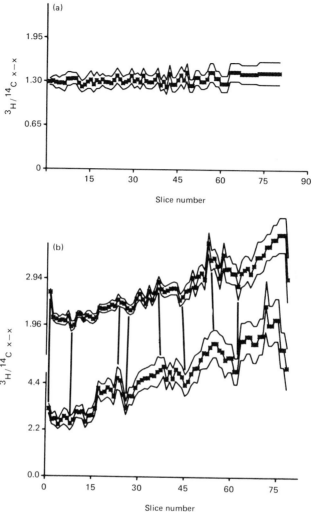

Fig. 7.12 (a) and (b). Comparisons of proteins synthesized by normal and by partial *Ilyanassa* embryos. (a) Control experiment comparing ^{3}H and ^{14}C proteins synthesized during a 100-minute period in early cleavage (second to third quartet stage). The proteins of 166 ^{3}H-leucine labeled embryos and 89 ^{14}C-leucine labeled embryos were extracted together and displayed by gel electrophoresis. The ratio of ^{3}H to ^{14}C radioactivity in each gel slice is shown. Solid lines represent ±15% error limits. (b) Differences in the proteins synthesized by AB-derived (^{14}C) and CD-derived (^{3}H) embryos, reflected in the ^{3}H/^{14}C ratio profiles. Results of two completely independent experiments are compared, using 146 CD and 104 AB fragments (top) or 99 CD and 118 AB fragments (bottom). Vertical lines indicate landmark differences which are repeatable and statistically significant in one or both experiments. Suggestions of repeatable although not statistically established differences can also be seen in the gel segments between vertical lines. Acceptable error limits are set as ±10% and ±15% of the ratio for top and bottom, respectively. From P. Donohoo and F. C. Kafatos (1973), *Dev. Biol.* **32**, 224.

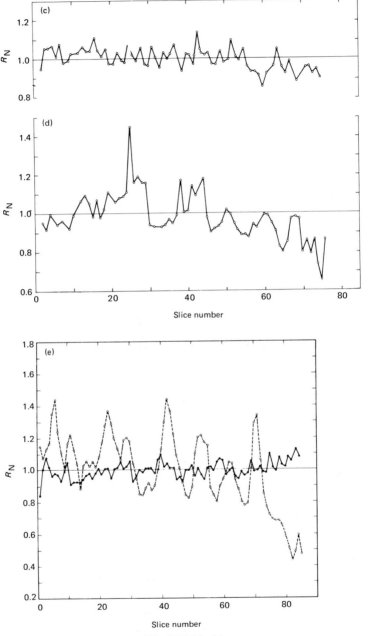

Fig. 7.12 (c)–(e)

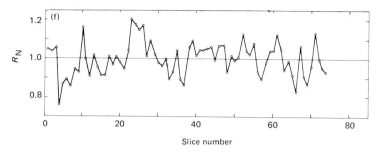

Fig. 7.12 (f) Proteins from ^{14}C-leucine labeled normal embryos which had been cultured 24 hours in actinomycin D coelectrophoresed with proteins from ^{3}H-leucine labeled lobeless embryos also cultured for 24 hours in actinomycin D. From K. M. Newrock and R. A. Raff (1975). *Dev. Biol.* **42**, 242.

hours of development. However, pool specific activity determinations could not be carried out, and a large but unknown fraction of the labeled RNA in these experiments was probably nuclear RNA. The relative amounts of precursor incorporated into the RNA's of lobeless and normal embryos were also compared by Koser and Collier (1976). They found that there is less incorporation in the largest RNA species in lobeless embryos than in normal embryos after gastrulation (i.e., beyond 1 day of development). Neither Davidson *et al.* (1965) nor Koser and Collier (1976) detected differences in incorporation between lobeless and normal embryos within the first day. Similarly, Collier (1975a) measured the fraction of poly(A)RNA which can be labeled in a 6.5 hour exposure to ^{3}H-uridine during cleavage, and found that the amount of labeled poly(A)RNA is the same in lobeless and normal embryos. Around 30% of this RNA was polyadenylated in both cases, and much of it may also belong to the heterogeneous nuclear RNA class. The significance of these observations is uncertain even if it is true that the absolute heterogeneous nuclear RNA synthesis rate is greater in normal than in lobeless embryos after gastrulation. Thus it is within the first day that crucial events of determination

---

Fig. 7.12 (c)–(e). (c) Control experiment similar to that in (a) comparing newly synthesized ^{14}C- and ^{3}H-leucine labeled proteins of 24 hour normal embryos. R_N is the normalized ratio of the coelectrophoresed ^{14}C- and ^{3}H-labeled proteins in each gel slice. (d) Comparison of proteins from ^{14}C-leucine labeled normal embryos coelectrophoresed with proteins from 24-hour ^{3}H-leucine labeled lobeless embryos. (e) Comparison of ^{3}H-leucine labeled proteins synthesized in normal 48-hour embryos with ^{14}C-leucine labeled proteins synthesized in 48-hour actinomycin-treated embryos (dashed line). Actinomycin was present from the beginning of development. The control (solid line) shows a comparison of proteins from 48-hour normal embryos labeled with ^{3}H-leucine coelectrophoresed with proteins from ^{14}C-leucine labeled normal embryos. No actinomycin was used in the control. From K. M. Newrock and R. A. Raff (1975). *Dev. Biol.* **42**, 242.

must occur, and as the studies reviewed above show, a normal pattern of protein synthesis during the first day does not occur if the polar lobe is removed.

No evidence yet exists which either demonstrates or excludes a transcription level effect of the polar lobe determinants. The reality of the morphogenetic determinants in the molluscan polar lobes cannot be questioned, however. As originally inferred by Wilson in 1904, they are required to set in train the events of coelomic mesoderm differentiation. It is clear that they affect protein synthesis in a major way from early in development. This provides a precise molecular index of their effects, an advantage offered by few other cases of cytoplasmic localization.

Localization of Germ Cell Determinants

CLASSIC EXPERIMENTS ON LOCALIZATION OF GERM CELL DETERMINANTS

An early embryonic cell type which seems generally to be specified through the action of cytoplasmic determinants is the primordial germ cells. In many organisms both primordial germ cells and the special regions of egg cytoplasm which induce their formation stand out cytologically. This facilitated study by early investigators, and several interesting and convincing cases of primordial germ cell determination had been worked out soon after the turn of the century. We begin with a brief review of two particularly striking examples from this period. These concern the determination of primordial germ cells in the eggs of certain beetles, and in the eggs of the nematode *Ascaris megalocephala* (*Parascaris equorum*).

Primordial germ cell differentiation was studied in chrysomelid beetles by Hegner (1911, 1914). In these eggs germ cell formation can be said to initiate when the cleaving nuclei reach the periphery of the egg and blastoderm organization begins. Nuclei arriving at the polar region of the oblong egg enter a special region of cytoplasm which appears to function as the germ cell determinant. That is, only the cells formed from these nuclei and their polar cytoplasm differentiate as germ-line stem cells. Hegner succeeded in selectively destroying the germ cell determinant cytoplasm with a hot needle before the peripheral movement of the nuclei had brought them to the polar region of the egg. His drawings of this experiment are reproduced in Fig. 7.13. The injury induced by the needle is rapidly walled off by the forming blastoderm, as illustrated, and normal development of a differentiated gastrula and ultimately a hatching insect takes place. Germ cells are visible in normal embryos as shown in Fig.

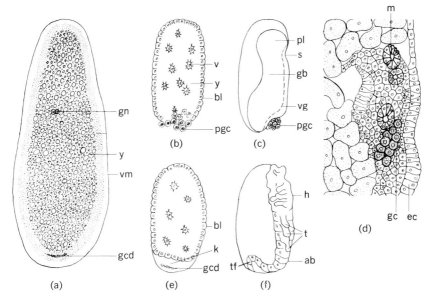

Fig. 7.13. Germ cell determination in beetle eggs. (a) A longitudinal section through an egg of *Calligrapha bigsbyana* 4 hours after deposition. (b)–(f) *Leptinotarsa decemlineata*. (b) A longitudinal section through an egg 1 day after deposition when in the blastoderm stage. (c) Superficial view of the right side of an egg 36 hours after deposition. Note mass of primordial germ cells. (d) Longitudinal section through the tail fold of a normal embryo 60 hours old, showing germ cells. (e) Longitudinal section through an egg, the posterior end of which was killed with a hot needle just after the egg was laid; the egg was then allowed to develop for 24 hours. (f) Side view of an egg similarly treated. The posterior end was killed with a hot needle just after deposition, and the egg was then allowed to develop for 60 hours. gc, germ cells; gcd, germ cell determinants; gn, germ nuclei undergoing fusion; vm, vitelline membrane; y, yolk; bl, blastoderm; pgc, primordial germ cells; v, vitellophage; gb, germ band; pl, procephalic lobes; s, stomodeum; vg, ventral groove; k, portion of egg killed; ab, abdomen; h, head; t, thoracic appendages; tf, tail fold; ec, ectoderm; m, malpighian tubules. From R. W. Hegner (1911). *Biol. Bull.* **20**, 237.

7.13d, but were found to be absent in embryos descended from the cauterized eggs. The adults developing from these eggs are consequently sterile. No genomic material is directly affected, since no nuclei are in the vicinity of the polar cytoplasm at the time of cauterization. However, Hegner's experiment shows that when the germ cell determinant cytoplasm is destroyed, the capacity of the embryo to elicit germ cell differentiation is lost.

Germ cell determinants were demonstrated in *Ascaris* eggs by Boveri and several of his associates. Boveri (1899) showed that in *Ascaris* the cell lineage which gives rise to the primordial germ cells undergoes an unequal division for the first five cleavages. At each cleavage one daughter cell retains the complete chromosome complement while the other eliminates a large fraction of the genome as deeply staining pycnotic granules. This process is called "chromosome diminution." At each division the cell not undergoing chromosome diminution is the germ-line stem cell, while the other becomes a somatic cell. After fifth cleavage, i.e., the 32-cell stage, there are 31 somatic cells and one germ-line stem cell. Figure 7.14 portrays the appearance of the early cleavages of the *Ascaris* embryo. From the beginning the germ-line stem cell possesses a distinct cytological appearance. Not only do chromosome diminutions occur in the other cells and not in this one, but in the early cleavages the germ cell mitotic spindles are oriented perpendicularly to those of the somatic cells. The orientation of cleavage-stage mitoses is generally found to depend on the cytoplasmic organization of the egg (recall, e.g., the fact that cleavage patterns are maintained in actinomycin-treated eggs).

Boveri and his associates induced abnormal cleavage either by polyspermy or by centrifugation and investigated germ cell determination in the treated eggs. These experiments showed that the number of germ-line stem cells depends on the number of cells into which is distributed the polar egg cytoplasm normally enclosed in the single initial germ-line stem cell. Examples are diagrammed in Fig. 7.15 (Hogue, 1910). These studies demonstrated that a component of the egg cytoplasm specifies the germ cells. The cytoplasmic germ cell determinants produce an immediate effect on the blastomere nuclei even when these are different nuclei from those normally in the polar region. Thus nuclei surrounded with this cytoplasm are protected from diminution, while all nuclei distributed into the remainder of the cytoplasm undergo chromosome diminution. The conclusion of this process is marked by the cessation of diminution mitoses at the 32-cell stage. From this point on no further somatic cells are given off from the germ cell stem line. In other experiments the polar germ cell determinants were destroyed by irradiation with ultraviolet light, resulting in failure of any primordial germ cells to develop (Stevens, 1909). Additional centrifugation experiments on *Ascaris* eggs were recently performed by Guerrier (1967) with results basically in agreement with Boveri's conception of the organization of this egg. Possible roles of the chromosomal DNA which is retained in the germ-line cells and absent from the somatic cells of *Ascaris* are discussed in Chapter 8.

Germ cell determinants are now known to exist in the egg cytoplasm of many creatures. Often the region of the cytoplasm containing the germ cell determinants is marked by special, easily visualized granules. Charac-

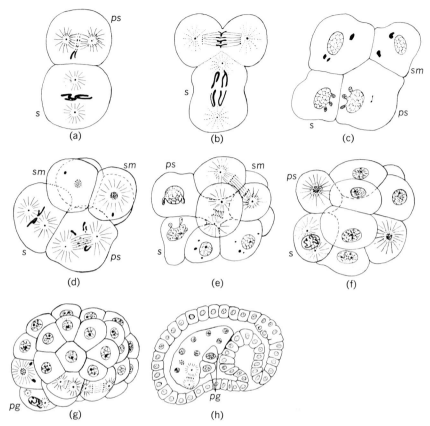

Fig. 7.14. Chromosomal diminution and determination of primordial germ cells in *Ascaris megalocephala*. ps, primordial somatic cell, yet to undergo diminution; sm, somatic cell; s, germ-line stem cell. (a) Second cleavage in progress. In the somatic cell chromosome diminution is in progress. (b) Later stage, elimination-chromatin at equator of upper spindle (T stage). (c) 4-cell stage showing eliminated chromatin in upper two cells. (d) Third cleavage in progress, second diminution at ps. (e) 10-cell stage showing mitosis of somatic cells with diminished nuclei each containing many small chromosomes rather than the four large chromosomes seen in the germ-line cell in (b). (f) 12-cell stage, third diminution in progress at ps. (g) About 32 cells, fourth diminution in progress, leaving primordial germ cell (pg) (in prophase). (h) Gastrula completed with two primordial germ cells. From T. Boveri (1899). Reproduced by E. B. Wilson (1925). "The Cell in Development and Heredity," pp. 322–324. MacMillan, New York.

teristically, these "polar granules" are distributed only to the primordial germ cells. Early cytologists observed polar granules in the primordial germ cells of crustaceans (Amma, 1911), chaetognaths, and several in-

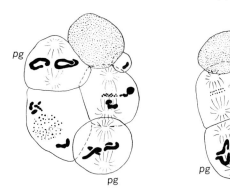

Fig. 7.15. Centrifuged *Ascaris* eggs with two rather than one primordial germ cell (pg) owing to redistribution of cytoplasmic germ cell determinants before cleavage. The stippled balls of cytoplasm at the top of the embryo are the result of centrifugation. Diminution is observed in the somatic cells. The drawings represent two of the forms which occur when the eggs are centrifuged. From M. J. Hogue (1910). *Arch. Entwicklungsmech. Organ.* **29**, 109.

sects (reviewed by Hegner, 1914). Polar granules have now been identified in the primordial germ cells of many other animals, among which are annelids, dipteran insects, ascidians, teleosts, both anuran and urodele amphibians, and various mammals including the human (reviewed by Kerr and Dixon, 1974; Eddy, 1975). The following discussion concerns the two cases of germ cell determination which are at present most extensively studied, viz., amphibian eggs and *Drosophila* eggs.

EVIDENCE FOR GERM CELL DETERMINANTS IN AMPHIBIAN EGGS

It has been known for many years that the amphibian germ cell stem line originates early in cleavage from blastomeres forming at the vegetal pole of the egg. Bounoure (1937) and Bounoure *et al.* (1954) showed that ultraviolet irradiation of the vegetal region of the cytoplasm of frog eggs results in a sharp decrease in the number of primordial germ cells, and frequently in sterility. Many other workers subsequently confirmed this finding in experiments in which the vegetal pole cytoplasm of uncleaved 2- or 4-cell eggs was damaged by ultraviolet irradiation or by physical means (reviewed by Blackler, 1970). The most direct and convincing demonstration of germ cell determinant cytoplasm in amphibian eggs was provided by Smith (1966). Smith showed that complete sterility can always be obtained when the vegetal pole of uncleaved *Rana pipiens* eggs is exposed to sufficient doses of ultraviolet light. Irradiation is also effective at the 2-cell stage, but no longer affects germ cell determination after the 8-cell stage. Parallel irradiation of the animal pole cytoplasm produces no visible defects, and certainly none in the germ cell line. The most important result obtained by Smith (1966) is that the deficiency in the cytoplasmic factors resulting from ultraviolet irradiation can be compensated by injection into the vegetal pole of cytoplasm from the vegetal pole of an unirradiated egg. In a significant fraction of cases the recipient eggs were able to give rise to embryos containing germ cells, while controls receiving no vegetal cytoplasm or cytoplasm from the animal pole never developed germ cells.

POLAR GRANULES IN AMPHIBIAN EGGS

The polar granules in the vegetal region of amphibian eggs have been much studied (reviewed by Smith and Williams, 1975). They are usually associated with high concentrations of mitochondria and have a distinctive morphology. These polar granules are about 0.2 μm in diameter and are surrounded by arrays of ribosomes. Their appearance at several early stages is shown in the electron micrographs reproduced in Fig. 7.16. Classical investigators such as Hegner (1914) believed that the polar granules are the actual site of the germ cell determinants. Attempts have been made to follow them in germ-line cells throughout the life cycle from egg to egg. The following account is derived mainly from the work of Mahowald and Hennen (1971) and Williams and Smith (1971) on *Rana pipiens* and Smith and Williams (1975) on *Xenopus*. Many other workers have also contributed to this field, and excellent reviews concerning the disposition of polar granules and related structures throughout the life cycle are provided by Kerr and Dixon (1974) and by Smith and Williams (1975). Polar granules are initially located just under the vegetal cortex of the fertilized egg (Fig. 7.16a). These structures are also found in unfertilized eggs and matured oocytes in *Xenopus* (Smith and Williams, 1975). At the 16-cell stage they appear to have undergone fusion or at any rate are somewhat enlarged (Fig. 7.16d). They are still present, with associated ribosomes, in the primordial germ cells of blastulae and gastrulae. By the time the germ cells are localized in the germinal ridges, however, they no longer contain identifiable polar granules. Instead a fibrous component is found, applied to the nuclear membrane. Though transition stages have not been convincingly described, most investigators believe these are descendants of the earlier polar granules. These fibrous structures (known as "nuage" in mammalian oogonia) remain present in young ovarian oogonia and persist through most of oogenesis. However they are absent in mature oocytes. Typical polar granules reappear during maturation, at about the time of germinal vesicle breakdown. We may conclude that the

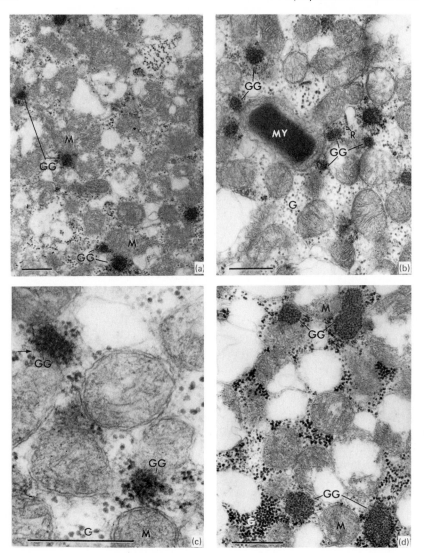

Fig. 7.16. Ultrastructure of region containing polar granules in fertilized amphibian eggs. (a) 1.5 hours after fertilization. (b) 2-cell stage. (c) 4-cell stage. (d) 16-cell stage. GG, polar or germinal granules; M, mitochondria; MY, mitochondria which contain yolk; R, ribosomes; and G, glycogen. Arrow in (c) points to fibril connecting germinal granule and a ribosome. Scale lines equal 0.5 μm. From M. A. Williams and L. D. Smith (1971). *Dev. Biol.* **25**, 568.

natural history of the polar granules is consistent with the proposition that they contain the germ cell determinants originally localized in the egg cytoplasm. It is to be stressed, however, that no direct evidence to this effect exists.

Several correlations seem to link the germ cell determinants with the polar granules. According to Blackler (1970) germ cell determination occurs before neurulation, and it is in the preneurula stages that the polar granules can be observed in their original form, surrounded by ribosomes. Persuasive arguments have been based on various studies using ultraviolet irradiation. As noted above this no longer causes sterility at the 8-cell stage or later. Interestingly, at just this point the polar granules move inward from the cortex of the egg so that they no longer lie within the shallow penetration range of the ultraviolet irradiation (Tanabe and Kotani, 1974). Furthermore, when *Xenopus* eggs are centrifuged so as to move the polar granules inward, irradiation even at the 2-cell stage fails to produce sterility. This result is significant in that it implies that the polar granules are not just associated with a special form of endoplasm containing the real localized determinants. Little change would be likely to result in the position of any but fairly large, dense particles under the centrifugation forces applied in this experiment, only 150 g for 60 seconds. That the germ cell determinants, whatever their form, are more or less cortically located was also shown by Tanabe and Kotani (1974). In this experiment the number of primordial germ cells per tadpole was demonstrated to vary almost exactly in inverse proportion to the *area* of the vegetal pole irradiated with ultraviolet light.

Attempts to isolate polar granules have so far not been successful. Almost nothing is known about their composition, though Smith (1966) reported that the "action spectrum" for induction of sterility by ultraviolet irradiation resembles that of nucleic acids. While it is clear that germ cell determinants are localized in the cytoplasm of amphibian eggs, and can be moved from one egg to another, their mode of action remains a matter for speculation. However, it may turn out that the implications of the above data are correct and that the polar granules are indeed the sites of the germ cell determinants. In this case their unique structure and high density should eventually provide an opportunity to isolate and analyze a known morphogenetic determinant.

EVIDENCE FOR GERM CELL DETERMINANTS IN
DROSOPHILA EGGS

Direct evidence for the determination of primordial germ cells by a similar localization mechanism now also exists for *Drosophila* eggs. This is an important point, since it means that germ cell determination occurs the same way in both the protostomial and deuterostomial branches of metazoan evolution. The primordial germ cells in *Drosophila* eggs arise at the posterior pole of the egg, where they can be easily distinguished, as shown in Fig. 7.17a. The "pole cells" are the first cells formed in the embryo and come to lie outside of the blastoderm wall. Geigy (1931) showed that ultraviolet irradiation of vegetal pole cytoplasm prior to the migration of the cleavage nuclei into this region results in otherwise normal, but agametic animals. Many subsequent experiments demonstrated that the same cleavage nuclei as give rise to pole cells when surrounded by polar cytoplasm will become somatic blastoderm nuclei if the polar cytoplasm is destroyed and they are surrounded by other cytoplasm (reviewed by Counce, 1973). Observations of this kind have also been made with a large number of other insect species. Recently, cytoplasmic transfer experiments similar to those carried out by Smith (1966) on amphibian eggs have been performed with *Drosophila* embryos. These studies prove the existence of the polar germ cell determinants in a particularly convincing way. Okada *et al.* (1974a) and Warn (1975) demonstrated that eggs irradiated in such a manner as to prevent pole cell formation can be rescued by injection of normal posterior pole cytoplasm. Figure 7.17 shows one such experiment (Okada *et al.*, 1974a). A section of the polar region of an irradiated egg lacking pole cells can be seen in Fig. 7.17b, which is to be compared to the normal egg shown in Fig. 7.17a. In Figure 7.17c is an egg which was irradiated and then injected with polar cytoplasm. Pole

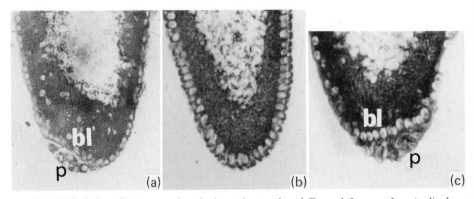

Fig. 7.17. Pole cells in normal and ultraviolet irradiated *Drosophila* eggs. Longitudinal sections of the posterior regions of eggs fixed at the blastoderm stage. (a) Normal unirradiated egg. Complete blastoderm and pole cells (P) are evident. × 300. (b) Egg was irradiated at cleavage stage. Blastoderm has formed over the entire egg, but no pole cells are found. × 300. (c) Eggs were irradiated at cleavage stage and subsequently injected with polar plasm. Blastoderm (bl) and pole cells (P) similar to those of normal eggs. Damage on right side is an artifact of sectioning. × 450. From M. Okada, I. A. Kleinman, and H. A. Schneiderman (1974a). *Dev. Biol.* **37**, 43.

cells similar to those in Fig. 7.17a can clearly be seen. A demonstration by
Illmensee and Mahowald (1974) takes this approach one step further. In
their experiments posterior pole cytoplasm was injected into the *anterior*
end of the recipient embryos, where pole cells normally never form. In
many of the injected embryos, cells having the cytological characteristics
of pole cells were discovered at the site of the injection. To prove that
these cells are functionally primordial germ cells, they were induced by
injection of polar cytoplasm in eggs of *mwh e* genotype. Adult flies bearing
these integumentary markers [multiple wing hairs (*mwh*), ebony (*e*)] can
be recognized by inspection. Illmensee and Mahowald (1974) then transplanted the anterior pole cells into the posterior region of eggs of *y w sn³*
genotype, i.e., bearing different integumentary markers [yellow (*y*), white
(*w*), singed (*sn³*)]. The flies developing from these eggs were mated to
other *y w sn³* flies and the progeny tested for *mwh e* heterozygotes. A
number of these were found, proving that the induced anterior pole cells
were functional primordial germ cells. Germ cell determinants are thus
present in the posterior polar cytoplasm of *Drosophila* eggs, and this
cytoplasm evidently possesses the capacity to alter the fate of normally
somatic nuclei so that they become primordial germ cells. Similar experiments of Illmensee and Mahowald (1976) show that functional primordial
germ cells can be induced to occur by injection of polar plasm into ventral
as well as anterior regions of the egg. Once induced, the ectopic primordial germ cells display the capacity to migrate to the forming gonad and to
complete gametogenesis despite their initially abnormal location in the
embryo. Illmensee *et al.* (1976) also tested the ability of polar plasm from
oocytes to induce functional pole cells when injected into the anterior
ends of *Drosophila* eggs. Polar plasm from unfertilized mature oocytes or
newly mature oocytes (stages 13–14 of oogenesis) induce the formation of
anterior pole cells. These were again genetically marked and when transplanted into the polar regions of eggs of different phenotype were shown
to give rise to functional gametes. However, injection of pole plasm
extracted from younger oocytes (stages 10–12) does not induce the formation of anterior pole cells. Pole plasm from these oocytes contains
morphologically identifiable polar granules, but evidently lacks some
necessary constituents. These experiments clearly show that the synthesis
and deposition of the localized germ cell determinants occur as a result
of ontogenic processes taking place during oogenesis.

POLAR GRANULES IN *DROSOPHILA* EGGS

Ultrastructural studies on *Drosophila* eggs and early primordial germ
cells (Mahowald, 1962) demonstrated polar granules almost indistinguishable from those later identified in amphibian eggs (see Fig. 7.16). Thus

these granules also consist of dense, non-membrane-bound particles surrounded by clouds of ribosomes (Mahowald, 1968). In various species of Drosophila these particles fuse or fragment during early embryogenesis. As with the primordial germ cells of amphibian eggs, the ribosomes associated with the Drosophila polar granules are no longer observed after the earliest stages of embryogenesis (Mahowald, 1971a). By the time the germ cells are located in the larval gonad, the polar granules are replaced by amorphous fibrillar structures applied to the nuclear membrane. This again is similar to what is observed in amphibian and mammalian oogonia. These structures remain throughout the oogonial stage, but are absent in oocytes. Polar granules can again be observed in Drosophila oocytes during vitellogenesis, though as noted above they, or the cytoplasm with which they are associated, are not competent to determine germ cells until the end of oogenesis (Illmensee et al., 1976). The life cycle of the polar granules and their apparent fibrous derivatives [see Mahowald (1971a) for evidence of continuity between these fibrous structures and the polar granules present earlier] is in outline the same in Drosophila as in the amphibia. However, as is also the case in the amphibia, direct evidence relating the Drosophila polar granules with germ cell determination is lacking and almost nothing is known of their molecular structure. Mahowald (1971b) claimed that the polar granules contain RNA immediately after fertilization, but lost this RNA by the blastoderm stage. These results were based on the indium trichloride staining procedure for electron microscope visualization of RNA, and the reliability of this method is open to some question. An interesting additional aspect is the presence of helical arrays of polyribosomes associated with the polar granules. An implication drawn by Mahowald (1968, 1971b) is that the polar granule is a site for localization of the maternal messenger RNA's which direct germ cell differentiation.

A Phylogenetic Survey: How Universal Is Cytoplasmic Localization?

If determination of early embryonic structures by means of cytoplasmic localization is a basic aspect of development from egg and sperm, this phenomenon should be widespread in its phylogenetic occurrence. We have already seen that the most distantly related metazoa utilize similar mechanisms of cytoplasmic localization in germ cell determination. Germ cells could represent a special case, however, and it is important to consider the extent to which localization mechanisms generally affect somatic determination, as in the molluscs and ctenophores discussed earlier. In this section we review several further examples of localization which are of interest with respect to their phylogenetic position, as well as to the problem of understanding the mechanism of localization. In Fig. 7.18 a phylogenetic tree is presented for reference in the following discussion. One issue of importance is whether localization occurs in species which also display strong regulative abilities early in development. Embryos are usually defined as regulative if they can compensate for missing parts or an altered alignment of blastomeres and still develop normally.

FAILURE TO DEMONSTRATE PRECISE CYTOPLASMIC LOCALIZATIONS IN EGGS OF A TURBELLARIAN FLATWORM

The simplest animals for which there is clear evidence of cytoplasmic localization in the egg are the ctenophores, as described earlier. These are extremely primitive, relative to most bilateral metazoa. They achieve only a tissue-level grade of organization and form no organs (Hyman, 1951).

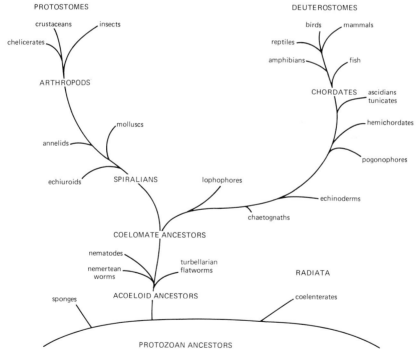

Fig. 7.18. Phylogenetic tree of the metazoa, essentially in accordance with the ideas of Hyman (1940). Phylogenetic groups referred to in this chapter are emphasized.

The lowest bilateral animals which have been studied from our present point of view are acoel turbellarians, the most primitive group of flatworms. Bresslau (1909) and Boyer (1971) showed that cleavage in the acoel *Childia groenlandica* is spiral like that of annelids and molluscs, except that it is based on *pairs* of micromeres rather than the micromere quartets typical of many higher protostomes. Boyer (1971) carried out blastomere deletion experiments in *Childia* eggs and was unable to find any particular morphogenetic value associated with any of the early blastomeres. Except when both of the two macromeres were deleted, complete embryos were formed by all remaining combinations. Only a small amount of macromere material, plus micromeres, is required for complete differentiation. Thus, while some very general kind of localization may exist in the macromeres, there are no cell types analogous to the D quadrant lineage of the *Ilyanassa* embryo. These experiments show that spiral cleavage itself cannot necessarily be equated with highly determinate development, as in the annelids and molluscs.

LOCALIZATION IN CEPHALOPOD MOLLUSCS

The cephalopod molluscs develop by a process which differs greatly from spiral cleavage. These embryos are telolecithal, and cleavage occurs only in a flat disc of cytoplasm at one end of the egg. During organogenesis this disc extends out over the surrounding yolk. It has been shown by Arnold (1968) that a prelocalized pattern of morphogenetic determinants exists in the peripheral egg cortex. These morphogenetic determinants act to determine the fate of the cells of the blastodisc as this forms from the cortical egg cytoplasm. If the blastoderm is prevented from contacting peripheral regions of the egg cortex by ligation, morphogenetic defects are observed which can be related to the particular region of the cortex affected. Similar results were obtained by ultraviolet irradiation of specific areas of the cortex and by disorganizing it by means of centrifugation (Arnold, 1968). Thus, despite the completely different geometry of the cephalopod embryo, cortical localization is likely to be as highly precise in these molluscs as in the gastropods.

LOCALIZATION IN DIPTERAN INSECTS

Another protostome group displaying a great variety of developmental processes is the insects. There is an enormous literature on determination and regulation in insect development. Certain regions of the insect egg appear to control developmental processes. Many insect eggs display potent regulatory capacities, so that whole mirror image embryos can be formed from bisected eggs [e.g., see the experiments of Sander (1971) on *Euscelis*]. This area of knowledge cannot be summarized here, and the interested reader is referred to the extensive review of Counce (1973). The discussion below is confined to two recent studies regarding localization in insect eggs, both of which provide insight into the nature of the localization process.

We first consider the localization of "anteriorizing" factors in the egg of the chironomid midge *Smittia*. In this organism ultraviolet irradiation of the anterior pole of the egg invariably produces a curious developmental abnormality called "double abdomen" (Kalthoff and Sander, 1968). Here the head, thorax, and anterior abdominal segments are replaced by posterior abdominal segments, in mirror image symmetry to the normal posterior end. A double abdomen larva of *Smittia* is shown in Fig. 7.19 (Kalth-

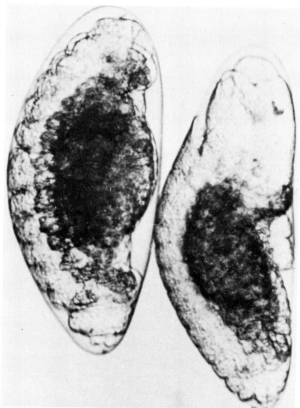

Fig. 7.19. "Double abdomen" larva of *Smittia*. Right: larva with normal body segment pattern. Left: double abdomen. Length of eggs: 250 µm. The photograph was taken by K. Sander. From K. Kalthoff (1971a). *Dev. Biol.* **25**, 119.

off, 1971a). This result indicates that the anterior end of this egg contains some positive control factor which elicits anterior segment differentiation. In its absence the anterior end differentiates in the same manner as does the posterior end. Puncturing the anterior pole with a needle also produces the double abdomen phenotype (Schmidt et al., 1975). No special cytological features can be seen in the cytoplasm of the anterior end of the egg (Zissler and Sander, 1973). However careful studies of Kalthoff (1971b) have indicated the exact area of the ultraviolet sensitive region of the egg. By shading different regions of the egg and studying the transmittance of the ultraviolet radiation as a function of depth from the egg surface, Kalthoff (1971b) showed that the sensitive target regions are symmetrically distributed around the anterior pole. The depth of penetration of the ultraviolet light is only about 5 μm or less (Kalthoff, 1973). An interesting aspect of the Smittia system is that the ultraviolet induction of double abdomen is photoreversible. As much as 60% increase in the frequency of normal embryos is reported by Kalthoff (1971a) when ultraviolet irradiated eggs were subsequently exposed to long wavelength ultraviolet and visible light. On this basis, and from measurements of the action spectrum of the UV irradiation, Kalthoff (1971a, 1973) suggested that the UV-sensitive targets localized in the egg cortex include nucleic acids, and in accord with this Kandler-Singer and Kalthoff (1976) showed that the anterior determinants are RNase sensitive. The RNase is admitted by a small puncture made in the anterior end of the egg, and up to 40% of eggs so treated display the double abdomen syndrome. Punctures made in other regions of the egg, denatured RNase, or other enzymes do not have this effect.

The morphogenetic targets of the ultraviolet irradiation in Smittia eggs are clearly present (i.e., destructible) long before nuclei arrive at the cortex to form the blastoderm. In other words, they are constituents of the original cortical egg cytoplasm (including the plasma membrane) and must have been synthesized during oogenesis. Strong support for this conclusion comes from studies of the "bicaudal" mutation in Drosophila, which also produces double abdomen embryos such as those in Fig. 7.19. Bull (1966) showed that bicaudal is a maternal effect mutant; that is, it affects synthesis of some component during oogenesis. Bicephalic maternal mutants are also known in Drosophila (Lohs-Schardin and Sander, 1976). In Chironomus both bicaudal and bicephalic embryos can be induced by centrifugation (Yajima, 1960). These forms were explained by Yajima as the result of the centrifugal movement of cytoplasm resulting in contact with the original posterior or anterior egg cortices. The cytoplasm is thus supposed to become determined to produce posterior or anterior cell lineages. After the centrifugation is terminated the determined cytoplasm spreads out symmetrically toward both poles of the egg. The result is the development of symmetrical embryos. Additional information derives from the experiments in which cellular blastoderm from *Drosophila* eggs, or preblastoderm cortex plus nuclei, were cultured *in vivo* in adult or larval fly abdomens. In this way their morphogenetic potentials could be assayed. Chan and Gehring (1971) carried out such experiments with groups of cells derived from anterior and posterior blastoderms. They found that anterior blastoderm cells give rise only to anterior and thoracic structures and posterior blastoderm cells give rise only to posterior structures. Similarly, Schubiger (1976) showed that partial embryos formed by ligation at the blastoderm stage can produce all of those anterior, thoracic, or caudal adult parts which would normally derive from them. Thus by the blastoderm stage determination is relatively advanced. When the same experiment was carried out with partial embryos which had been ligated in the preblastoderm stage, anterior and caudal structures were still formed, but some thoracic structures were missing. The latter appear to require blastoderm formation for their determination. From all these experiments the main conclusion pertinent to our present subject is that some dipteran eggs contain localized cortical factors for anterior and posterior determination from the beginning of development (see also Weischaus and Gehring, 1976). These factors appear to affect the contiguous cytoplasm, or are released into this cytoplasm. Their effects are observed following migration of the nuclei into the polar areas of the egg, and ultimately the formation of the cellular blastoderm.

If the localized factors in preblastoderm dipteran eggs are the product of synthesis during oogenesis, maternal effect mutations should be recoverable in which early determination occurs improperly or not at all. One example, the "bicaudal" mutation, was already mentioned. Rice and Garen (1975) have described three other maternal effect mutations which interfere with normal blastoderm formation in *Drosophila*. Eggs bearing the first of these, termed *mat(3)1*, form no somatic blastoderm at all, though the pole cells develop normally. Therefore, the polar determinants responsible for primordial germ cell differentiation result from the activity of different maternal gene(s) than do the cytoplasmic determinants involved in blastoderm formation. In mutation *mat(3)3* a region including about 30% of the normal blastoderm and located on the posterior dorsal surface fails to become cellularized. In mutation *mat(3)6* about 70% of the blastoderm is noncellularized. *Mat(3)3* is a temperature-sensitive mutant, and by measuring the period of temperature sensitivity Rice and Garen (1975) showed that the locus involved is functional during the last 12 hours of oogenesis [the total duration of the egg chamber in wild-type *Drosophila* is about 8 days (Grell and Chandley, 1965)].

The importance of blastoderm formation in *Drosophila* eggs has been emphasized by nuclear transplantation experiments. These demonstrate that prior to blastoderm cellularization the nuclei are totipotent. For example, Illmensee (1972) showed that nuclei from polar regions, cleavage nuclei, or syncytial blastoderm lateral nuclei can all support development of unfertilized eggs at least through the early embryo stage and into the larval instars. Tissues displaying many forms of differentiation derive from these nuclei. Similarly, Okada *et al.* (1974c) demonstrated that anterior nuclei injected into the posterior pole region of *Drosophila* eggs can participate in the formation of a number of posterior differentiated structures. Clearly in Diptera blastoderm cellularization is a critical feature of the determination process, for it is then that the nuclei are brought in contact with the morphogenetic substances which appear to be localized in the cortical region of the egg cytoplasm.

Annelids, molluscs, and arthropods all belong to the same great branch of animal evolution. These are the major protostomial phyla [see Fig. 7.18, and Hyman (1951) for a general discussion of protostome and deuterostome phylogeny]. The basic plan according to which protostome embryos are constructed differs from that of the deuterostomes. In protostomes the embryonic blastopore becomes the mouth, and coelomic mesoderm is generally formed from a solid column of stem cells proliferated from the endoderm. In the deuterostomes coelomic mesoderm arises by outfolding or delamination from the endodermal regions of the embryo, the blastopore becomes the anus, and the mouth is a new formation. These differences are so basic that they undoubtedly affect the initial organization of the egg. If we exclude germ cell determination, among the examples we have so far considered, only the studies of Conklin and his followers on the ascidian *Styela* indicate that cytoplasmic localization may be significant in deuterostome development. The remainder of this section is devoted to a brief review of further evidence regarding localization in deuterostomial animals.

CYTOPLASMIC LOCALIZATION IN SEA URCHIN EGGS

The sea urchin egg provides an important example. Detailed observations indicating that morphogenetic determinants exist in the cytoplasm of sea urchin eggs were made by Boveri (1901). This study was carried out with *Paracentrotus lividus*, a species in which a median band of pigment granules marks the location of the cytoplasm which will be needed for the formation of the archenteron. By shaking the eggs into pieces and fertilizing the fragments, Boveri showed that only fragments containing the pigment-marked cytoplasm are able to gastrulate and form archenteron. These results were confirmed by Hörstadius (1928). Like Boveri, Hörstadius found that the pigment layer marks the eventual axial orientation of the embryo. Thus, animal half-egg fragments develop in a manner similar to the animal halves of 16-cell cleavage-stage embryos, and the morphogenesis carried out by vegetal egg fragments corresponds to that observed in 8- or 16-cell vegetal half-embryos. Therefore the potentials for micromere, skeleton, and archenteron formation are localized in the vegetal region of the egg cytoplasm even before fertilization. The orientation and distribution of these materials remain unchanged as the egg cytoplasm begins to be divided up among the blastomeres. Hörstadius (1937a) extended these findings to *Arbacia punctulata*, orienting the eggs individually soon after fertilization, and cutting them in half at the pronuclear fusion stage. Haploid and diploid animal and vegetal halves are all obtained in this manner. The animal halves cleave equally, producing no micromeres, and form spherical structures with enlarged apical tufts. The vegetal halves, however, form micromeres and gastrulate, with archenteron and skeleton formation ensuing. Like lobeless *Dentalium* and *Ilyanassa* eggs, the animal egg fragments possess complete genomes, but fail to differentiate several important cell lineages. Hörstadius' experiments illustrate localization of morphogenetic factors in sea urchin egg cytoplasm and prove that this localization is normally definitive.

LOCALIZATION OF CAPACITY TO FORM HISTOSPECIFIC ENZYMES IN ASCIDIAN EMBRYOS

We now return to the ascidian embryo and a review of some recent studies in which localization can be followed at the molecular level. Whittaker (1973a) showed that the tail muscle cells of the postgastrular larva are marked by concentrations of acetylcholinesterase high enough to be easily detected with cytochemical methods. Similarly, tyrosinase can be observed in the larval brain cells (Minganti, 1951; Whittaker, 1973b). Normally, acetylcholinesterase is first observed at 8 hours of development in the presumptive muscle cells of the early neurula, and the stain becomes more intense thereafter. Whittaker (1973a) treated embryos of *Ciona intestinalis* with cytochalasin and other mitotic inhibitors, thereby arresting their morphological development at various cleavage stages (i.e., at 1–5 hours of development). The treated embryos remain alive and at least some of their molecular functions continue. In embryos arrested during cleavage and maintained for 12–14 hours, acetylcholinesterase activity develops in certain of the blastomeres. The maximum number of blastomeres displaying this enzyme at each cleavage-arrested stage is one at the 1-cell stage, two at the 2-cell stage, two at the 8-cell stage, four at the 16-cell stage, six at the 32-cell stage, and eight at the 64-cell stage. From these data and the location of the stained blastomeres, it is clear that the cells which contain acetylcholinesterase are precisely those comprising the muscle cell lineage. This lineage can be seen in Conklin's drawings in Fig. 7.1. Similar results were obtained in cytochalasin-blocked eggs stained for tyrosinase. Tyrosinase normally begins to develop at 9 hours (tailbud larva, similar to that portrayed in 1 of Fig. 7.1). In cytochalasin-treated 32-cell stages, tyrosinase also appears at 9 hours, confined to two blastomeres known to be the progenitors of larval brain pigment cells. It follows from these and similar observations that tyrosinase and acetylcholinesterase are synthesized only in cells containing the cytoplasmic constituents normally partitioned into the muscle and neural stem lines. Though the appearance of these enzymes requires a certain number of hours, it is not affected by the number of cells containing this cytoplasm, their morphological form or location, or their environment with respect to other cells. Three possible interpretations of this result are that the enzymes are stored in inactive form in the localized egg cytoplasm and are activated after some time by a protease; that maternal messages for these enzymes are sequestered in the localized egg cytoplasms and are released and translated after some time; or that regulatory signals are localized in the egg cytoplasms which cause transcription of the appropriate structural genes in the blastomeres. The appearance of both enzymes seems to require protein synthesis, since it is blocked by puromycin treatment. Whittaker (1973a) also found that actinomycin blocks the appearance of both enzymes, either in cytochalasin-treated or normal cells, if it is given between 5 and 7 hours for acetylcholinesterase and 6 to 7 hours for tyrosinase. Subsequent treatment with actinomycin does not prevent enzyme formation. These results argue against the first and second of the possibilities mentioned above, and in favor of new structural gene transcription. However, this argument is clearly not a strong one, since the actinomycin-sensitive transcripts could code for translational control factors or enzyme processing factors. Another histospecific enzyme, alkaline phosphatase, has also been studied by Whittaker (1973c). This enzyme is normally localized to endoderm cells, and also appears in appropriate blastomeres of cleavage-arrested embryos at the correct time, about 12 hours following fertilization. In contrast to the case with tyrosinase and acetylcholinesterase, however, the appearance of alkaline phosphatase activity is not actinomycin sensitive. Thus here the localization of a maternal messenger RNA would seem indicated, rather than new structural gene transcription.

LOCALIZATION OF AXIAL DETERMINANTS IN AMPHIBIAN EGG CYTOPLASM

Much evidence indicates that the cortex of the amphibian egg contains localized components which determine axial organization. Classical studies dating back to Spemann (1914) suggested that the gray crescent area of the egg cytoplasm contains factors controlling chordamesoderm and neural development (see Holtfreter and Hamburger, 1956; Briggs and King, 1959, for summaries of this evidence). Curtis (1962) developed a method for transplanting small (150 × 150 μm) sections of egg cortex to other eggs and embryos in order to investigate localization in these eggs. He reported that transplantation of a cortical graft from the gray crescent area of an 8-cell embryo to the ventral side of a fertilized *Xenopus* egg results in double gastrulation and axiation. This experiment is diagrammed in Fig. 7.20c. Figures 7.20a and d show that excision of the gray crescent cortical cytoplasm before (Fig. 7.20d), but not after (Fig. 7.20a),

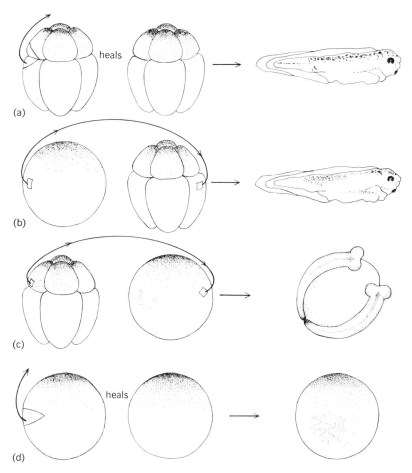

Fig. 7.20. Cortical transplants in the egg of *Xenopus*. (a) Excision of the gray crescent cortex from a stage 4 (8-cell) embryo results in a normal embryo being formed [compared with (d)]. (b) Grafting gray crescent cortex from stage 1 to the ventral margin of stage 4 does not result in the induction of a second embryonic axis. (c) Grafts of gray crescent cortex from stage 4 embryos to the ventral margins of stage 1 embryos induce secondary embryonic axes. (d) Excision of gray crescent cortex from stage 1 embryos prevents axial morphogenesis though cleavage and mitosis continue. From A. S. G. Curtis (1962). *J. Embryol. Exp. Morphol.* **10**, 410.

the 8-cell stage results in failure of gastrulation. One explanation is that the cortical cytoplasm of the crescent in some way effects blastomere determination as early as third cleavage. The experiment shown in Fig. 7.20b is consistent with this interpretation. Here it is found that double axiation can no longer be induced by gray crescent cytoplasm from unfertilized eggs when this is implanted after third cleavage. However, this could also be due to other causes, such as changes in the state of the recipient egg cortex after third cleavage. As shown in Fig. 7.20c the responsible cytoplasmic determinants are still present in the cortex of 8-cell stage embryos, since an implant of this cytoplasm is able to induce double axiation when transplanted back to an unfertilized egg. It may be significant that the third cleavage is the first transverse cleavage to occur.

The special interest of the experiments shown in Fig. 7.20 is that they clearly trace the source of the axiation factors eventually functional in the dorsal lip of the gastrula back to the cortical cytoplasm of the fertilized egg. The main results of Curtis have been confirmed by different methods in experiments of Tompkins and Rodman (1971). Rather than implant cortical segments into the cortex of a recipient embryo, these workers injected pieces of cortex into the blastocoel of the recipient embryos. Pieces of gray crescent cortex from either unfertilized eggs or 8-cell embryos which were implanted in this way gave rise to doubly axiated embryos. A cross section of the dorsal region of such an embryo is shown in Fig. 7.21. Tompkins and Rodman (1971) also concluded that the cortex of the egg in the gray crescent region contains factors which lead to axial differentiation. However, it is not clear by any means that these factors are confined to the gray crescent itself, or that the gray crescent alone suffices for axial determination (e.g., see Nieuwkoop, 1969, 1973).

Another approach to the localization of axiation factors in amphibian eggs has been the use of UV irradiation (Baldwin, 1915; Grant, 1969; Grant and Wacaster, 1972; Malacinski, 1972; Malacinski et al., 1974; Chung and Malacinski, 1975). The UV-sensitive region was localized temporally and spatially in both *Rana pipiens* and *Ambystoma mexicanum* eggs by Malacinski et al. (1974). In *Rana* eggs the most sensitive period was found to be before first cleavage, at 0–90 minutes after fertilization. Malacinski et al. (1974) showed that the effect of UV irradiation decreases sharply after 90 minutes. The physiological result of irradiation is faulty differentiation of chordamesoderm, and microcephaly, acephaly, or other aneural morphologies, even though gastrulation appears to occur normally. By marking the irradiated areas, and using albino strains of axolotl which lack egg pigment, Malacinski et al. (1974) were able to demonstrate that the UV-sensitive neuralizing factors are localized mainly in the future dorsal region of the vegetal egg cortex. This region

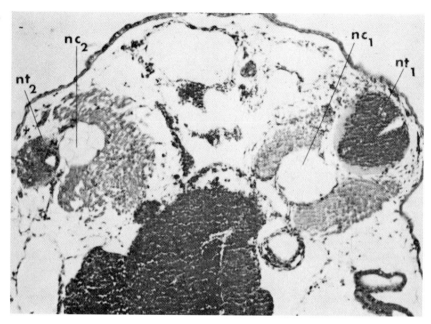

Fig. 7.21. Secondary axiation in a *Xenopus* egg induced by injection of a cortical implant into the blastocoel. Section through a recipient of a stage 4 gray crescent cortex implant, showing a secondary axis. nc_1, notochord of the primary axis; nt_1, neural tube of the primary axis; nc_2, notochord of the secondary axis; nt_2, neural tube of the secondary axis. From R. Tompkins and W. P. Rodman (1971). *Proc. Natl. Acad. Sci. U.S.A.* **68**, 2921.

usually but not invariably includes the gray crescent, and occasionally slight sensitivity was noted in the dorsal animal hemisphere of the egg. The UV irradiation has a purely cytoplasmic effect, since nuclei from irradiated eggs are able to support development when injected into enucleated eggs as well as control nuclei (Grant and Wacaster, 1972). When dorsal lips from normal gastrulae were grafted into gastrulae derived from irradiated eggs, normal embryos were obtained in a high percentage of cases (Chung and Malacinski, 1975). It follows that the UV irradiation of the precleavage egg cortex destroys factors required for dorsal lip differentiation. The dorsal lip, of course, is a major axial inducing center in the amphibian egg, and control experiments of Chung and Malacinski (1975) showed that over 90% of normal embryos receiving a second normal dorsal lip developed secondary axial structures similar to that shown in Fig. 7.21. In other experiments Malacinski (1974) showed that the deficiency caused by UV irradiation of the fertilized egg can be corrected by microinjection of germinal vesicle nucleoplasm at the blastula stage. Extracts of whole ovaries, or of germinal vesicles from oocytes of any stage were effective. In normal unirradiated embryos these extracts resulted in excessive head region development. Grant and Wacaster (1972) showed in addition that the irradiation effect could be reversed by injection of cytoplasm from the marginal region of unirradiated eggs. The simplest interpretation of all these experiments is that the morphogenetic egg substances required for axial differentiation are synthesized during oogenesis. These determinants are present in germinal vesicles in ovarian oocytes, but they are located in the cortical or subcortical structures of the marginal region of the egg after fertilization. This follows from the cortical implantation experiments and from their sensitivity to UV irradiation. As noted above, measurements of transmittance of UV light in egg cytoplasm show that the depth of penetration from the surface is only 5–10 μm or less (Kalthoff, 1971b, 1973; Grant and Wacaster, 1972). Almost no information regarding the chemical nature of the responsible cortical agents is available. However, electron microscope observations have shown that the cortical layer of the *Xenopus* egg is characterized by a special structure. This was observed by Hebard and Herold (1967). A dense layer of about the same thickness as the cortical grafts transplanted by Curtis lies directly beneath the plasma membrane in this egg. This layer forms immediately after fertilization, shortly before the dorsoventral localization patterns of the amphibian egg are established. In the absence of direct evidence, however, it is also possible that the essential structure in the cortical region is the egg membrane itself.

ABSENCE OF EVIDENCE FOR CYTOPLASMIC LOCALIZATION IN MAMMALIAN EGGS

A great deal of effort has been expended in the attempt to determine whether cytoplasmic localization affects development in mammalian eggs (see reviews of Gardner and Papaioannou, 1975; Kelly, 1975; Wilson and Stern, 1975). Most experiments have been carried out on the eggs of rat, mouse, and rabbit. In preimplantation mammalian embryos only two forms of differentiation are evident. The cells of the early blastula (morula) may differentiate either into the extraembryonic trophoblast or into the inner cell mass. Only the latter gives rise to the actual embryo. The experiments relevant to our present topic fall into two general classes: those in which the capacities of isolated blastomeres, fused embryos, and partial embryos are tested and those which seek to determine what controls the choice between inner cell mass and trophoblast differentiation. It may be observed that the latter question is not exactly analogous to the problems we have so far been considering, such as what controls the differentiation of mesoderm or germ cells. That is, the mechanism(s) which determine whether given cells will be part of the embryo proper are not necessarily expected to provide insight into the mechanism(s) which determine the specificity of cells within the embryo.

It is clear that individual blastomeres from 2- to 4-cell rodent and rabbit embryos can develop into complete blastocysts (Nicholas and Hall, 1942; Seidel, 1960; Tarkowski, 1959a,b; Mulnard, 1965; Tarkowski and Wrób-lewska, 1967; reviewed by Kelly, 1975). Furthermore, blastocysts and normal embryos can be formed from fused embryos up to the late morula stage (Tarkowski, 1961, 1963; Mintz, 1962, 1965; Gardner, 1968). Stern and Wilson (1972) showed that normal blastocysts could even be formed from 8-cell eggs fused with late morulae. Either inner cell mass or trophoblast tissue can be removed from the blastocyst, without affecting its ability to form a whole embryo (Lin, 1969; Gardner, 1971). Similarly, disaggregated mouse embryos of all stages from 8-cell to late blastocyst will reaggregate to form normal blastocysts, even if the disaggregated cell mixture contains blastomeres derived from different stages (Stern, 1972). These experiments show that the rodent and lagomorph eggs have an impressive ability to regulate. Even if specific morphogenetic assignments or polarities were initially present in parts of these eggs, they can easily be overridden under circumstances of blastomere isolation or fusion. Furthermore, no evidence for the existence of such assignments, or for fixed polarity or bilateral symmetry in the fertilized egg can be derived from these studies. Wilson *et al.* (1972) showed, however, that cleavage in the mouse egg *normally* occurs without any disturbance of the cytoplasmic architecture. Thus whatever its degree of lability, if developmental information were localized in the cytoplasm, it would be transmitted in a spatially organized way to the blastomeres.

The differentiation of a blastomere into trophoblasts as opposed to elements of the internal cell mass turns out to depend on whether it is on the inside or outside of the morula. Among the experiments on which this statement is based are those of Hillman *et al.* (1972), who produced chimeric combinations of marked cells and correlated their fates with their position in the early embryo. For example, though each blastomere of a 4-cell embryo can give rise to both trophoblast and inner cell mass, when these blastomeres are placed on the outside of other embryos they always formed trophoblast and did not contribute to the embryo proper. Conversely, blastomeres completely surrounded by other cells produced inner cell mass cells. Tarkowski and Wróblewska (1967) also used this explanation in interpreting the high frequency with which isolated one-eighth blastomeres give rise only to trophoblast vesicles compared to one-fourth and one-half embryos. The smaller one-eighth blastomeres produced smaller morulae, and therefore when cavitation occurs, fewer cells can be completely enveloped by other cells, which is the necessary requirement for inner cell mass differentiation. Along the same lines, alkaline phosphatase, an inner cell mass marker, is reported to appear in chimeric aggregates only when these include blastomeres totally surrounded by other blastomeres (Izquierdo and Ortiz, 1975). Stern (1973) flattened mouse embryos under a glass plate and showed that in these embryos trophoblast cells appear even in internal regions where the blastomeres abut the glass interface rather than other cells. Finally, in the mouse egg it is the *outer* regions of the egg cytoplasm which normally end up in trophoblast cells. Wilson *et al.* (1972) showed this by microinjecting droplets of silicone into the egg at the 2-cell stage and later determining the nature of the cells which contain these markers. Droplets injected peripherally appear only in trophoblast cells, while those injected centrally show up either in trophoblast or inner cell mass cells. Once determined, inner cell mass cells can no longer give rise to trophoblast, even when exposed to "outside" conditions (Rossant, 1975a,b). Nor can isolated inner cell masses implant (Gardner, 1972). The inner cell mass, which is very different in its properties from the trophoblast, appears to exercise control over cell division in the trophoblast (Gardner, 1972).

In the only mammalian eggs which have been well studied, the main conclusions with respect to localization might be summarized as follows. Inner cell mass trophoblast are probably determined by internal or peripheral position, but this form of localization is of a type not obviously related to what we have considered earlier in this chapter. Cytoplasm on the outside of the egg and blastomeres on the outside of the morula give rise to trophoblast. While the cleavage pattern may be fixed with reference to the architecture of the egg, no evidence for polarity or other forms of morphogenetic localization exists. However, the meaning of this result remains obscure. Cytoplasmic localization may be completely absent in the mammalian developmental scheme. Alternatively it could play an important role, but be so easily disrupted that the manipulations of the researcher reveal only the regulative response to experimental perturbation.

PHYLOGENETIC SUMMARY

The evidence reviewed in this section shows that morphogenetic determinants are localized in the cytoplasm of both protostome and deuterostome eggs, with the possible exception of mammals and acoel turbellarians. Other cases may also come to light. Examples have been considered from almost every major animal phylum. Though it may not be universal, and clearly is of varying significance depending on the species, cytoplasmic localization is obviously widespread in phylogenetic terms. The nearest common ancestor of protostome and deuterostome organisms was probably a precoelomate worm, and these major branches of metazoan evolution apparently had already diverged by the early Cambrian. Furthermore, cytoplasmic localization exists even in eggs of the Radiata, such as the ctenophore. Localization is thus evolutionarily more ancient than the coelomic grade of organization, and perhaps is even older than the Bilateria. It seems likely that localization is among the basic mechanisms of metazoan embryogenesis and evolved along with the processes of development from sperm and egg.

Interpretations of the Localization Phenomenon

CLASSIC THEORIES OF CYTOPLASMIC LOCALIZATION IN EGGS

The study of cytoplasmic localization has an interesting intellectual history. We have seen that our current view of localization derives in large part from the researches of classical cell biologists. Wilson (1925) pointed out the close relation between these investigations of the localization phenomenon and the late nineteenth century controversy over performationist as opposed to epigenetic theories of development. The participants in this controversy included many eminent biologists engaged personally in research on the localization problem. Huxley (1878), Hertwig (1894), Bourne (1894), and Whitman (1895a) all published discussions of the theoretical implications of localization. It is worthwhile to begin a current discussion of this subject by considering the insights developed by its early students.

By the last third of the nineteenth century it seemed clear that embryos develop by epigenetic mechanisms. The theory of epigenetic development held that organisms increase in absolute complexity of organization as they develop, rather than simply increasing in size without further "differentiation or essential modification." Preformationist eighteenth century philosophers and scientists such as Bonnet (1762) had believed that little change in complexity actually occurs in development, and that fully "organized bodies preexist from the beginning" (Bonnet, translated by Whitman, 1895b). The earliest investigations which directly supported epigenetic interpretations of development were the demonstrations by Wolff in 1759 that leaves and flowers develop from undifferentiated tissues, Panter's (1817) description of epigenetic development in the chick from primitive germ layers, and the studies of von Baer (1828–1837), as a result of which the germ layer theory was generalized to other animals. Von Baer showed that skin develops epigenetically from ectoderm, muscular and skeletal systems from mesoderm, etc. In 1867 Kowalewsky demonstrated that the germ layers themselves are formed epigenetically.

The first observations on the importance of the cytoarchitecture of the egg gave rise to what was regarded for a time as a new form of preformationism. This was developed by men such as His, the teacher of Miescher, and a proponent of the view that satisfactory explanations of biological phenomena can only be obtained at the molecular level. His suggested in 1874 that the epigenetic character of early chick development is only apparent, the underlying phenomenon being the "coalescence of preformed germs":

> It is clear on the one hand that every point in the embryonic region of the blastoderm must represent a later organ or part of an organ, and on the other hand, that every organ developed from the blastoderm has its preformed germ in a definitely localized region. *The material of the germ is already present, but is not yet . . . directly recognizable* (His, 1874).

A farsighted statement of this school was made by Lankester (1877); which extended the "molecular preformation" hypothesis further:

> Though the substance of an egg cell may appear homogeneous under the most powerful microscope, excepting the fine granular matter suspended in it, it is quite possible, indeed certain, that it may contain *already formed and individualized*, various kinds of physiological molecules. The visible process of segregation is only the sequel of a differentiation already established. . . . Thus, since the fertilized egg already contains hereditarily acquired molecules, . . . invisible though differentiated, there would be a possibility that these molecules should part company, *not* after the egg-cell had broken up into many cells as a morula, but at the very first step in the multiplication of the egg-cell. . . . We should not be able to recognize these molecules by sight; the two cleavage cells would present an identical appearance, and yet the segregation . . . has already taken place. This hypothesis may be called that of *Precocious Segregation* (Lankester, 1877).

For some years after 1900 a theory espoused by Boveri, Loeb, and Morgan (see Wilson, 1925) proposed that the egg cytoplasm and its regions of localized determinants are responsible for the form of the "embryo in the rough." It was proposed that the role of the embryo genomes is to determine the details of the *individual's* development, such as its color, size, and the exact shape of its skeleton. For example, Boveri (1903) listed among the "preformed" characters determined by the structure of the cytoplasm the following: the tempo of development, polarity, axis of symmetry, pattern of cleavage, and the crude areas of morphogenetic localization, e.g., the location of the mesoderm stem cells. He proposed that those characters which are determined by the nuclear genes, e.g., the skeleton of the sea urchin pluteus, are the ones which develop *epigenetically*, in contrast to the "embryo in the rough" characters already preformed in the cytoplasm. Thus, on the one hand, those developmental processes stemming from the action of the embryo genome were considered epigenetic, and, on the other, maternally programmed development independent of embryo genome action was considered preformational, the whole pregastrular period of development in the sea urchin belonging to the latter category.

The appearance of a more modern resolution of the localization problem dates to about 1895–1896, and the writings of Whitman, Wilson, and Driesch. Whitman (1895a,b) stated that the real problem posed by the discovery of cytoplasmic localization was to preserve the idea that the nucleus contains genetic determinants through whose action epigenetic development must take place. The ideas stemming from this resolution of the problem led to the modern view of the localization problem. Extending Whitman's basic views, Driesch and Wilson suggested that even apparently preformed characters can only be regarded as the product of an *earlier epigenetic process originating in the oocyte nucleus* during oogenesis. Thus in an appendix to Crampton's study of localization in *Ilyanassa*, Wilson (1896) wrote:

> Cytoplasmic organization, while affording the immediate conditions for development, is itself a result in the last analysis of the nature of the nuclear substance which represents by its inherent composition the totality of heritable potence. Logically carried out this view inevitably involves the conclusion that the specific plasma structure of the egg is acquired during its ovarian maturation (Wilson, 1896).

This view abolishes the dichotomy between epigenetic, genome-directed morphogenesis and the preformational morphogenesis of the "embryo in the rough" mediated by the egg cytoplasm. Wilson (1925) considered that the only real preformation is that of the genome itself: "Heredity is effected by the transmission of a *nuclear preformation* which in the course of development finds expression in a process of cytoplasmic epigenesis." This insight provided a fruitful basis for further advances in the treatment of the localization problem, but further than this classical writers did not go. The morphogenetic determinants present in the egg cytoplasm were always referred to simply as "organ-forming substances."

"REGULATIVE" AND "MOSAIC" EGGS

A basic problem in interpreting localization, and one which arose early in the history of this subject, is the fact that in some eggs individual blastomeres display great regulatory powers, while in others cleavage seems completely determinate. This dichotomy led to the classical view that eggs can be classified as either "mosaic" or "regulative" in their developmental character. Localization was supposed to be important in mosaic eggs but not in regulative eggs. It has become clear that this exclusive classification is basically wrong. As has been repeatedly demonstrated by the data reviewed in this chapter, localization occurs in eggs which are also well known for their regulatory powers. Examples are the eggs of amphibia and sea urchins. In both of these groups partial embryos and fused embryos can regulate to form whole embryos. This was shown first for amphibian eggs by Spemann (1903), McClendon (1910), Mangold (1920), and Ruud (1925) and has been demonstrated in many ways since. Similarly, Driesch (1891) showed that a single sea urchin blastomere possesses the capability of forming a complete embryo. Subsequently, Driesch (1900) demonstrated that the first two blastomeres and some of the individual blastomeres of the 4-cell stage sea urchin embryo possess the capability of forming a complete embryo. Even some of the 8-cell stage blastomeres can develop far enough to produce a pluteus. One of the most impressive demonstrations of regulative ability in the literature is to be found in the sea urchin blastomere recombination experiments carried out by Hörstadius (1939). These experiments proved that the morphogenetic fate of given tiers of blastomeres can be completely changed depending on which other blastomeres are present. Other examples of blastomeres which can form complete embryos include the CD cell (but not the AB cell) of *Tubifex* (Penners, 1926), either of the first 2 cells and some of the first 4 cells of the cephalochordate *Amphioxus* (Wilson, 1893), and any of the first 2 or 4 blastomeres in some teleosts (Morgan, 1927). In all of these cases it is clear that the developing blastomere lineage performs activities not normally assigned to it, e.g., the formation from a right-hand blastomere of both a left and a right side of the embryo.

These "regulative" organisms contrast sharply with those in which isolated blastomeres produce only the fractional embryonic structures to which their descendants normally give rise. We have already considered many such examples, for instance, *Styela* in which only left or right half embryos are derived from isolated blastomeres (Fig. 7.5) and *Sabellaria* in which fused embryos give rise to double monsters (Hatt, 1931). Striking though these differences may be, it is clear that this criterion cannot be relied upon to provide an index of whether localization occurs in normal development of an egg. Various factors contribute to the outcome of these tests and complicate their interpretation. The following discussion deals with some of these factors, in particular (a) with the geometric relation between the early cleavage planes and the topographical distribution of the localized determinants in the egg, (b) with the point in development at which the localization patterns actually become established, (c) with the extent to which the events of early embryogenesis depend on cellular interactions, and (d) with the physical lability of the cytological structures on which the localization pattern may depend. Classical knowledge with regard to the first two of these factors is reviewed by Wilson (1925); for emphasis on the third, see Watterson (1956).

Wilson (1925) pointed out with great clarity that blastomere totipotency and cytoplasmic localization are not exclusive. In Fig. 7.22 is reproduced a heuristic diagram which illustrates this point (Wilson, 1925). Here localized areas of morphogenetic potential are mapped out for the eggs of a hydromedusa, a gastropod or annelid, an ascidian, and an echinoderm. The diagram shows how mesodermal and other cytoplasmic determinants are partitioned among the blastomeres. The developmental potency of the early blastomeres of each type of egg can be understood in terms of the relation between the planes of cleavage and the distribution of the morphogenetically significant regions in the egg. For example, if the mesodermal determinants are asymmetrically distributed with respect to the cleavage planes, all of the blastomeres cannot be totipotent (Fig. 7.22d–f).

To quote Wilson (1925):

> Totipotence on the part of the early blastomeres is dependent primarily on a symmetrical or merely quantitative distribution of the protoplasmic stuffs of the cleavage. In the hydromedusa (Fig. 7.22j–l) the original grouping of these materials is, broadly speaking, concentric about the center of the egg, and all of the radial cleavages accordingly are quantitative. . . . Since the first five cleavages are of this type, complete dwarfs may be produced from any of the blastomeres up to the 16-cell stage . . . when the first qualitative divisions begin by the delamination-cleavages parallel to the surface. In the sea urchin the ooplasmic stuffs are polarized, displaying a symmetrical horizontal stratification at right angles to the axis of the egg. Since the first two cleavages pass exactly through the axis and cut all the strata symmetrically (Fig. 7.22a–c) the first four or two blastomeres receive equal allotments of these strata in their normal proportions and hence remain totipotent . . . we should expect the third cleavage to be qualitative; this is borne out by both observation and experiment.

Similarly, as shown in Fig. 7.22d–f, in annelids, molluscs, and other creatures undergoing spiral cleavage, the mesoderm determinants initially located in the polar region of the egg are distributed only to the D blasto-

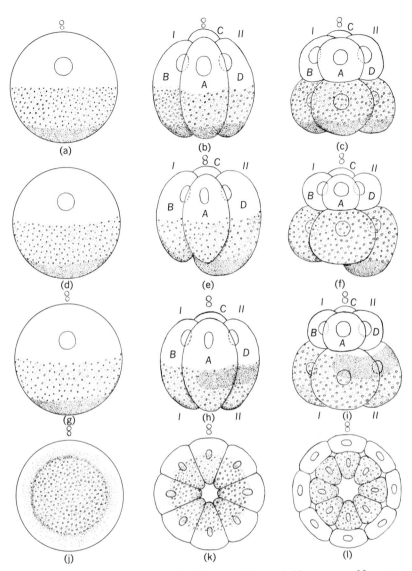

Fig. 7.22. Localization of morphogenetic substances in the early blastomeres of four types of egg. Diagrams of the primary stratification in the eggs of the sea urchin (a)–(c) and the annelid or gastropod (d)–(f). The first two cleavage planes are designated as I or II. The upper or white zone of each egg is ectoblastic, the middle or granular one the entoblastic, the lower or stippled one the mesoblastic. In (a)–(c) all the zones are equally divided; in (d)–(f) only the two upper zones are thus divided, the lower one passing entirely into the D quadrant. (g)–(i) primary stratification in the ascidian and (j)–(l) in the hydromedusa. In the ascidian the lower mesoblastic stratum is equally divided between A and D quadrants. In the hydromedusa this stratum is absent, and the remaining two are equally distributed up to the time of delamination (l). From E. B. Wilson (1925). "The Cell in Development and Heredity," pp. 1072–1076. MacMillan, New York.

mere. Consequently it is this cell alone which retains the capacity to develop into a qualitatively complete dwarf embryo, for example, the D cell of *Ilyanassa*. In the ascidian egg (Fig. 7.22g–l) only the posterior blastomeres A and D retain the mesodermal determinants, but other essential determinants not figured (e.g., those for neural ectoderm) are missing from these cells (see Figs. 7.1 and 7.2). As Conklin showed in *Styela*, no one blastomere gives rise to a complete embryo in this form (Fig. 7.5). The potentialities of blastomeres isolated from embryos of the types shown in Fig. 7.22 would depend in part on these cytoplasmic localization patterns. Because of the relation between these patterns and the cleavage planes, the sea urchin and hydromedusa embryo would be characterized as "regulative," while the annelid would be characterized as "mosaic," if the criterion were the developmental potentialities of isolated blastomeres. Yet the development of all four of these embryos depends on the same kind of localized cytoplasmic determinants.

GRADUAL ESTABLISHMENT OF LOCALIZATION PATTERNS

Another variable which affects the outcome of isolated blastomere experiments is the point when the definitive localization patterns are established in different organisms. An example is the visible process by which localization patterns are set up between fertilization and the initiation of cleavage in *Styela* (Fig. 7.2). In many experiments precleavage cytoplasmic movements have been correlated with the appearance of localization by sectioning eggs under the microscope at various times between release from the ovary and cleavage, fertilizing the fragments, and measuring the ability of the egg fragment to develop. An excellent example is the egg of the nemertean worm *Cerebratulus lacteus*. When tested at the 8-cell stage, the blastomeres of this egg behave in a perfectly determined fashion (Wilson, 1903; Zeleny, 1904; Yatsu, 1910a,b). An exhaustive series of experiments by Hörstadius (1937b) shows that these embryos can be dissected and recombined in any way, but irrespective of its new context each blastomere lineage in the recombinants differentiates as it would have in the intact egg. This is the behavior expected of a perfect mosaic embryo. On the other hand, removal of large nonnucleated sectors of egg cytoplasm before cleavage does not interfere qualitatively with development (Wilson, 1903). Egg fragments as small as one-fourth the original mass of the egg can be fertilized and induced to develop after the germinal vesicle has broken down, and the plane of the section makes little difference. Yatsu (1903) carried out similar operations on *Cerebratulus* eggs, extirpating portions of the egg cytoplasm at three successive points in the precleavage maturation process: before germinal vesicle breakdown, at the metaphase of the first reduction division, and at pronuclear fusion. These operations do not prevent qualitatively normal development if performed before germinal vesicle breakdown, but almost all embryos resulting from eggs cut at the pronuclear fusion stage are defective. Yatsu concluded that localization begins with germinal vesicle dissolution and becomes definitive in the period between fertilization and fusion. In the sea urchin, while the future animal–vegetal blastomere values are determined before fertilization, localization of the micromere determinants takes place later (Hörstadius, 1939, 1973; Morgan, 1927). Many additional demonstrations of progressive localization are cited by classical writers (Wilson, 1925; Morgan, 1927).

Attention has recently been focused on the role of cleavage itself in setting up the localization patterns. Guerrier (1968, 1970a–d) investigated the effect on later development of altering cleavage planes by compression, an approach first used by Driesch (1892) (see Chapter 1). In the eggs of Limax maximus, a gastropod, the orientation of the first cleavage spindle can be shifted without effect on later morphogenesis. Therefore, in this egg, the dorsoventral polarity is not established until after the second maturation division. Similarly, in the annelid Sabellaria the position of the plane of bilateral symmetry depends on the position of the first cleavage spindle and cannot be explained on the basis of a preformed dorsoventral polarity originally fixed in the egg (Guerrier, 1970b). In this latter egg, which has a polar lobe bearing the usual D quadrant determinants, a basic change in localization can be detected between first and second cleavage. If the first cleavage is made to occur equatorially so that the vegetal polar lobe determinants are separated by the cleavage plane from the future micromeres, subsequent development is completely abnormal, but if this is done at second cleavage, normal larvae are derived. Since spiral cleavage proceeds after the egg is released from compression in either case, this experiment suggests that the first cleavage itself is involved in the proper distribution of the polar lobe materials, an idea which is particularly interesting if it is considered that these substances may be cortical in location. In Sabellaria, as in Ilyanassa, the polar lobe determinants cannot be moved by low centrifugal forces which effectively displace the endoplasm (Guerrier, 1970b). Other experiments of Guerrier (1970c) on the eggs of lamellibranch molluscs indicate that the undivided egg contains a polar organization which affects the position of the first cleavage spindle. However, in normal development this also depends in part on the point of entry of the sperm, as in so many other eggs (see, e.g., Fig. 7.2). According to Guerrier (1968, 1970a,b,c) the cytoplasm of spiralian eggs contains a mechanism which sets up the alternating spindle orientations characteristic of spiral cleavage. Since this mechanism functions even when the initial orientation of the whole cleavage process is changed experimentally, it cannot be "imprinted" on the egg surface from the beginning. The definition of the dorsoventral axis of the embryo and the plane of bilateral symmetry occur by means of the interaction between the spiral cleavage pattern and the cortical determinants in the egg. These epigenetic processes are not completed in the spiralian eggs studied by Guerrier until the second or third cleavage. It is interesting in this connection that the second cleavage in eggs of the gastropod *Lymnaea* is a period when the eggs are extremely sensitive to pulse treatments of heat, cold, lithium chloride and sodium azide (Geilenkirchen, 1966, 1967; Camey and Geilenkirchen, 1970). Embryos treated during the metaphase–anaphase period develop into normal blastulae, but gastrular differentiation fails and subsequent organogenesis is abnormal.

These observations all show that the morphogenetic determinants present in the cytoplasm of the mature oocyte are not yet topographically localized. The process by which the definitive localization patterns are established occurs in stages. Cytoplasmic movements *prior* to first cleavage are clearly involved in some organisms, such as ascidians and amphibians. In other organisms cytoplasmic movements *accompanying* the first cleavages are also significant, as in the spiralian embryos just discussed and the ctenophores. As we have seen, however, there are cases in which cleavage is arrested and localization still takes place (e.g., Lillie, 1902; Whittaker, 1973a). Thus Boveri (1905) and Hörstadius (1928) showed that if cleavage is delayed in the sea urchin, micromere formation occurs earlier than the fourth division. It follows that the progressive localization of cytoplasmic determinants is at least partially governed by some form of internal "clock" mechanism which controls both cleavage and the processes of topographical localization [see Hörstadius (1973) for a discussion of the change in determination with time in sea urchin embryos]. In any case it is clear that the extent to which an embryo seems to conform to the model of a "mosaic of self-differentiating parts" depends partly on when the test is made, for organisms differ in when and how the cytoplasmic determinants are distributed in their eggs.

LABILITY OF CYTOPLASMIC LOCALIZATION PATTERNS

Mention has already been made of the fact that localization patterns are in some cases highly labile (see Watterson, 1956). An example is the frog egg, in which determinants responsible for the axial organization of the embryo are clearly localized, as discussed earlier. Nonetheless, Landström and Løvtrup (1975) reported that a blastomere from a 2-cell embryo which has not inherited gray crescent cytoplasm and would normally not gastrulate can be induced to do so by anaerobic treatment of one side. The posterior end always appears at the anaerobic side. Drastic alterations in the morphogenetic values of the early blastomeres have been experimentally induced in a number of ways and in a variety of other animals. Among the best known examples are the effects of lithium chloride and sodium thiocyanate on the course of differentiation in early sea urchin embryos, though many other agents have similar effects (Child, 1940; see review in Hörstadius, 1973). Lithium chloride treatment induces the development of a disproportionately enormous gut. This effect is known as "vegetalization." Thiocyanate causes the embryos to become ciliated balls and is known as an "animalizing" agent. The mechanism by which these agents act is unknown. Runnström and Markman (1966) showed that the effect of LiCl treatment is blocked by actinomycin treatment of the embryos, which suggests that transcription may be involved.

Experimental treatments which display the lability of the primary localization patterns often display the regulative abilities of the egg as well. A basic point is that regulative development in experimentally altered embryos or parts of embryos depends on cell–cell interactions. The example which most clearly demonstrates this is Hörstadius' (1939) experiments on chimeric sea urchin eggs mentioned above. Hörstadius showed, for instance, that in an artificial combination of animal pole blastomeres plus micromeres the central tiers of cells function as endodermal and equatorial blastomeres, rather than continuing to develop as animal hemisphere cells. The formation of giant though otherwise normal single embryos from fused eggs (Mangold, 1920) is another example. In order for these results to occur it is necessary that intercellular interactions, including of course the induction of gastrular structures, must have exercised influence over the behavior of each cell lineage. Interaction between blastomere lineages also functions in the development of embryos displaying relatively little regulative ability, but in these embryos inductive processes seem to become important only at later stages. Several examples have been mentioned earlier, including shell gland induction in gastropod molluscs (Raven, 1958; Cather, 1967; Hess, 1971) and induction of anterior neural structures in ascidians (Reverberi and Minganti, 1947).

We conclude that all embryos (with the possible exception of the mammals) rely on *both* blastomere self-differentiation occurring as a result of localized cytoplasmic determinants and on intercellular interactions. In some organisms, such as ctenophores, annelids, molluscs, and ascidians, the capacities for self-differentiation are dominant early in development and the localization patterns are relatively nonlabile. In others, such as sea urchins and some insects, the initial patterns of localization can easily be overridden, and we infer from this that cell–cell interaction normally plays a more important role in the early development of these creatures. These differences in the significance of blastomere interaction define the regulative ability of each embryonic system, but not the presence or absence of cytoplasmic localization.

TENTATIVE CONCLUSIONS AND SPECULATIONS

Despite the large array of particular characteristics we have touched on, several generalities about localization stand out. Whatever the nature of the morphogenetic determinants in egg cytoplasm, they must be synthesized during oogenesis. At least in many cases their final topographical distribution does not occur until after germinal vesicle breakdown, and often not until the first cleavages. Therefore, the topographical organization of the egg cytoplasm is not in general a direct product of the processes of oogenesis, that is, a *preformation*, in classical terms, but is rather an epigenetic or developmental construction. What is preformed is the morphogenetic determinants themselves. Evidence from many systems suggests that when finally localized these are fixed in the cortex or subcortical regions of the egg. In some cases this may mean the egg membrane itself.

There is no convincing evidence as to the molecular mechanism by which the localized determinants of the egg function, only that they exist. In many cases they seem to be responsible for large programs of differentiation, such as the whole set of functions required for the differentiation of primordial germ cells, or coelomic mesoderm, or cephalic plate cells. It is possible that sets of maternal messenger RNA's end up being sequestered in different regions of the egg, particularly in view of the results of Donohoo and Kafatos (1973). As will be recalled, their study showed that the CD and AB blastomeres of *Ilyanassa* eggs synthesized different sets of proteins (Fig. 7.12). Though these could be the products of newly transcribed messenger RNA's, this seems improbable, considering that the vast majority of messages being translated during early cleavage are maternal. However, in order to account for the eventual differentiation of the blastomere lineages, that is, for organogenesis, specific sets of embryo genes clearly must be activated.

Davidson *et al.* (1965) and Davidson and Britten (1971) proposed explanations of the localization phenomenon based on embryo gene activation. According to these interpretations, regulatory molecules or messenger RNA's synthesizing such molecules are topographically localized in the egg cytoplasm. If the conclusions of Kalthoff and his colleagues are correct, the head determinants of *Smittia* eggs would represent a model for localized RNA's which serve as morphogenetic determinants. Other examples which might fall into this class have been reviewed in this chapter, such as germ cell determinants. In such cases it could be hypothesized that the pattern of gene activity in the early blastomeres is controlled by these RNA's. Such arguments indicate that a basic understanding of localization depends in the end on knowledge of the molecular mechanism of gene regulation. Localization seems likely to have evolved very early in the history of the metazoa, along with other basic mechanisms of development. Thus, if egg cytoplasmic determinants control gene activity, they must be similar in composition and function to other gene regulatory molecules. From this point of view the cortex (or membrane) of the early embryo could be regarded as a two-dimensional array of stored gene regulatory agents.

8

Lampbrush Chromosomes and the Synthesis of Heterogeneous Nuclear and Messenger RNA's during Oogenesis

The structure of lampbrush chromosomes is discussed, and the traditional conclusion that these are sites of active synthesis is reviewed in light of ultrastructural evidence. The lampbrush chromosome matrices are composed of ribonucleoprotein granules arranged in loop-specific aggregates. Each loop contains a specific, fixed region of the DNA. Organisms with larger genomes tend to have larger chromosome loops, leading to a pardox, since the genetic function of the DNA in the loop should be the same in related animals. One or more transcription units may exist per loop, as visualized in the electron microscope, and these are maximally packed with active polymerases. Estimates of lampbrush chromosome RNA complexity and synthesis rates are derived for *Xenopus* oocytes from structural evidence. The occurrence of lampbrush chromosomes in the chordate life cycle is considered, and it is concluded that the lampbrush stage always requires many weeks or months, if not longer. A detailed staging system for *Xenopus* oogenesis is summarized. Available data regarding the duration of the lampbrush phase and the phylogenetic occurrence of lampbrush chromosomes are collated in Table 8.1. In meroistic oogenesis, as studied in holometabolous insects, lampbrush chromosomes are absent or little developed, and instead the oocyte is supplied by nurse cells. Nurse cell–oocyte complexes originate through a programmed series of incomplete oogonial divisions. Nurse

319

cells and oocytes are joined by open "ring canals," and through these canals RNA's, including ribosomal RNA, flow from their sites of synthesis in the nurse cells to the oocyte. In several invertebrates, special portions of the genome are required only for gametogenesis and are discarded after gametogenesis is complete by processes of chromosome diminution. Germ-line-specific DNA in *Ascaris* is largely satellite DNA, but in some insects germ-line-specific DNA is transcribed during oogenesis and is probably not satellite DNA. A few lampbrush loops are present in *Drosophila* spermatocyte Y chromosomes and may play a role similar to that of oocyte lampbrush chromosome loops. Molecular evidence regarding transcription of transfer RNA and 5 S, 18 S, and 28 S ribosomal RNA's during oogenesis is reviewed. In previtellogenic amphibian and teleost oocytes these low molecular weight RNA's are synthesized actively and stored for later use. Genes for these RNA species are highly repetitive and are not amplified in oocytes, but special subsets of these genes may be transcribed in oogenesis. Genes for the high molecular weight ribosomal RNA's are amplified extensively in many organisms, though not in those which utilize meroistic oogenesis. The replicated copies of the ribosomal genes are present in extrachromosomal nucleoli. Early in oogenesis these genes are not transcribed extensively, but they are almost completely occupied with closely packed transcripts at midvitellogenesis growth stages. In some amphibians these genes are again less extensively transcribed in mature oocytes. Estimates and measurements of ribosomal RNA synthesis rates in lampbrush stage and in mature oocytes are collated in Table 8.2. The lampbrush chromosome RNA's are of the heterogeneous nuclear type, but no direct measurements of their turnover or synthesis rates exist. Measurements of synthesis rates in mature *Xenopus* oocytes for a rapidly turning over heterogeneous nuclear RNA class, and a more slowly decaying RNA class which may be messenger RNA, are also collated in Table 8.2. In addition this table contains estimates of synthesis rates for poly(A)RNA in mature and lampbrush stage oocytes. During the lampbrush stage a relatively complex set of repetitive sequence transcripts is synthesized and stored in the amphibian oocyte. The poly(A)RNA stored in the mature oocyte is largely maternal message, and it is present at its final levels before the lampbrush stage begins. Messenger RNA is translated during oogenesis, and studies with maternal mutants in *Drosophila* and axolotl show that some of the proteins then synthesized are required for early development. The functional role of lampbrush chromosomes is not understood. Since the lampbrush stage lasts a long time and the active transcription units are transcribed with maximum intensity, the function of these structures appears to be the accretion of heterogeneous RNA species. Difficulties with the conventional proposal that the transcripts of lampbrush chromosomes are maternal messenger RNA precursors are considered. These problems arise from the quantitative evidence reviewed regarding synthesis and accumulation of various heterogeneous classes of RNA during oogenesis. Insufficient data are yet available to reach a firm conclusion, and an alternative speculation is mentioned.

The mature oocyte is a remarkable cell. It contains a tremendous diversity of structural gene transcripts, which are almost certainly destined to serve as maternal messenger RNA's. Their complexity, as we saw in Chapter 6, is as high as 2×10^7 to 4×10^7 nucleotides. The data reviewed in Chapter 7 show that the oocyte also contains morphogenetic determinants, or messenger RNA's which can be translated into morphogenetic determinants. In addition, a great deal of organized cytoarchitecture obviously exists in the mature oocyte, which together with epigenetic processes occurring after fertilization gives rise to the definitive localization patterns. The maternal messenger RNA's, possibly other heterogeneous oocyte RNA's of yet unknown function, and many developmentally important proteins must arise through the activity of genes in the oocyte chromosomes. Alternatively, in some animals these constituents may derive from nurse cell chromosomes. In most phyletic groups, however, nurse cells are lacking, and the site of synthesis of heterogeneous RNA's is clearly the oocyte itself. At certain stages the oocyte chromosomes assume the laterally looped "lampbrush" form unique to germ-line cells. The distribution and structure of lampbrush chromosomes is reviewed in this chapter, and their function in transcription is considered from comparative, cytological, genetic, and molecular viewpoints. Many other important aspects of oogenesis are not discussed here or are dealt with only in summary. These include the amplification of ribosomal DNA in the oocyte nucleoli, the activities of the oocyte mitochondria, the events taking place in oogonia and in very early meiotic oocytes, and the biosynthetic processes occurring in maturing oocytes between ovulation and fertilization. A comprehensive treatment of the latter subject is to be found in a recent review of Smith (1975), and references for others of these topics are provided below.

Structure and Transcription in the Lampbrush Chromosomes of Amphibian Oocytes

DNA CONTENT AND STRUCTURE OF LAMPBRUSH CHROMOSOMES

Lampbrush chromosome structure has always been considered suggestive of intense transcriptional activity. The chromosomes attain a high degree of extension at the midlampbrush stage and the linear loop-chromomere arrangement appears to reflect the linear arrangement of genetic sites in the DNA. It is important to note that lampbrush chromosomes are diplotene meiotic prophase structures, so that the complete set of these chromosomes contains a 4C genome. Each lampbrush chromosome bivalent contains the four homologous chromatids, joined at about 2–4 chiasmata (Callan and Lloyd, 1960; Mancino *et al.*, 1969; Giorgi and Galleni, 1972; Müller, 1974).

The amount of nuclear DNA in meiotic prophase oocyte nuclei has been measured spectrophotometrically in a number of organisms in which the nucleus is small enough to provide the necessary DNA concentrations. As expected, oocytes of mammals (Alfert, 1950; Van de Kerckhove, 1959), insects (Swift and Kleinfeld, 1953), and various worms (Mulnard, 1954; Govaert, 1957) contain either the 4C nuclear DNA content or an amount slightly greater than this, usually attributed to excess nucleolar DNA. In organisms whose oocytes contain very large nuclei, such as the amphibia, the chromosomal apparatus becomes so extended that its ability to stain for DNA tends to disappear by the maximum growth stage. The significance of this was discussed long ago by Maréchal (1907). The absence during oogenesis of those chromatin elements staining with basic dyes, while the chromosomal structures themselves persist, was for years regarded as a strong argument against the idea that DNA could be the genetic material in the chromosomes [see, for example, Wilson's (1925) discussion of this problem]. However, Brachet (1940) showed that if the chromatin of the lampbrush-stage amphibian oocyte nucleus is concentrated at one pole of the nucleus by centrifugation, a Feulgen-positive reaction can be easily demonstrated. The DNA content of manually isolated lampbrush chromosomes from the newt *Triturus* was measured by Izawa *et al.* (1963), and about four times the 4C value was obtained. This was probably due to contamination of the chromosomes by extrachromosomal nucleoli, though efforts were made to avoid this. Perkowska *et al.* (1968) were able to account quantitatively for the total DNA of the *Xenopus* oocyte nucleus, about 42 pg, by adding the amount of nucleolar

DNA, 30 pg, to the 4C quantity expected to be present in the chromosomes, or 12 pg.

Gall (1954) showed that concentrations of DNA sufficiently high to be stained by the Feulgen reaction are confined to the chromomeres (see Fig. 8.1). This finding suggested that the chromomeres contain densely packed

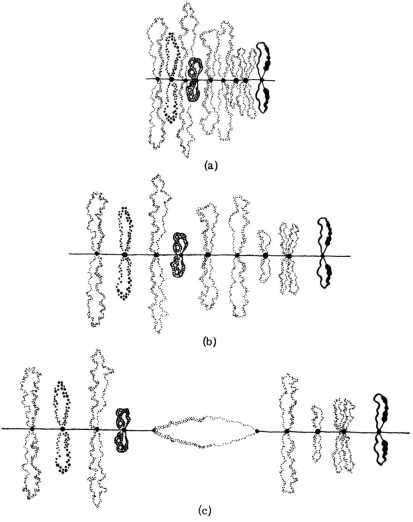

(a)

(b)

(c)

Fig. 8.1. Diagrams to illustrate what happens when parts of a lampbrush chromosome are stretched. (a) Unstretched; solid dots represent condensed chromomeric DNA. (b) Stretched within the elastic limit. (c) Stretched beyond the elastic limit—one chromomere has broken and a pair of lateral loops span the break. From H. G. Callan (1963). *Int. Rev. Cytol.* **15,** 1.

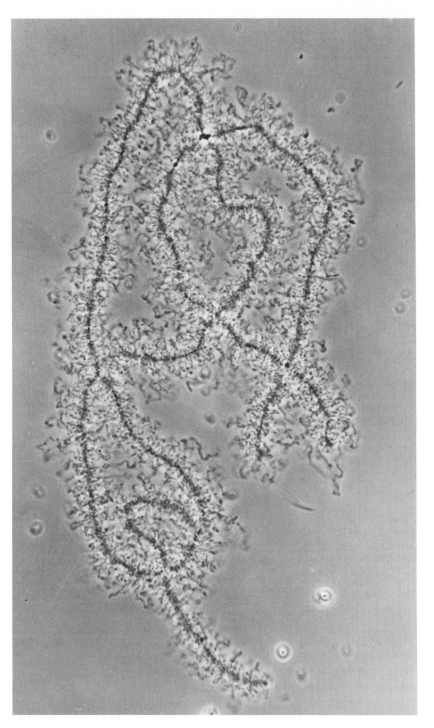

DNA. Callan (1955) demonstrated the paired structure of the loops by stretching individual chromosome regions with needles to the point where the chromomeres separate transversely into their separate strands. A drawing showing this experiment is reproduced in Fig. 8.1 (Callan, 1963). The general appearance of newt lampbrush chromosomes in the phase microscope can be seen in Fig. 8.2. Each lateral loop contains a single DNA duplex, and each pair of loops and each chromomere contains two such duplexes, as suggested by the diagram in Fig. 8.1. These facts were deduced by Gall (1963) from a kinetic study of lampbrush chromosome breakage by DNase. The single DNA nucleoprotein fibers present in the loops have subsequently been visualized in the electron microscope by Miller (1965) and others (e.g., Ullerich, 1970; Angelier and Lacroix, 1975). There are about 20,000 loops in the whole chromosome set of *Triturus*, some of which are very large, over 200 μm in axial length (Callan, 1963). Thus, there are about 5000 loops per haploid set. Estimates as high as 10,000 were made by Vlad and MacGregor (1975) for another urodele, of the genus *Plethodon*.

Specific loops are easily recognizable in lampbrush chromosomes, both by their size and their morphology (e.g., see Fig. 8.1). A "map" of the lampbrush chromosomes of the newt was constructed by Callan and Lloyd (1960), and similar analyses have been carried out for many other amphibia (e.g., see Mancino and Barsacchi, 1969; Mancino *et al.*, 1969), including *Xenopus* (Müller, 1974). Specific loop morphologies apparently derive from the loop size and the characteristics of the ribonucleoprotein matrix of each loop. The RNA of the lampbrush chromosome matrix is DNA-like in base composition. This was established by Edström and Gall (1963) who carried out microanalyses on manually isolated lampbrush chromosomes. Both the amount of RNA and the amount of protein in these chromosomes is unusually large. RNA and protein contents of *Triturus* lampbrush chromosomes were measured by Izawa *et al.* (1963). Their data show that the complete chromosomal set contains about 0.4 μg of protein and about 0.007 μg of RNA. They reported the mass ratio of RNA to DNA in the lampbrush chromosomes to be about 9, but since their value for DNA is probably too high (see above), the chromosomal RNA to DNA ratio is even greater. In comparison, a typical somatic cell chromatin contains about 0.05 of the DNA mass as RNA. Similarly, the ratio of protein to DNA mass in lampbrush chromosomes is extremely

---

Fig. 8.2. Lampbrush chromosomes of *Triturus viridescens* unfixed. The loops can be seen projecting laterally from the main chromosomal axis. Note the chiasmata. ×525. From J. G. Gall (1966). *In* "Methods in Cell Physiology" (D. M. Prescott, ed.), Vol. II, p. 37. Academic Press, New York.

high, at least 200 times greater than this ratio in typical somatic chroma-tin. More recent measurements (Malcolm and Sommerville, 1974) indi-cate that the RNA to protein ratio in the loop matrix is about 0.03, similar to the value of about 0.02 reported by Izawa et al. (1963).

SPECIFICITY OF LAMPBRUSH CHROMOSOME LOOPS

Malcolm and Sommerville (1974) showed by high voltage electron mi-croscopy that the organization of the ribonucleoprotein in the various loops is distinct. This was also the conclusion drawn by Angelier and Lacroix (1975) from an electron microscopic study, and it explains the heterogeneous appearance of the loops in the phase microscope (e.g., Fig. 8.1). Examples of particular loop matrices are shown in Fig. 8.3 (Malcolm and Sommerville, 1974). The ribonucleoprotein particles of which the loop matrix consists appear to be assembled from 20 nm subparticles (Fig. 8.3f) which aggregate in different ways in different loops. A possible impli-cation of this result is that different loops contain different species of nonhistone chromosomal protein. Scott and Sommerville (1974) demon-strated this to be the case in *Triturus* lampbrush chromosomes at least for several loops, by using fluorescein-linked antibodies against nuclear ribonucleoprotein. Remarkably, some fractions of these antibodies reacted only with about 10 loops. Some of these are shown in Fig. 8.4. It can be seen that the whole of each loop reacts with the antibody. The important conclusion from these studies is that both the morphology of the ribonucleoprotein granules and its protein constituents are particular to given loops. These loops therefore each represent a *unit* with respect to the nature of the accreted ribonucleoprotein which they contain.

The individual loops of the lampbrush chromosome have a genetic significance. Thus in their morphological characteristics they behave as Mendelian markers. Particular loop morphologies are the property of species, subspecies, or individuals (Callan, 1963). Heterozygotes, or hy-brids between related species, generally produce heterozygous sets of lampbrush chromosomes which display the loop morphologies charac-teristic of each parent. Many cases are known in which allelic alternatives exist for given loops within various *Triturus* subspecies, and the frequen-cies at which these alternatives appear are distributed in wild populations as predicted by a Hardy–Weinberg calculation. These observations all show that the loops are specific manifestations of the DNA sequences in the genome. In *Triturus*, furthermore, certain sets of loops have been shown by *in situ* hybridization to be the location of the 5 S ribosomal RNA genes (Pukkila, 1975). A general conclusion is that each loop in-cludes a specific region of the DNA. The existence of morphologically heterozygous lampbrush chromosome bivalents shows that the chromosome structure is determined by the DNA which it contains, rather, for instance than by its position on the chromosome, conditions in the nuclear sap, or biosynthetic activities occurring elsewhere in the nucleus.

DNA CONTENT OF LAMPBRUSH CHROMOSOME LOOPS

Among the features of lampbrush chromosome structure which are most puzzling is the relation between lampbrush chromosome size and genome size. This subject has two aspects, the number of chromomeres (or loops) and the size of the loops (or length of DNA per loop). Comparisons can easily be made in the amphibia because of the large differences in genome size found within this class. The oocytes of species with large genomes are found to contain longer lampbrush chromosomes, with bigger loops. This is shown dramatically in Fig. 8.5, where the lampbrush chromosomes of *Triturus* and *Xenopus* are compared to the same scale (Müller, 1974). Depending on the species the genomic sizes of *Triturus* species are about 7–15 times larger than that of *Xenopus*. In *Xenopus* the loops average only a few micrometers and large loops are about 10–15 μm in length (Callan, 1963; Müller, 1974), while in *Triturus* species they average 50 μm and some are as large as 200 μm (Gall, 1955). Since the structure of the loops is probably similar, there is much more DNA in a typical loop of a *Triturus* lampbrush chromosome than in a typical loop of a *Xenopus* lampbrush chromosome. This seems paradoxical since the loops contain specific genetic sequences, and since their physiological function must be similar in the oocytes of these two species. Furthermore, this paradox extends to the number of loops present. Vlad and MacGregor (1975) compared the size of lampbrush chromosomes among three species of the salamander genus *Plethodon*, varying in genome size from 20 to 38.8 pg. The *relative* dimensions of the individual chromosomes were about the same in the three species, while the absolute size was greater in species with larger genomes. However, the number of chromomeres or loops is about the same per unit length of chromosome. The total number of loops is 60–70% greater in the chromosomes of the species with the larger genome. Therefore, though the loops represent specific genetic regions, the number of loops cannot be equated with the number of active structural genes, nor is it proportional to the number of active structural genes. Obviously, there are not 60–70% more structural genes active in one species of salamander than in another, unless the structural genes are repetitive in one species and not in the other. Rosbash *et al.* (1974) showed, however, that most of the structural genes active in oogenesis and which produce poly(A)RNA's are single copy sequences,

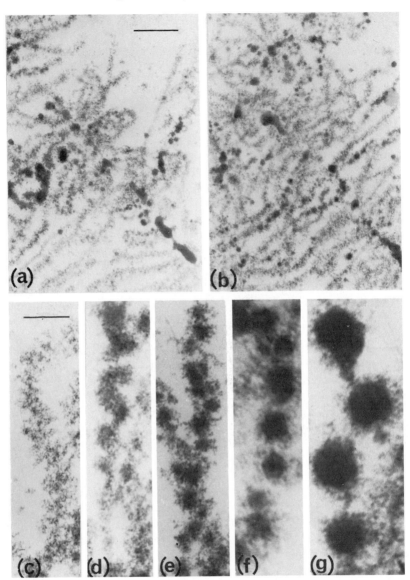

Fig. 8.3. High voltage electron micrographs showing lampbrush chromosomes of *Triturus* oocytes and the arrangement of the ribonucleoproteins which constitute the matrix of the lateral loops. (a) and (b) Low magnification micrographs of chromosomal regions including several loop pairs. (c)–(g) Various morphological types are shown. The scale lines in (a) and (b) represent 2 μm and 0.5 μm in (c)–(g). (h) Part of a "fuzzy" lateral loop taken at high magnification showing 20 nm subparticles. The scale line represents 0.1 μm. From D. B. Malcolm and J. Sommerville (1974). *Chromosoma* **47**, 137.

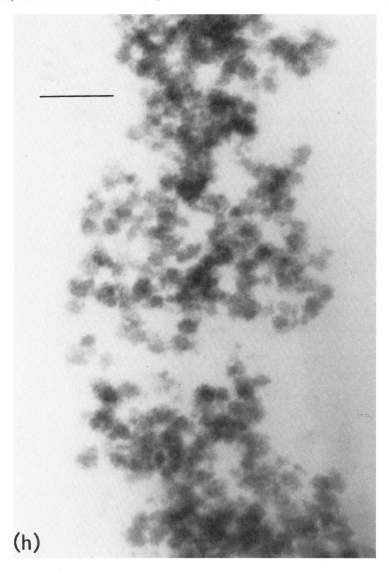

(h)

Fig. 8.3(h)

both in *Triturus* and in *Xenopus*. As will be recalled from Chapter 6, these authors also found that the complexity of oocyte RNA is about the same in *Xenopus* and *Triturus* oocytes. That is, about the same number of structural genes appear to be active in the synthesis of at least those messenger RNA's which are stored in the mature oocytes of these two species. It

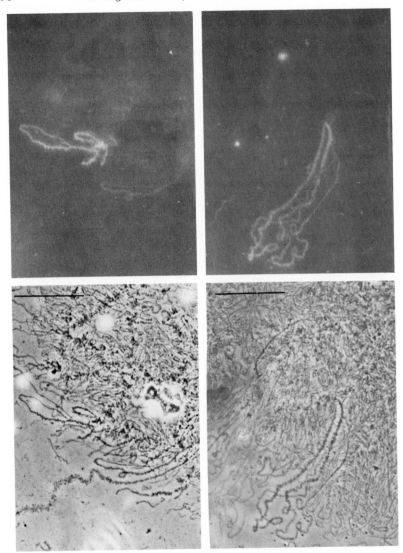

Fig. 8.4. Phase contrast and fluorescence photomicrographs showing location and specificity of reaction of fluorescein-conjugated antibodies with *Triturus* oocyte lampbrush chromosomes. The photomicrographs were taken on Ilford FP4 using a 1-second exposure for phase (top); and 2.5-minute exposure for fluorescence (bottom). The scale lines represent 10 μm. From S. E. M. Scott and J. Sommerville (1974). *Nature (London)* **250,** 680.

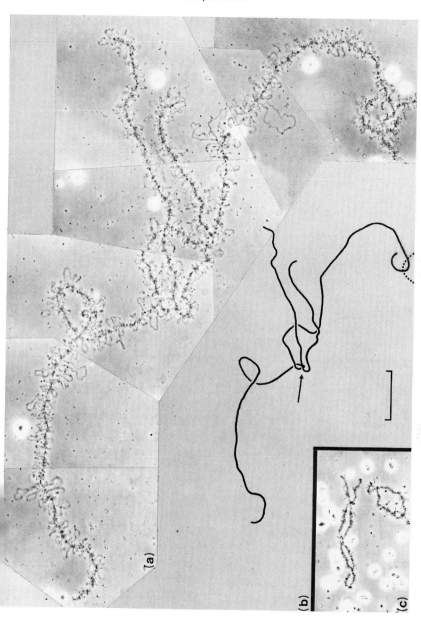

Fig. 8.5. Comparison between lampbrush chromosomes of *Triturus cristatus* and *Xenopus laevis*. (a) Bivalent I of *Triturus cristatus*. (b) Schematic drawing of the same chromosome. The arrow indicates the position of the chiasma. (c) Bivalents I and XIII of *Xenopus laevis*. The scale line represents 50 μm for (a) and (c). From W. P. Müller (1974). *Chromosoma* 47, 283.

follows that the absolute length of the DNA included in the lampbrush chromosome loops and the way this DNA is divided up among the loops are independent of the number of nonrepetitive structural genes active in the preparation of maternal messenger RNA.

Even in the small lampbrush chromosome loops of *Xenopus*, the length of the DNA is at least several thousand nucleotide pairs, and it is tens or hundreds of thousands of nucleotide pairs in the loops of urodele lampbrush chromosomes. These are of course minimal estimates, since they are based on the contour length of the loops, and the DNA is probably not fully extended. Callan (1955) found that the loops can be stretched by a factor of about 2.5. In any case it is probable that most loops contain both repetitive and nonrepetitive DNA sequence, since these are interspersed in amphibian genomes on a much finer scale than that defined by the length of the loops. As reviewed in Chapter 1, most of the DNA consists of single copy sequences one to a few thousand nucleotides long interspersed with repetitive sequence elements. Sequence organization has been studied in detail in *Xenopus* DNA (Davidson *et al.*, 1973; Chamberlin *et al.*, 1975), and it is also known that short repetitive sequences are interspersed with single copy sequences in the DNA of *Plethodon* (MacGregor *et al.*, 1976) and of *Triturus* (Sommerville and Malcolm, 1976).

TRANSCRIPTION UNITS IN LAMPBRUSH CHROMOSOMES

Radioautograph experiments showed some years ago that newly synthesized RNA can be observed along the whole length of most lampbrush chromosome loops. This suggests that most of the DNA fibril in the loops is being transcribed, though other interpretations are possible. A representative radioautograph is shown in Fig. 8.6. The loops are the main site of labeling (Gall, 1958; Gall and Callan, 1962; Izawa *et al.*, 1963) though RNA could be labeled in the chromosomal axis as well (see Fig. 8.6). Newly synthesized proteins as well as newly synthesized RNA are found all over the chromosome loops (Gall and Callan, 1962). The loops, however, are polarized in their structure. That is, the matrix of loop material is thicker at one end of the loop than at the other (Callan and Lloyd, 1960; Callan, 1963; Malcolm and Sommerville, 1974). There are at least two pairs of giant loops (out of the hundreds sufficiently prominent to be observable) which display an interesting deviation from the uniform pattern of RNA synthesis portrayed in Fig. 8.6, in that these loops *label* in a polarized fashion (Gall and Callan, 1962). Only the areas toward the thinner insertion of the loop appear to serve as sites of synthesis, since in radioautographic experiments only this region of the loop shows uptake into RNA after 1-day exposure to labeled precursor. At 4 days half of the loop is labeled, at 7 days two-thirds is labeled, and at 14 days labeled RNA is present all over the loop. Gall and Callan (1962) and Callan (1967)

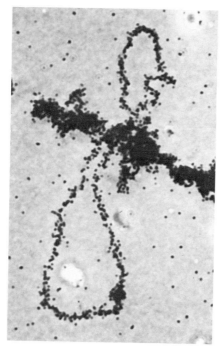

Fig. 8.6. Radioautograph of a single *Triturus* lampbrush chromosome loop pair exposed to ³H-uridine. Labeling along the chromosome axis could result from smaller, heavily labeled loops which have collapsed against the axis during fixation. ×1250. From J. Gall (1966). *In* "Methods in Cell Physiology" (D. M. Prescott, ed.), Vol. II, p. 37. Academic Press, New York.

interpreted this as movement of the RNA gene product together with movement of the DNA template. It was supposed that the DNA present in the chromomeres at any one time is spun out into the loops at a later time, and at a still later time, having traversed the whole loop structure, it is again found in the chromomere. The RNA is postulated to move along with the DNA, and Callan (1967) proposed that this mechanism pertains to all the chromosome loops. Much evidence of different kinds, to be reviewed below, now seems to exclude this idea at least as a general feature of lampbrush chromosomes. The alternative is that in the few loops which label sequentially, the ribonucleoprotein matrix itself spreads over these loops in a polarized fashion.

The polarity of the loop matrices can be understood on the basis that the loops contain one or a few very large transcriptional units. This is clear from electron micrographs in which the transcriptional complexes in the chromosome loops are visualized. Some striking examples are presented in Fig. 8.7. Figure 8.7a shows a region of the loop matrix in *Triturus*

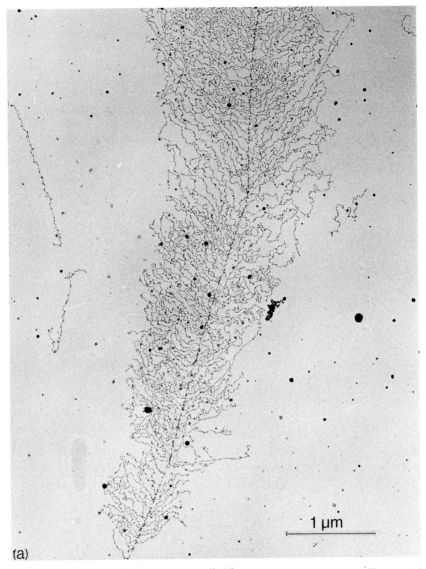

(a)

Fig. 8.7. Visualization of transcription units in lampbrush chromosomes. (a) Electron micrograph of a portion of a *Triturus viridescens* lampbrush loop near its thin insertion end. From O. L. Miller, Jr. and A. H. Bakken (1972). *Acta Endocrinol. (Copenhagen) Suppl.* **68,** 155. (b) A complete transcription unit at least 40 μm in length, from a lampbrush chromosome loop of *Pleurodeles poireti*. (c) A region of a loop showing parts of two transcriptional units of opposite polarity separated by a long nontranscribed region. (b) and (c) from N. Angelier and J. C. Lacroix (1975). *Chromosoma* **51,** 323.

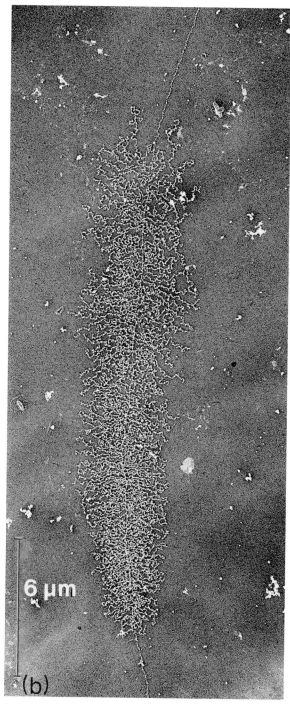

6 µm

(b)

Fig. 8.7(b)

335

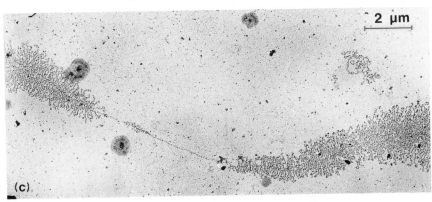

Fig. 8.7(c)

lampbrush chromosomes (after solubilization or disaggregation of some of its constituents) which is more than 30,000 nucleotide pairs long (Miller and Bakken, 1972). The transcription unit begins below the lower left corner and continues beyond the upper boundary of the photograph. The transcripts increase in length as one proceeds upward along the central DNA fibril. The transcripts appear to be spaced only about 100 nucleotides apart. Some of the transcription units seen in such preparations are over 10^5 nucleotides in length. These length estimates are of course only approximate since all the nucleoprotein fibrils are likely to be less than maximally extended. Figure 8.7b and c, from the work of Angelier and Lacroix (1975), shows several more transcription units from urodele lampbrush chromosome loops. In Fig. 8.7b a unit over 10^5 nucleotides in length can be seen. Again the transcripts appear to originate only one to a few hundred nucleotide pairs apart. In Fig. 8.7c parts of two such units are shown. This structure is particularly interesting because there is a "spacer" or nontranscribed region several thousand nucleotide pairs in length separating the two transcriptional units and because these have opposing polarity. These two features indicate convincingly that a given loop may include more than a single transcriptional unit. It is likely from the work of Sommerville and his colleagues (see Fig. 8.3) that the ribonucleoprotein fibrils in all of the preparations shown in Fig. 8.7 have lost some of their folded structures. Malcolm and Sommerville (1974) have convincingly interpreted the thick matrix and large granules found accumulating toward the distal side of each loop as aggregates of the 20 nm particles shown in Fig. 8.3f. According to this interpretation, proteins are added to the RNA as it is synthesized, so that the size of the ribonucleoprotein aggregate increases with the length of the transcript. These observations provide an excellent explanation of loop polarity without requiring that the DNA template be constantly moving out of the chromomere and into the loops.

A main conclusion from these fine-structure studies on loop polarity is that in urodeles the lampbrush chromosome transcripts are very large, at least 5×10^4 to 10×10^4 nucleotides in length. Furthermore, there is no obligatory relation between the number of transcription units and the number of loops, since more than one transcription unit is often observed associated with a given DNA region (Angelier and Lacroix, 1975). While the number of loops in the lampbrush chromosomes of the *Plethodon* species studied by Vlad and MacGregor (1975) differs, the number of transcription units could still be the same. However, since a loop is mainly composed of one or a few transcription units, the same data show that the length of the individual transcription units in *Xenopus* lampbrush chromosome loops must be significantly lower than that found in the loops of the urodele lampbrush chromosomes.

ESTIMATES OF COMPLEXITY AND SYNTHESIS RATE FOR LAMPBRUSH CHROMOSOME RNA FROM STRUCTURAL EVIDENCE

Since each loop represents a given region of the DNA, it should be possible to calculate very roughly the sequence complexity of the total loop transcripts. Suppose there are 20,000 loops in a set of *Triturus* chromosomes, each containing an average of two transcription units of 10^5 nucleotides. In this case 4×10^9 nucleotides or about 5% of the 4C genome would be transcribed. In accord with this value, Gall (1955) estimated that at least 5% of the total DNA is in the loops of *Triturus* lampbrush chromosomes, and Vlad and MacGregor (1975) state that about 10% of the genome is extended in the loops of *Plethodon* lampbrush chromosomes. This suggests that at least 5–10% of the DNA length is represented in the lampbrush chromosome transcripts. Another rough calculation possible is the rate of RNA synthesis. If 4×10^9 nucleotides are being transcribed by polymerases 100 nucleotides apart (Fig. 8.7), and the transcription rate is 15 nucleotides sec^{-1} (D. M. Anderson and L. D. Smith, personal communication), the total RNA synthesis rate would be about 6×10^8 nucleotides sec^{-1} or about 20 pg min^{-1} per oocyte nucleus for *Triturus*. Since the length of the loops seems to vary roughly with genome size, the equivalent value might be one-seventh as great for *Xenopus*, or about 3 pg min^{-1} per (4C) oocyte nucleus. This estimate may be compared to the rate of heterogeneous nuclear RNA synthesis in the postgastrular *Xenopus* embryo cells, which we calculated in Chapter 5 to be about 0.01–0.02 pg min^{-1} per (2C) nucleus. Despite the crudeness of these calculations it is clear that the oocyte nucleus is by far the more active, by more than two orders of magnitude. The difference must lie mainly in the number of polymerases transcribing each region, i.e., in the close packing of the polymerases in the lampbrush chromosomes compared to their distribution in the active regions of embryo nuclei. As calculated in Chapter 6, the density of nascent heterogeneous nuclear RNA molecules in (sea urchin) embryo nuclei is only about one per 10^4 nucleotides. If the DNA in the lampbrush loops is typical interspersed repetitive and nonrepetitive sequence, the complexity of heterogeneous nuclear RNA in somatic cells and in lampbrush chromosomes might not be so different, since complexities ⩾10% of the single copy sequence have been observed in various embryonic and other somatic cells (reviewed by Davidson and Britten, 1973; see Chapter 6). We conclude, on the basis of the structural evidence so far reviewed, that the lampbrush chromosomes are a device for extraordinarily *rapid* synthesis of a class of RNA's which has many of the characteristics of heterogeneous nuclear RNA. Thus the RNA synthesized in lampbrush chromosomes is DNA-like in base composition; it is very large in size; it may have an interspersed sequence organization, and if so, its complexity is likely to be similar to that known for nuclear RNA in other systems. In all of these characteristics it differs sharply from messenger RNA, including that stored in the oocytes of the same species. As we recall from Chapter 6, the complexity of the putative maternal messenger RNA's in *Xenopus* oocytes is only about 1% of the single copy sequence, an order of magnitude lower than that estimated here for the lampbrush chromosome transcripts.

The Occurrence of Lampbrush Chromosomes

Lampbrush chromosomes develop after the termination of premeiotic DNA synthesis and the completion of the leptotene, zygotene and pachytene stages of meiotic prophase. As noted above they are diplotene structures. A long period may separate lampbrush stage oocytes from ovulation and maturity. As documented below the lampbrush phase itself typically requires a considerable length of time. In some organisms the process of vitellogenesis or yolk deposition follows the lampbrush phase, and in others a lengthy period of oocyte "storage" intervenes prior to ovulation. As an aid in appreciating the meaning of the synthetic activity of lampbrush chromosomes we begin with a brief review of the course of oogenesis in the context of the chordate life cycle.

Following their early determination (Chapter 7) the primordial germ cells in chordate embryos migrate to the position of the forming gonad (reviewed by Franchi *et al.*, 1962). During larval or embryonic life a definitive ovary with nests of proliferating oogonia is established. In amphibians and some teleosts, oogonial divisions appear to continue even after sexual maturity is attained. The events of meiotic prophase recur cyclically with each breeding season in these animals and in reptiles (Franchi *et al.*, 1962). On the other hand in mammals, birds, cyclostomes, elasmobranchs, and other teleosts the total number of oogonia ever possessed by an individual is already formed long before sexual maturity has been attained. In these animals premeiotic DNA synthesis and the first stages of meiotic prophase are completed early in the life cycle, before or soon after hatching, metamorphosis, or birth. In the mouse, for example, premeiotic DNA synthesis occurs during fetal life (Borum, 1967); in the rabbit this synthesis takes place within the first few days after birth (Kennelly *et al.*, 1970). After this no further *de novo* production of oocytes or oogonia can take place. Understanding of this aspect of germ cell differentiation in higher chordates dates back to the studies of Waldever (1870), who observed that mitotic figures are absent in the ovarian tissues of birds and mammals during neonatal life.

Chromosomal condensation and synapsis occur soon after premeiotic DNA synthesis in mammals. For example, the data of Kennelly *et al.* (1970) show that in the rabbit premeiotic DNA synthesis begins within 6 hours of the final oogonial division and requires about 9 hours. The initial stages of the meiotic prophase, leptotene, zygotene and pachytene, are completed within about 11 days. The oocytes then enter diplotene but within a few days the chromosomes become too diffuse to be observed by conventional methods in fixed sections. The oocytes remain in this condition, which is termed the dictyate stage, until sexual maturity and ovulation. The same course of events is observed in rodents, where again a diplotene stage lasting only a few days is observed, followed by a prolonged dictyate stage (Franchi and Mandl, 1963). Lampbrush chromosomes have not been observed in rodent oocytes, but they are reported in human oocytes (Baker and Franchi, 1966, 1967). Like those of amphibians the lampbrush chromosomes of primate oocytes are the sites of RNA synthesis (Baker *et al.*, 1969). In the human the pachytene stage is completed during fetal life, by about 7 months of pregnancy, and the oocytes are in diplotene by birth. Thus in a species as long-lived as ourselves more than 40 years may separate the initiation and the termination of meiotic prophase.

An interesting feature of chordate oogenesis illustrated by studies on human oocytes is the progressive decline in the number of available germ cells. At 5 months of pregnancy human ovaries contain about 6.8×10^6 oocytes; by 7 months, at termination of pachytene, they contain about 2×10^6 oocytes; and at 7 years only 0.3×10^6 oocytes, or under 5% of the initial population (Baker, 1963). Much the same has been found to be true in the rat (Franchi and Mandl, 1963) and in lower chordates such as the lamprey (Hardisty and Cosh, 1966). In the lamprey the oogonial divisions are also completed during larval life and meiotic oocytes appear before metamorphosis (Okkelberg, 1921). By metamorphosis the oocytes are in the diplotene stage, and this condition lasts for several years (Okkelberg, 1921; Lewis and McMillan, 1965). About two-thirds of the oocytes which initiate meiosis in lampreys are discarded during the diplotene phase, and those actually completing diplotene represent less than one-fifth of the original number of oogonia. One interpretation is that the mature oocyte population in chordates is the product of a stringent selective process of some kind which is survived by only a small minority of the starting oocytes.

STAGES OF OOGENESIS AND DURATION OF THE LAMPBRUSH PHASE IN AMPHIBIANS

The most detailed studies on chromosomal changes during oogenesis have been carried out on amphibian oocytes. A diagram summarizing these changes is shown in Fig. 8.8 (Duryee, 1950). The earliest oocytes figured are already in meiotic prophase and contain a 4C genome. The chromosomes assume the lampbrush form at Duryee's stage 3 and are maximally extended throughout stage 4. After this they retract. At the end of oogenesis in stage 6, when animal–vegetal polarity has been established and the eggs are ready to be ovulated, the chromosomes are once again condensed in preparation for the first meiotic metaphase. A useful classification for the stages of oogenesis in *Xenopus* has been prepared by Dumont (1972), and we refer to this staging system in the following discussions. Dumont's stage 1 oocytes are previtellogenic and include zygotene, pachytene, and very early diplotene oocytes. According to Coggins and Gall (1972) zygotene and pachytene stages require about 23 days in young female *Xenopus*. A very interesting aspect of stage 1 *Xenopus* oocytes is that they are found in complexes of 16 cells connected by intercellular bridges (Coggins, 1973). The adjacent oocytes of each such nest develop synchronously. This type of structure is also known in the rabbit (Zamboni and Gondos, 1968) and in the early oocytes of other organisms (see, e.g., Fawcett *et al.*, 1959). In Dumont's classification vitellogenesis is divided into stages 2, 3, 4, and 5. Stage 2 of Dumont is the early lampbrush stage (equivalent to Duryee's stage 3, in Fig. 8.8). Pigment deposition

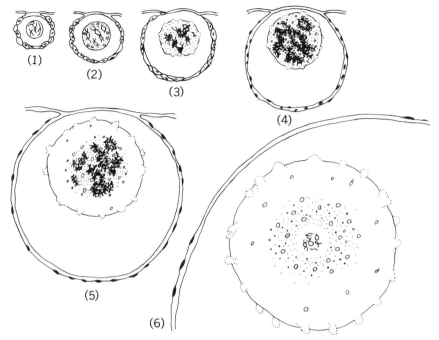

(1)

(2)

(3)

(4)

(5)

(6)

Fig. 8.8. Schematic diagram of nuclear growth stages during the later development of frog eggs. Modified from W. Duryee (1950). *Ann. N.Y. Acad. Sci.* **50**, 920.

begins in stage 3 of Dumont, which is the maximum lampbrush (i.e., stage 4 in Fig. 8.8). During stage 4 of Dumont the lampbrush structures are retracting and the oocyte grows rapidly, from 600 to 1000 μm in diameter. Dumont divides Duryee's stage 6 (Fig. 8.8) into two stages. By Dumont stage 5 the lampbrush chromosomes have been condensed, and during this stage yolk deposition is completed. Stage 6 of Dumont is the definitive mature oocyte of 1200–1300 μm diameter, marked by an equatorial white band which separates the animal and vegetal hemispheres. A section displaying the condensed lampbrush chromosomes of a stage 6 oocyte is reproduced in Fig. 8.9a (Dumont, 1972). These chromosomes may be compared to the extended lampbrush structures in a stage 3 oocyte photographed at the same magnification (Fig. 8.9b). Stage 6 oocytes may be held in the ovary for some time and may eventually become atretic and be resorbed (Dumont, 1972). About 45% of the stage 2–6 oocytes are at stage 2, and about equal quantities of each of the following stages were present in the animals studied by Dumont (1972). This suggests that the ovary contains a "reserve" of oocytes held at stage 2 from which groups of oocytes are selected to undergo further oogenesis.

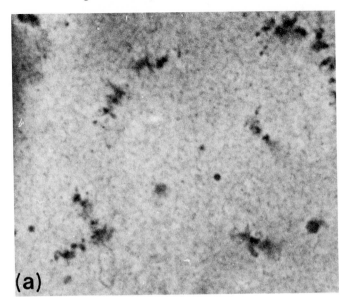

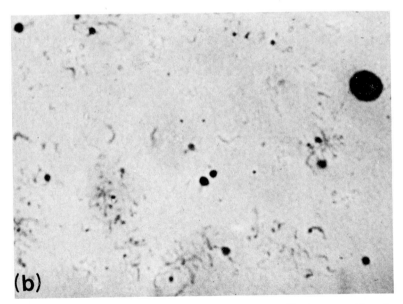

Fig. 8.9. Comparison of maximum lampbrush chromosomes (Dumont stage 3) and condensed lampbrush chromosomes (Dumont stage 6) from *Xenopus* oocytes. Phase contrast micrographs of fixed, sectioned materials (×1050). (a) Stage 6 oocyte nucleus showing condensed lampbrush chromosomes. (b) Stage 3 oocyte nucleus showing lampbrush chromosomes and associated small, dense nucleoli. From J. N. Dumont (1972). *J. Morphol.* **136,** 153.

Of particular interest for our present subject is the length of time the lampbrush chromosome stages persist. Unfortunately, little useful information is available. Davidson (1968) cited a duration of 3 months for the lampbrush stage in *Xenopus*. However, Scheer (1973) found that the duration of the lampbrush stage may be as short as about 1.3 months under conditions which might be expected to stimulate growth maximally. A measurement of the minimum lampbrush phase under laboratory conditions was obtained for a neotropic anuran, *Engystomops pustulosus* by Davidson and Hough (1969a). This species is a temporary water breeder able to shed several clutches of eggs during each rainy season. Under laboratory conditions which mimic the onset of the dry season, oogenesis occurs synchronously and a large clutch of oocytes enters the lampbrush phase of oogenesis together. The total time required for Dumont stages (2+3+4+5+6) is $2\frac{1}{2}$ to 3 months. Of this about 30 days are occupied with stages (4+5+6), meaning that the lampbrush stages (2+3) occupy $1\frac{1}{2}$ to 2 months in this species.

PHYLOGENETIC OCCURRENCE OF LAMPBRUSH CHROMOSOMES AND GENERAL CONCLUSIONS

In Table 8.1 data are collected regarding the distribution of lampbrush chromosomes in animals and where possible the duration of the lampbrush stage. It is clear that lampbrush chromosomes occur in animals of many major groups, both deuterostome and protostome (see Fig. 7.18). Thus, like the process of oogenesis itself, lampbrush chromosomes have an ancient evolutionary history antedating the appearance of coelomate organisms. Lampbrush chromosomes must play some fundamental role in oogenesis since they have been retained throughout most of metazoan evolution.

The list of organisms in which lampbrush chromosomes have been reported is of course limited by the choices made by investigators and the difficulty of observing lampbrush chromosomes in organisms with very small genomes. It is clear nonetheless that lampbrush chromosomes are not ubiquitous. This is shown by their absence in certain insect groups. Here the oocyte chromosomes remain visibly condensed throughout oogenesis. In these insects nurse cells play a dominant role in oogenesis as described in the following section of this chapter. The similarity in structure between the lampbrush chromosomes of the most distantly related creatures is remarkable. Examples are shown in Fig. 8.10 in which lampbrush chromosomes are displayed from an orthopteran insect, *Decticus albifrons* (Kunz, 1967b); a squid, *Sepia officinalis* (Ribbert and Kunz, 1969); a snail, *Bithynia tentaculata* (Bottke, 1973); and a starfish, *Echinaster sepositus* (Delobel, 1971). These may be compared with the

TABLE 8.1. Occurrence of Lampbrush Chromosomes in Oocytes and Duration of the Lampbrush Stage

Taxonomic affiliations of animals in which lampbrush chromosomes have been reported	Reference	Estimated duration of lampbrush stage where available	Reference
Deuterostomes			
Chaetognaths			
Arrow worm	Benoit (1930)		
Echinoderms			
Starfish	Delobel (1971)		
Sea urchins	Jörgenssen (1913); Davidson (1968)		
Chordates			
Cyclostomes	Okkelberg (1921)	Several weeks in lamprey *Petromyzon*	Lewis and McMillan (1965)
Elasmobranchs	Rückert (1892); Maréchal (1907)		
Teleosts	Rückert (1892); Maréchal (1907)		
Amphibians	Callan (1957); Duryee (1950)	Urodele about 7 months in *Triturus*	Callan (1963)
		Anurans ≥3 months in *Xenopus*	Davidson (1968)
		1.3 months in *Xenopus*	Scheer (1973)
		1–1.5 months in *Engystomops*	Davidson and Hough (1969a)
Reptiles	Loyez (1905)	Some months in lizards	Loyez (1905); Boyd (1941)
Birds	D'Hollander (1904); Romanoff (1960)	3 weeks in chick	D'Hollander (1904)
Mammals	Baker and Franchi (1966)	Perhaps years in man	Baker and Franchi (1966, 1967)
Protostomes			
Molluscs			
Gastropods	Davidson (1968); Bottke (1973)		
Cephalopods	Callan (1957); Ribbert and Kunz (1969)		
Insects			
Orthopterans	Kunz (1967a,b); Bier *et al.* (1969)	3 months in cricket	Ribbert and Bier (1969)

amphibian lampbrush chromosomes shown under similar conditions in Fig. 8.2. Delobel (1971) prepared a "map" of the individual bivalents in the nucleus of *Echinaster* oocytes and found that their detailed cytological features are similar to those observed in amphibian lampbrush chromosomes. The typical loops are 1–2 μm long in maximum lampbrush oocytes of *Echinaster sepositus*, and the largest loops are 15–20 μm. The genome size of this species is not known, but another starfish species of this genus, *Echinaster echinophorus*, has a genome size of 0.96 pg (Hinegardner, 1974). Thus, the loop dimensions are in accord with expectation. Lampbrush chromosomes were reported in sea urchin oocytes by Jörgenssen (1913) and by Davidson (1968). The duration of the lampbrush stage in echinoderm oocytes is not known. However, they are present in the small previtellogenesis oocytes of sea urchins between annual breeding cycles (Davidson, 1968). Diplotene chromosomes were observed in such oocytes many years ago by Tennent and Ito (1941), but their cytological form could not be distinguished with the methods then in use. Vitellogenesis in sea urchins is a relatively rapid process requiring only several weeks (analogous to Dumont stages 4–6 in amphibian oocytes). Thus, since the period between breeding seasons is many months long, it is likely that in sea urchins the smaller previtellogenesis oocytes bearing lampbrush chromosomes persist for a long time, as in the amphibia. However, further information on this point is clearly required. Except for the case of a cricket, where the lampbrush stage lasts for about 3 months (Bier, 1967; Ribbert and Bier, 1969), almost no data on the duration of the lampbrush stages in invertebrate oocytes are available. The remaining estimates of the duration of the lampbrush stages in Table 8.1 refer to chordates, and it can be seen that in each case the lampbrush phase is measured in terms of weeks or months, if not years.

Though other interpretations are of course possible, these data suggest that a prolonged period of lampbrush chromosome activity is required to complete the preparation of the oocyte. The leptotene, zygotene, pachytene, and diplotene stages can be regarded as essential to the mechanics of the meiotic process, but this is not so of the lampbrush structures themselves. The long duration of the lampbrush stage, the dense packing of polymerases in the lampbrush transcription units, the unusual ribonucleoprotein matrices on the loops, and the fact that there are four rather than two copies of the active sequences per nucleus all support the view that the lampbrush chromosomes are generating transcription products, some fraction of which are accumulated in the oocyte. The length of time taken by oogenesis is of obvious adaptive significance, and in many animals the lampbrush structures appear to be functional during a large fraction of this period. An extreme example is found in the tailed frog

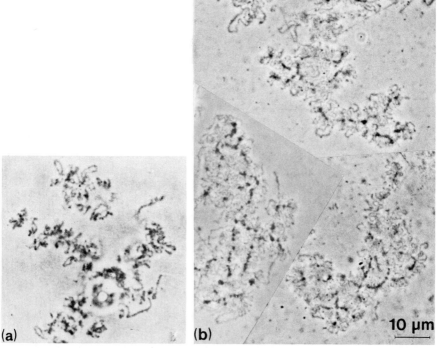

Fig. 8.10 (a) and (b). Lampbrush chromosomes of invertebrates. Chromosomes are isolated in salt solutions and photographed through the phase microscope. (a) Lampbrush chromosomes from the oocyte of an orthopteran insect, *Decticus albifrons*. From W. Kunz (1967b). *Chromosoma* **21**, 446. (b) Lampbrush chromosomes from the oocyte of a cephalopod mollusc, the squid *Sepia officinalis*. From D. Ribbert and W. Kunz (1969). *Chromosoma* **28**, 93.

Ascaphus (MacGregor and Kezer, 1970). In this organism the oocyte contains eight nuclei rather than one lampbrush stage nucleus. At the end of oogenesis seven of these disappear. MacGregor and Kezer (1970) pointed out that during the whole of its growth phase this oocyte has no less than 352 lampbrush chromosomes, containing 32 copies of each active locus.

Meroistic Oogenesis in Insects and the Role of Germ-Line Gene Expression

ACCESSORY CELL FUNCTIONS IN MEROISTIC OOGENESIS

In meroistic oogenesis oocyte lampbrush chromosomes are absent or are only very slightly developed. Their role seems to be taken over by the

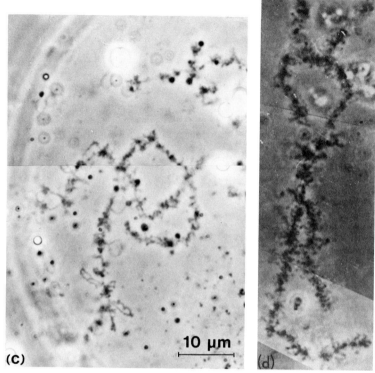

Fig. 8.10 (c) and (d). (c) Lampbrush chromosomes from the oocyte of a gastropod mollusc, *Bithynia tentaculata*. From W. Bottke (1973). *Chromosoma* **42**, 175. (d) Lampbrush chromosomes from the oocyte of an echinoderm, the starfish *Echinaster sepositus*. From N. Delobel (1971). *Ann. Embryol. Morphol.* **4**, 383.

nurse cell nuclei. This form of oogenesis is confined to certain holometabolous insect orders, the most prominent of which are Lepidoptera, Diptera, and Coleoptera, and may exist in other protostome phyla as well. As noted in the previous section, however, nurse cells are absent and lampbrush chromosomes are functional in the oocytes of other insect orders, such as the Orthoptera. The same end result is of course required of both types of oogenesis, viz., the production of an oocyte. It is illuminating to consider meroistic oogenesis from the point of view that the nurse cells must carry out many of the same functions for meroistic oocytes as do the lampbrush chromosomes in organisms lacking nurse cells.

In meroistic oogenesis the oocyte is fed through large open channels which link it with the nurse cells. These are always descendants of the same oogonial stem cell as has given rise to the oocyte. Two types of meroistic oogenesis are diagrammed in Fig. 8.11a and b (Bier, 1967).

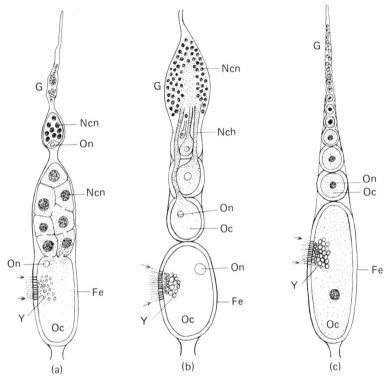

Fig. 8.11. Diagram of the three types of insect ovary. Polytrophic and telotrophic meroistic oogenesis are portrayed in (a) and (b), respectively, and panoistic oogenesis (i.e., with lampbrush chromosomes) is portrayed in (c). In polytrophic meroistic oogenesis the oocyte is fed directly by the nurse cells via individual junctions linking the nurse cells to each other and to the oocyte. In telotrophic meroistic oogenesis the nurse cells communicate with the oocyte via a common "nutritive cord." Cell nuclei which synthesize RNA are shown in black, and those which are inactive are shown as open circles. Concentrations of labeled RNA in the cytoplasm are represented by fine black dots. Yolk proteins (Y), derived originally from the blood, are pictured entering the oocyte (Oc) via the follicular epithelium (Fe). Ncn, nurse cell nucleus; Nch, nutritive chord; G, germarium; On, oocyte nucleus. After K. Bier (1967). *Naturwissenschaften* **54**, 189.

Oogenesis of the lampbrush chromosome type, termed "panoistic" oogenesis in insects, is illustrated in Fig. 8.11c. Nurse cells are absent in panoistic oogenesis. Follicle cells play a role in both meroistic and panoistic oogenesis, however. Figure 8.11 shows that in all three forms of insect oogenesis the follicle cells are involved in the uptake of yolk protein from external medium. Follicle cells are nearly ubiquitous accessories of oogenesis in metazoa. They are involved in yolk transport in the oocytes of many animals other than insects [earlier references are reviewed by Telfer (1965) and Davidson (1968); for the most extensive current studies on yolk protein incorporation in an amphibian, see, e.g., Wallace *et al.*

(1972) and Bergink and Wallace (1974)]. Follicle cells also synthesize and secrete various components such as chorion proteins (e.g., Paul *et al.*, 1972). Follicle cell function lies outside the scope of this essay, except to make the point that these cells are to be clearly distinguished from nurse cells. They do not contribute nucleic acids to the oocyte, and unlike nurse cells they are not of germ-line origin. The condensed chromosomes of meroistic insect oocytes either do not synthesize RNA during oocyte growth, or synthesize it at a very low rate. Instead, as indicated in Fig. 8.11a and b, the oocyte RNA is synthesized in the nurse cells and is transported into the oocyte via the cytoplasmic bridges linking the latter with the nurse cells. We now review briefly the ontogeny of the nurse cell–oocyte complex, and consider some of the evidence relating to the nurse cell functions suggested in Fig. 8.11.

Nurse cells are fairly common among the protostomes, though little molecular evidence exists as to their function except in insects. The role of nurse cells in feeding the oocyte was remarked upon by classical writers [see, e.g., Wilson (1925) for references]. Early observers claimed that in several species mitochondria pass from nurse cells to oocyte. An interesting variation exists in turbellarian flatworms, where the nurse cells, filled with yolk, are encapsulated in a cocoon along with the oocytes *after* oogenesis is completed, and the nurse cell contents are used to sustain the growth of the embryos just as is *intracellular* yolk in other eggs. This unusual course of events draws attention to the essential aspect of nurse cell function, that of providing the oocyte with materials it will require for development. Wherever nurse cells are found, the physiological nature of the oocyte–nurse cell interaction is evident. In certain annelids, for example, one or two nurse cells with large polyploid nuclei are applied to each oocyte, and the oocyte–nurse cell complex is released into the lumen of the ovary relatively early in oogenesis. Oocyte growth then occurs at the expense of the nurse cells, which shrink progressively until they become small compared to the relatively enormous oocytes. A general survey of accessory cell–oocyte arrangements in invertebrate oogenesis is given by Raven (1961).

STRUCTURE AND ORIGIN OF NURSE CELL–OOCYTE COMPLEXES IN HOLOMETABOLOUS INSECTS

In Fig. 8.12a nurse cell–oocyte and nurse cell–nurse cell junctions in the *Drosophila* germarium are shown as they appear in the light microscope (Koch *et al.*, 1967), and in Fig. 8.12b an electron micrograph of a junction between the oocyte and a nurse cell is reproduced (Brown and King, 1964). The process by which the polytrophic egg chamber is formed was worked out for *Drosophila* by Koch *et al.* (1967), for the moth *Hyalophora cecropia* by King and Aggarwal (1965), and for the wasp

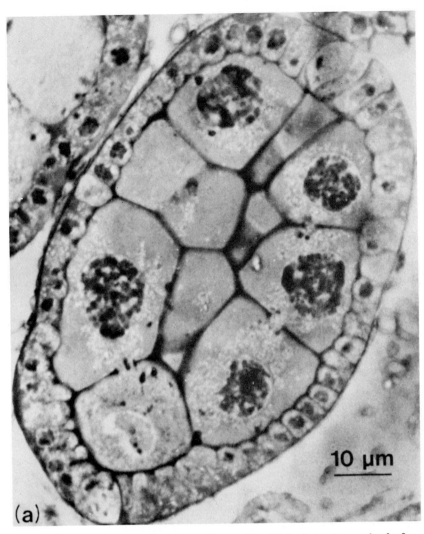

Fig. 8.12 (a). Polytrophic egg chambers of *Drosophila*. (a) A photomicrograph of a 2 μm
section (stained by the periodic acid–Schiff procedure) through a stage 6 *Drosophila*
egg chamber. At this stage a single layer of cuboidal follicle cells surrounds the oocyte,
which occupies a position at the lower left corner of the chamber, and the 15 nurse cells,
9 of which can be seen in this section. Three ring canals are evident, one connecting the
oocyte with a nurse cell and two interconnecting nurse cells. From E. A. Koch, P. A. Smith,
and R. C. King (1967). *J. Morphol*. **121**, 55.

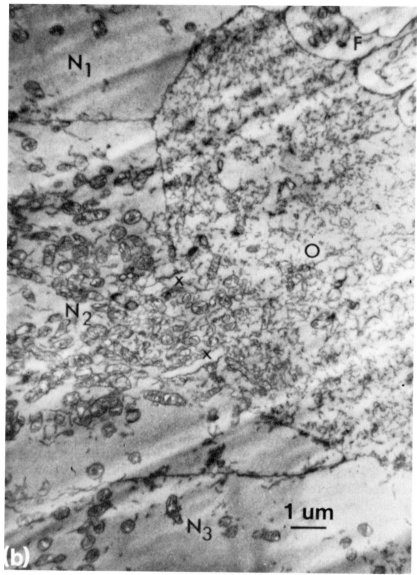

Fig. 8.12 (b). An electron micrograph of a section through a ring canal connecting a nurse cell (N_2) to the oocyte (K).Two other nurse cells (N_1 and N_3) and portions of several follicle cells (F) are evident. An electron-pale material ($\times$) can be seen. Mitochondria which seem to have been fixed while entering the oocyte are evident within the ring canal. The higher concentration of particulate material in the oocyte suggests that materials contributed by the nurse cells are accumulating in the oocyte. ($KMnO_4$–OsO_4 fixation, embedded in Epon.) From E. H. Brown and R. C. King (1964). *Growth* **28**, 41.

Habobracon juglandis by Cassidy and King (1972), and this area is reviewed by King (1970). The oocyte–nurse cell complex in Hyalophora includes 7 nurse cells and in Drosophila 15 nurse cells. These complexes are constructed in the terminal three and four oogonial divisions, respectively. With the first of these divisions the future oocyte is clearly differentiated from the nurse cell. Each nurse cell is connected to other nurse cell and/or to the oocyte by the open cytoplasmic bridges shown in Fig. 8.11. These are termed "ring canals" (Brown and King, 1964) or "fusomes" (Bier, 1963). The ring canals are very highly organized membranous structures (Cassidy and King, 1972; Kinderman and King, 1973) which originate when the daughter cells are incompletely walled off after each oogonial mitosis. The disposition of the ring canals therefore indicates the order of appearance of the nurse cells and the sequence of steps by which the egg chamber is constructed. Reconstructions of this process as it occurs in Drosophila and in Hyalophora cecropia are shown in Fig. 8.13a (Koch et al., 1967) and Fig. 8.13b (King and Aggarwal, 1965). It can be seen that except for that nurse cell which is formed first, the oocyte is the cell with the largest number of intercellular ring canals, and it is significant that both the first nurse cell and the oocyte initially form synaptinemal chromosomal complexes. The synaptinemal complexes developing in the nurse cell nucleus later disappear, whereas in the oocyte the usual meiotic prophase movements continue. An interesting female sterile mutation, fes, has been studied by Johnson and King (1972) in which the incomplete cytokinesis responsible for egg chamber formation is disturbed. In Drosophila homozygous for fes cytokinesis is often complete rather than incomplete, with the result that large numbers of abnormal cell clusters containing less than 16 interconnected cells are formed. Johnson and King (1972) suggest that the signal which normally stops further division at the 16-cell stage emanates from the differentiating oocyte, and that this signal diffuses to the other cells through the ring canals. This investigation shows the importance of the special form of cytokinesis involved in the construction of the meroistic egg chamber. Normally the interconnected oogonia all develop synchronously, dividing at the same time, and this form of coordination is absent in fes mutants. It is interesting to consider the function of the interconnected oogonia observed in other organisms, such as mouse and Xenopus, in light of this analysis of ring canal function. An ancient evolutionary origin is suggested for this aspect of oogenesis. The oogonial interconnections probably provide means of controlling divisions (Coggins, 1973), but conceivably some of the interconnected cells also provide macromolecular constituents to the one which will become the oocyte. Thus, it is possible that even oogenesis based on lampbrush chromosome function may involve cooperative behavior of cells in ways which resemble insect nurse cell function.

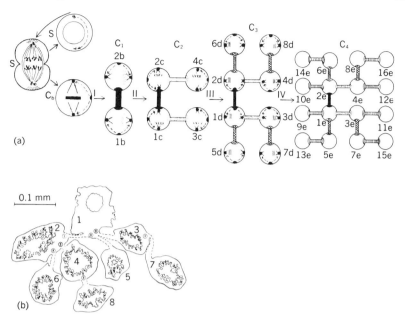

Fig. 8.13. Diagrammatic models showing steps in the production of polytrophic oocyte-nurse cell complexes. (a) The 16 interconnected cystocytes in *Drosophila*. In this drawing the cells are represented by circles lying in a single plane, and the ring canals have been lengthened for clarity. The area of each circle is proportional to the volume of the cell. The stem cell (S) divides into two daughters, one of which behaves like its parent. The other differentiates into a cystoblast (C_b) which by a series of four divisions (I–IV) produces 16 interconnected cystocytes: C_1, first; C_2, second; C_3, third; and C_4, fourth generation cystocyte. The original germ-line stem cell is shown at early anaphase. Each parent–daughter pair of centrioles is attached to the plasma membrane by astral rays. The daughter stem cell receives one pair of centrioles. One remains in place while the other moves to the opposite pole. This movement is represented by the broken arrows. In the daughter cystoblast and all cystocytes the initial position of the original centriole pair is represented by a solid half-circle, whereas their final positions are represented by solid circles. The position of the future cleavage furrow is drawn as a strip of defined texture. The canal derived from the furrow is coded similarly. The future oocyte is cell 1. From E. A. Koch, P. A. Smith, and R. C. King (1967). *J. Morphol.* **121**, 55. (b) A diagram showing the way in which the 8 cells of a stage 3 egg chamber of *Hyalophora cecropia* are interconnected by 7 canals. Each cell is traced from a magnified image of a section passing through its center, and its nucleus is outlined. Since the cells are represented as lying in one plane, the canals have been lengthened. Note the large, deeply crenelated polytene nuclei of the nurse cells. The number of the division at which each ring canal is formed is given in the numbers in the small circles. From R. C. King and S. K. Aggarwal (1965). *Growth* **29**, 17.

SYNTHESIS OF OOCYTE RNA IN NURSE CELLS

In Fig. 8.12b mitochondria can be observed densely packed in the ring canal, justifying the earlier claims based on light microscopy. If mitochondria can pass between nurse cell and oocyte, macromolecules could obviously do so as well, as suggested in Fig. 8.11. Radioautographic evidence that RNA synthesized in the nurse cells is fed into the oocyte is shown in Fig. 8.14 (Bier, 1963). Here newly synthesized RNA can be seen localized over the polytene nurse cell nuclei after a 30-minute labeling period (Fig. 8.14a). Five hours later (Fig. 8.14b) the labeled RNA has moved into the nurse cell cytoplasm and is apparently pouring through a ring canal into the cytoplasm of the oocyte. Note that no RNA synthesis can be observed over the oocyte at 30 minutes, even though the film is clearly overexposed with respect to the amount of incorporation in the nurse cell nuclei. This cannot be due to inavailability of precursor, considering the open channels between nurse cell and oocyte. Nor is it a reflection of the fact that the oocyte nucleus contains only the 4C amount of DNA, while the nurse cell nuclei are polytene and contain a quantity of DNA which is several hundred times greater. Labeled RNA can be seen

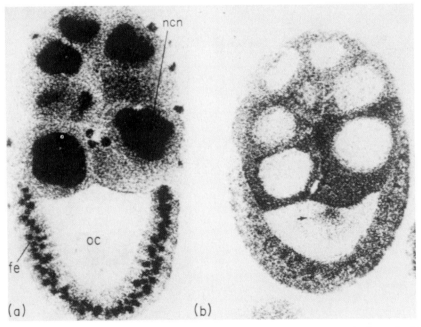

Fig. 8.14. Synthesis of RNA in nurse cell nuclei and transfer to the oocyte in the housefly. (a) Radioautograph of an oocyte (OC), its nurse cells, and follicular epithelium (fe) of *Musca domestica* incubated for 30 minutes with ^{3}H-cytidine. (b) The same, 5 hours later. Labeled RNA can be seen entering the oocyte from an adjacent nurse cell (arrow). The densely labeled RNA originally present in the nurse cell nuclei (ncn) in (a) is now mainly localized in the nurse cell cytoplasm. From K. Bier (1963). *J. Cell Biol.* **16**, 436.

clearly over the follicle cell nuclei at 30 minutes (Fig. 8.14a), and these are diploid or at least not highly polyploid. Results similar to those shown in Fig. 8.14 were also reported by Bier (1965) and Bier *et al.* (1969) for various other dipteran egg chambers. In some Coleoptera, however, where the oocyte chromosomes are somewhat less condensed, a small amount of labeling is noted over the oocyte nuclei (Bier *et al.*, 1969). The amount of RNA synthesis in the oocyte nuclei remains quantitatively trivial compared to the nurse cell activity, though of course it might be of great importance qualitatively.

Little is known about the species of RNA transferred from nurse cells to oocytes, except that most of this RNA is ribosomal. Pollack and Telfer (1969) concluded that the ribosomal RNA of the large oocytes of the moth *Hyalophora cecropia* derives from nurse cells. These oocytes contain 3 μg of RNA, most of which is ribosomal. The ribosomal RNA of *Drosophila* oocytes also is synthesized in nurse cells (Klug *et al.*, 1970; Dapples and King, 1970). In ovarioles of the moth *Antheraea polyphemus* Hughes and Berry (1970) demonstrated that ribosomes are transferred from the nurse cells into the oocyte. Oocytes in intact ovarioles which had been incubated for 6 hours following a 1-hour labeling period contained labeled ribosomes. If the nurse cells were removed after labeling but prior to the 6-hour incubation period, however, no labeled ribosomes appeared in the oocytes. In *Calliphora*, it has been shown that most ribosomal RNA synthesis occurs on extrachromosomal nucleoli in the nurse cells. Ribbert and Bier (1969) calculated that these nucleoli are responsible for at least 72% of the total nurse cell RNA synthesis. However, this function does not require significant amplification of the ribosomal gene sets beyond the complete genomic multiplication occurring when the nurse cells become polytene. Renkawitz and Kunz (1975) showed that the fraction of mature ovariole DNA which is ribosomal in *Calliphora* is only 1.35 times that measured in diploid brain cells. In two dipteran species, *Drosophila hydei* and *Sarcophaga barbata*, the ribosomal DNA is actually underreplicated by about a factor of two during nurse cell polytenization (Renkawitz and Kunz, 1975). Nor does ribosomal DNA amplification occur in *Oncopeltus fasciatus*, the milkweed bug (Cave, 1975). In this organism, which carries out oogenesis in a telotrophic meroistic manner similar to that illustrated in Fig. 8.11b, the fraction of DNA which is ribosomal is the same in nurse cells as in somatic cells.

Meroistic oogenesis is very rapid, compared to panoistic oogenesis in insects or to the lampbrush chromosome type of oogenesis in other organisms. Ribbert and Bier (1969) pointed out that it takes 100 days to carry out oogenesis in the cricket *Acheta domestica*, in which oogenesis is of the panoistic type, and only 6 days in *Calliphora*. In *Drosophila* the total period between premeiotic DNA replication and the production of mature oocytes is about 8 days (Grell and Chandley, 1965). On the other hand the nurse cells in the *Drosophila* egg chamber are 1024-fold polytene (Klug *et al.*, 1970), and in *Calliphora* the nurse cells are 256-fold polytene (Ribbert and Bier, 1969). There are 15 nurse cells feeding each oocyte in *Drosophila*, and thus several thousandfold more genomic DNA is active in the preparation of oocyte RNA in a meroistic oocyte than in the panoistic oocyte. This is the probable explanation for the short duration of meroistic oogenesis and also indicates its great adaptive value. Evidently oogenesis involves the accumulation of transcription products, the synthesis of which is accomplished in only a few days through the cooperative effort of hundreds of genomes in meroistic oogenesis. Alternatively, if this synthesis must be carried out on the 4C oocyte genome alone, weeks or months are required, as in the panoistic oogenesis of orthopteran insects and similar types of oogenesis in other animals. The significance of the fact that in meroistic oogenesis the nurse cells are of germ-line origin remains obscure. One possible implication would be that a specific set of genomic loci is required for oogenesis and that their expression occurs early in germ line ontogeny.

CHROMOSOMAL ELEMENTS CONTAINING DNA USED ONLY FOR GAMETOGENESIS

One component of the genome which may be specially required for meiotic pairing in germ-line cells is satellite DNA (Walker, 1971; Moritz and Roth, 1976). This is suggested by some observations regarding chromosome diminution in the nematode *Ascaris* (discussed in Chapter 7). As shown in Fig. 7.14 in *Ascaris* all somatic cells, but not germ-line cells, undergo diminution early in cleavage. Moritz and Roth (1976) showed that germ-line (spermatocyte and sperm) DNA includes satellite sequences, while the somatic DNA does not. These satellites are located in heterochromatic regions of the chromosomes. They are easily detectable by equilibrium banding in isopycnic CsCl gradients and also by renaturation kinetic analyses, since they represent as much as 85% of the total germ line DNA in *Parascaris equorum* and about 22% in *Ascaris lumbricoides*. The complexity of the single copy sequences is about the same in the germ line and the somatic DNA's. In other words the DNA extruded during chromosome diminution in somatic cells appears to consist largely of satellite sequences required only in germ-line cells. As first shown by Boveri, this DNA is essential for germ cell differentiation, and a role in the organization of meiotic chromosomal structures therefore seems possible for the germ-line satellite sequences.

Chromosome diminution also occurs in sciarid insects. Here again chromosomes are eliminated early in cleavage from somatic cells but not from germ cells. However, in these organisms chromosome elimination occurs later in the germ line as well. Germ-line diminution takes place after pole cell division is complete and the germ cells have migrated into the gonad, but before the germ cells enter meiotic prophase (Berry, 1941). Thus, if the last eliminated chromosomal DNA is utilized at all it must be in the course of the premeiotic development of the germ line, rather than during meiotic pairing.

Another group in which chromosomal diminution occurs during germ-line differentiation is the gall midges (Cecidomyidae). Here again the somatic stem cells are marked by elimination of chromosomes early in cleavage. As a result they retain only 6–12 chromosomes, this number depending on the sex of the organisms and on the species, while the germ-line cells retain the full complement of over 40 chromosomes. The function of the discarded portion of the genome has been studied experimentally in *Wachtliella persicariae* by Geyer-Duszyńska (1966) and Kunz *et al.* (1970) and in *Mayetiola destructor* by Bantock (1970). Germ-line stem cells which have undergone diminution in the same way as somatic cells can be produced in *Wachtliella* and *Mayetiola* by ligaturing or centrifuging the embryo in such a way as to prevent the descent of the nuclei into the germ cell determinant cytoplasm. When the ligature is removed after chromosome elimination has taken place, nuclei containing 8 rather than 40 chromosomes move into the presumptive germ cell region and give rise to the presumptive germ-line stem cells. Both Geyer-Duszyńska (1966) and Bantock (1970) showed that females developing from ligatured eggs, while normal in other aspects, are unable to carry out oogenesis. The fault appears to lie early in the process of oogenesis in the development of the premeiotic meroistic ovarian structures. Thus, the ovaries of females whose germ cells contain only the 8 somatic chromosomes lack both oocytes and nurse cells. It is interesting that at least in *Wachtliella* the structure of the reproductive system in males developing from ligatured eggs is more normal, and such males even produce sperm. During pupation, however, the larval sperm in these males degenerates. These experiments show directly that the portion of the genome normally confined physically to the germ-line cells contains genetic information needed in oogenesis. The peculiarity of these organisms is that the genetic elements bearing this information are packaged together. Kunz *et al.* (1970) showed that the special germ-line chromosomes are present in an extended form throughout oogenesis in *Wachtliella*. Throughout this period they synthesize RNA in the oocyte nucleus. Though oogenesis in these dipterans is meroistic, and most of the RNA derives from the nurse cells, this example proves that some necessary constituents come from gene activity in the oocyte nuclei as well. In any case these examples demonstrate the existence of special subsets of the genome which are needed only in oocytes and oogonia. In all probability these are not satellite DNA since the DNA specific to the germ cells is not confined to heterochromatic regions and is actively transcribed during oogenesis.

ANALOGIES BETWEEN OOCYTE LAMPBRUSH CHROMOSOME LOOPS AND LAMPBRUSH LOOPS IN *DROSOPHILA* SPERMATOCYTES

An interesting example relevant to the possible functions of oocyte lampbrush chromosomes is to be found in *Drosophila* spermatocytes. These cells contain a small number of lampbrush-like loops, probably about 6 in *Drosophila hydei*. These have been mapped by cytogenetic methods to specific loci on the Y chromosome (Hess, 1966). Several differences exist between the lampbrush structure of these chromosomes and those of oocytes, including the extremely small number of loops, compared to oocyte lampbrush chromosomes, and the fact that the loops do not originate from a common paired axis (Hennig, 1967). Nonetheless, they may be homologous in function as suggested by their ribonucleoprotein matrices. Electron micrographs show that the transcription units on the Y chromosome loops are at least 3×10^4 nucleotides in length, and probably more (Hennig *et al.*, 1974), just as in amphibian lampbrush loops. Also reminiscent of the latter, several transcription units separated by nontranscribed "spacers" have been visualized on single loops. In many regions, though not all, the transcripts are less densely packed along the axis of the transcription unit than in amphibian lampbrush loops. The presence of the Y chromosome loops is essential to the production of viable sperm. Thus, deficiencies in the Y chromosome involving individual loops produce specific defects in spermiogenesis (Hess, 1965, 1966; Meyer, 1968). Since after meiosis only one-half of the spermatids possess a Y chromosome, genetic information in the Y chromosome loops must be expressed during the spermatocyte stage, though its effect is seen only during spermatid maturation. Protein synthesis occurs in spermatids, but RNA synthesis does not. Thus, spermatid protein synthesis requires stored messenger RNA's. Hennig (1968) showed that RNA could be recovered from spermatids which hybridize with Y chromosome DNA in a filter system. Whether this RNA was messenger RNA is of course unknown. However, it is clear that messenger RNA derived from synthesis on the Y chromosome loops in spermatocytes is present in postmeiotic spermatids. This has been shown in two ways. For one thing temperature sensitive male sterile mutants have been found which at nonpermissive temperatures develop the same sperm defects as caused by Y chromosome deficiencies. More direct evidence is the identification of particular proteins synthesized only in postmeiotic stages with particular Y chromosome loops (Hennig *et al.*, 1974). In addition interspecific hybrids have been constructed which display both parental varieties of one particular loop and both parental varieties of at least one postmeiotic protein. One may conclude, as did Hess (1966), that the spermatocyte lampbrush loops include among their functions the accumulation of various species of stored messenger RNA's required for later sperm cell development. It is to be noted that in contrast to oogenesis in amphibians, the time between synthesis on the loops and utilization is in this case a matter of only a few days.

Synthesis of Ribosomal and Transfer RNA's in the Oocyte Nucleus

All of the classes of RNA which are synthesized in somatic cells are synthesized in oocytes, including mitochondrial RNA, transfer RNA, ribosomal RNA's, and complex RNA's of heterogeneous sequence. Synthesis of mitochondrial RNA in mature amphibian oocytes (Dawid, 1972; Webb *et al.*, 1975) was discussed briefly in Chapter 5 and is not further considered here. Synthesis and storage of the stable low molecular weight RNA's is interesting from our present point of view as a possible model for more complex heterogeneous RNA species, and we begin with a summary of current information on this subject.

LOW MOLECULAR WEIGHT STABLE RNA's

Transfer RNA and 5 S RNA are synthesized during oogenesis and stored for use during embryogenesis (see Chapter 4). Both species are coded by highly repetitive genes. There are about 50,000 copies of the 5 S RNA genes in the *Xenopus* genome (Brown and Weber, 1968) and an average of 200 copies of the genes for each transfer RNA (Clarkson *et al*, 1973a). These may exist in isocoding clusters (Clarkson *et al*, 1973b). Unlike the 18 S and 28 S ribosomal RNA genes, neither transfer RNA genes nor the 5 S RNA genes are amplified in oocytes (Wegnez and Denis, 1972). Transfer RNA and 5 S RNA are the major transcription product of very young *Xenopus* oocytes (Dumont stages 1–2). According to Thomas (1974), after 24 hours of labeling about 24% of the radioactivity in the cytoplasmic RNA of these oocytes is in transfer RNA and 39% is in 5 S RNA. Ford (1971) reported that in oocytes of this stage the molar ratio of newly synthesized transfer to ribosomal RNA is as high as 25, while that for 5 S RNA is over 100. Both transfer RNA and 5 S RNA continue to be synthesized throughout oogenesis. The previtellogenic oocytes are unusual in that these stable low molecular weight RNA's constitute a major fraction of their total accumulated RNA. This condition lasts until the onset of the massive ribosomal RNA synthesis which occurs after the beginning of Dumont stage 2. Both 5 S and transfer RNA are stored in a 42 S ribonucleoprotein particle present in great abundance in previtellogenic *Xenopus* oocytes (Ford, 1971; Denis and Mairy, 1972). The same seems to be true of the previtellogenesis oocytes of various teleosts (Mazabraud *et al.*, 1975). When the oocytes begin vitellogenesis these particles release the 5 S RNA, which is then incorporated into the oocyte ribosomes (Mairy and Denis, 1972).

Differences exist in the chromatographic behavior of the transfer RNA's present in *Xenopus* oocytes and those present in somatic cells (Denis *et al.*, 1975). Whether these differences stem from modification of the transfer RNA's or distinctions between the sets of genes in oocytes and those used in somatic cells is not known. However, the 5 S RNA of *Xenopus* oocytes has been sequenced and has been found to differ from somatic cell 5 S RNA by about 6 of the 120 or so nucleotides in this molecule (Denis *et al.*, 1972; Wegnez *et al.*, 1972; Denis and Wegnez, 1973; Ford and Southern, 1973; Brownlee *et al.*, 1974). Furthermore, several nonidentical sets of 5 S RNA genes are active in oocytes. The finding that special sets of "oogenesis genes" exist among the 5 S gene sequences is reminiscent of the examples considered in the last section in which a special portion of the genome is set aside for use in oogenesis.

AMPLIFICATION OF GENES FOR 18 S AND 28 S RIBOSOMAL RNA's

As is now well known in many organisms, the ribosomal RNA genes are amplified during early oogenesis. This phenomenon has been extensively reviewed elsewhere (e.g., see Brown and Dawid, 1968; Davidson, 1968; Gall, 1969; Hourcade *et al.*, 1974), and the more important facts can only be summarized here. Replication of the ribosomal DNA occurs mainly at the pachytene stage in amphibian oocytes (Gall, 1968; Pardue and Gall, 1969; Van Gansen and Schramm, 1974), though the replication process begins even earlier in premeiotic oogonia (Kalt and Gall, 1974). Similar observations have been made in a variety of phyletic groups, including some of the most distantly related. An example is the cricket, where amplification takes place during the pachytene stage (Cave, 1973). The extrachromosomal nucleolar DNA functional during oogenesis in both amphibians and insect oocytes is circular in form (Peacock, 1965; Miller, 1966; Lane, 1967; Hourcade *et al.*, 1973; Gall and Rochaix, 1974). It has been shown that the original copies used for ribosomal DNA replication are of chromosomal origin (Brown and Blackler, 1972), but the bulk of this DNA is synthesized by rolling circle replication of preexistent extrachromosomal ribosomal DNA (Hourcade *et al.*, 1973, 1974; Bird *et al.*, 1973; Rochaix *et al.*, 1974). The ribosomal DNA of the extrachromosomal nucleoli is somewhat heterogeneous in that the spacers differ detectably in length (Wellauer *et al.*, 1975).

In *Xenopus* the end result of ribosomal DNA amplification is the production of about 1500 extrachromosomal nucleoli, each containing several sets of ribosomal DNA genes. The total number of these gene sets is 3×10^3 to 5×10^3 (Perkowska *et al.*, 1968; Brown and Dawid, 1968; Gall, 1969). Since each haploid ribosomal gene set includes about 450 copies of the individual ribosomal RNA genes (Brown and Weber, 1968), the total number of these genes in the oocyte is 1.5×10^6 to 2.5×10^6, and their mass is about 30 pg per nucleus (Perkowska *et al.*, 1968).

SYNTHESIS RATES FOR RIBOSOMAL RNA IN AMPHIBIAN OOCYTES

In the earliest previtellogenic diplotene stages there is little ribosomal RNA synthesis, though extrachromosomal replication of the ribosomal DNA is by now complete. Scheer *et al.* (1976) showed that in the newt *Triton alpestris* previtellogenic oocytes (equivalent to Dumont stage 1 in *Xenopus*) synthesize RNA at only about 0.01–0.5% of the rate measured in vitellogenic oocytes. Electron micrographs show that this relatively low synthetic rate is correlated with the sparse packing of transcripts in the extrachromosomal nucleolar gene regions. In the previtellogenic oocytes the density of transcripts per ribosomal gene region is less than 3% of that in the nucleoli of midvitellogenic oocytes, and some gene regions appear totally inactive. In contrast, 90–95% of the nucleolar ribosomal genes are being transcribed in midvitellogenic oocytes, and the transcripts visible on these are tightly packed, with as many as 130 transcripts per gene region (Scheer *et al.*, 1976). Figure 8.15 displays comparable transcription units in the nucleolar ribosomal DNA of midvitellogenic *Triturus* oocytes (Miller and Beatty, 1969a). Structures exactly the same as those shown in Fig. 8.15 are found in *Triton* oocytes (Scheer *et al.*, 1976) and in *Xenopus* oocytes (Miller and Beatty, 1969b), where transcription of nucleolar genes occurs at high rates after Dumont stage 2. Except for variations in matrix length, structures such as those reproduced in Fig. 8.15 have been observed by other workers in a variety of nonamphibian oocytes as well.

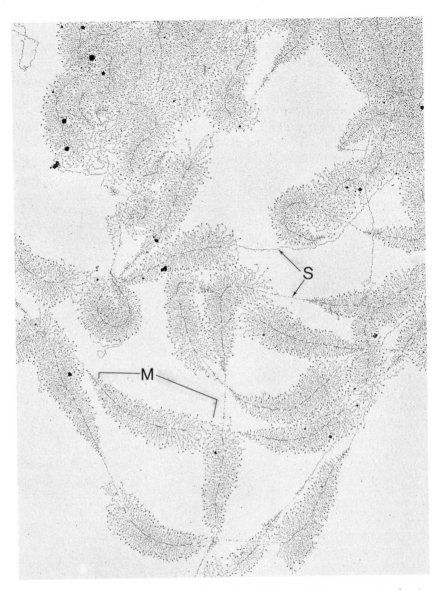

Fig. 8.15. Portion of nucleolar core isolated from *Triturus viridescens* oocyte showing ribosomal RNA transcription units. Matrix-covered axis segments (M) are separated by matrix-free axis segments (S). Matrix units are ~2.5 μm long and the matrix-free segments range from ⅓ to 10 times or more the length of the matrix unit. There are ~100 fibrils of increasing lengths attached to each matrix unit axis. Nucleoli were isolated into distilled water at pH 8.5, centrifuged through 0.1 M sucrose + 10% formalin (pH 8.5) onto carbon-coated grids, rinsed in 0.4% Kodak Photo-flo, dried, and stained in 1% phosphotungstic acid in 50% ethanol. ×15,125. From O. L. Miller Jr. and B. R. Beatty (1969a). *J. Cell Physiol.* 74, *Suppl.* 1, 225.

From the known length of the amphibian 40 S ribosomal RNA precursor and the observation that about 100 transcripts are present in each ribosomal transcription unit, the spacing between polymerases is calculated to be only about 100 nucleotide pairs (Miller and Beatty, 1969a–c). Thus, in *Xenopus* oocytes there would be about 2×10^8 polymerase molecules transcribing ribosomal RNA simultaneously, assuming that all ribosomal RNA genes are active as in *Triton* oocytes.

D. M. Anderson and L. D. Smith (personal communication) have measured the RNA "step time," i.e., the transcription rate per polymerase, at about 15 nucleotides sec^{-1} in stage 6 *Xenopus* oocytes. If this rate applies generally, the rate of ribosomal RNA synthesis in midvitellogenic oocytes would be 1.8×10^{11} nucleotides min^{-1}, or about 99 pg min^{-1} per oocyte if all the ribosomal genes are active. In comparison, Scheer (1973) measured a rate of *accumulation* of ribosomal RNA in growing *Xenopus* oocytes of about 76 pg min^{-1}. His experiments were carried out under conditions which stimulated maximum growth and may represent a synthesis rate which is close to the highest attainable. This synthesis rate and the others discussed here are collated in Table 8.2, where they can be compared with synthesis rates for various other classes of RNA.

Anderson and Smith (1976) carried out kinetic experiments on mature (stage 6) *Xenopus* oocytes in which the absolute synthesis rate for ribosomal RNA was measured on the basis of the rate of GTP incorporation and the GTP pool specific activity. These measurements yielded a rate of about 12 pg min^{-1}. Ribosomal RNA constitutes most of the stable RNA synthesized in midvitellogenic and mature oocytes as shown earlier by Davidson *et al.* (1964), Ford (1971), LaMarca *et al.* (1973), and Colman (1974), among others. The rate of total stable RNA synthesis in mature oocytes varies severalfold according to the individual females from which the oocytes are extracted, possibly depending on the hormonal state of the animal. Nonetheless, these data suggest that ribosomal RNA synthesis rates decrease between stage 3 and stage 6 in *Xenopus* (see Table 8.2), and this is clearly the case in *Triton alpestris* (Scheer *et al.*, 1976). In mature oocytes of the latter species the nucleolar genes contain only about 15% as many transcripts per ribosomal gene region as do growing oocytes, and the rate of synthesis is about 13% of that observed in midoogenesis. Thus, during oogenesis the rate of ribosomal RNA synthesis on the extrachromosomal nucleolar genes increases from a low initial level to a maximum at midvitellogenesis and then falls again at maturity. These changes appear to reflect large differences in the frequency of chain initiation and in the packing of active polymerases on the ribosomal genes.

The measurements of Anderson and Smith (1976) on stage 6 *Xenopus* oocytes, as well as other observations (La Marca *et al.*, 1973, 1975), show

TABLE 8.2. Calculated and Measured Rates of RNA Synthesis in *Xenopus* Oocytes

Species of RNA	Synthesis rate (pg min^{-1}) at oocyte stage[a]	
	3	6
Ribosomal RNA	99[b]	12[c]
	76[d]	—
Heterogeneous nuclear RNA $\geq$ 40 S with $t_{1/2}$ of 30 minutes	Existence not yet demonstrated	11[e]
Stored maternal poly(A) messenger (calculated maximum synthesis rate)	$\leq$0.8–1.4[f]	—
Lampbrush loop RNA	3[g]	—
4 S to 40 S heterogeneous RNA (relatively stable cytoplasmic component)	—	~1[h]
Total poly(A)RNA	1.3–2.8[i]	2.3–4.6[i]

[a] Oocyte stage according to Dumont (1972).

[b] Calculated from the number of polymerase molecules transcribing ribosomal RNA in the extrachromosomal nucleoli if all of the ribosomal genes are active, or about 2×10^8 (see text) and the step time for transcription, about 15 nucleotides sec^{-1} per polymerase (D. M. Anderson and L. D. Smith, personal communication).

[c] Data from Anderson and Smith (1976). These authors report about 0.85 pmoles GMP hr^{-1} incorporated in ribosomal precursor of which about 0.65 pmoles GMP hr^{-1} is incorporated in the mature ribosomal RNA's of stage 6 oocytes. This is 12 pg min^{-1} assuming 30% guanosine (see Fig. 8.17b).

[d] Calculated from data of Scheer (1973), who observed 3.8 μg of rRNA to be synthesized in 38 days in rapidly growing *Xenopus* oocytes. Scheer (1973) thus calculated that the average rate of synthesis of rRNA precursor molecules is 2.3×10^9 nucleotides sec^{-1}. This is 76 pg min^{-1}.

[e] Calculated from data of Anderson and Smith (1976). These investigators report that about 2.11 pmoles GMP hr^{-1} is incorporated in RNA by stage 6 oocytes, and of this 1.26 pmoles GMP hr^{-1} is incorporated in heterogeneous RNA's, the remainder being ribosomal precursor. Approximately 35% of the GMP incorporated into heterogeneous RNA is in >40 S RNA which turns over rapidly. Thus, about 0.45 pmole GMP hr^{-1} is incorporated in this RNA (see Fig. 8.17a). This is equivalent to 11 pg min^{-1} per nucleus, assuming 25% GMP in the RNA. The rest of the newly synthesized heterogeneous RNA is 4 S to 40 S RNA, some of which appears to decay more slowly (see footnote h and Fig. 8.17b). Since some of the 4 S to 40 S heterogeneous RNA may also belong to the rapid turnover class, the rate of synthesis of the latter is probably greater than 11 pg min^{-1}.

[f] Calculated on the basis that synthesis of the poly(A)RNA message stockpile *requires* $\geq$35 days (see text) and that the total maternal message mass is $\geq$40-70 ng (Table 4.2 and 6.1). Thus the synthesis rate is

$$\frac{4 \times 10^4 \text{ pg}}{\geq(35 \times 60 \times 24)\text{min}} \leq 0.8 \text{ pg min}^{-1}$$

[g] Calculated from the number of loops per nucleus, about 2×10^4, and the assumptions that in *Triturus* each contains about 2×10^5 nucleotide pairs in transcription units (see text). If the polymerases are separated by 100 nucleotides and transcribe at 15 nucleotides sec^{-1} per polymerase (D. M. Anderson and L. D. Smith, personal communication), the

chromosomal rate of RNA synthesis would be 3.6×10^{10} nucleotides or 20 pg min^{-1} in *Triturus*. Since the lampbrush chromosomes seem to be proportionally smaller in *Xenopus*, whose genome size is about $\frac{1}{7}$ that of *Triturus*, it is assumed for this rough calculation that the synthesis rate would be about 20/7 or about 3 pg min^{-1} in *Xenopus* lampbrush chromosomes.

[h] Calculated from data of Anderson and Smith (1976). The rate of synthesis of this set of components is difficult to estimate, since the 4 S to 40 S heterogeneous RNA includes kinetic fractions which turn over slowly, perhaps at several rates, as well as rapidly decaying species. The rate given is a conservative estimate for RNA's whose half-life is >4 hours, but which may not be completely stable. This RNA is indicated by the dashed line in Fig. 8.17b.

[i] Calculated from data of G. J. Dolecki and L. D. Smith (personal communication) who report for Dumont stage 3 *Xenopus* oocytes absolute synthesis rates of 0.055–0.12 pmole GMP hr^{-1} and for stage 6 oocytes rates of 0.1–0.2 pmole GMP hr^{-1}.

that ribosomal RNA synthesis continues at a significant rate even after oogenesis is complete. However, at least according to some authors, the total amount of ribosomal RNA ceases to increase in *Xenopus* oocytes after the beginning of Dumont stage 5 (Davidson *et al.*, 1964; Scheer, 1973; Rosbash and Ford, 1974). This may again depend on the hormonal state of the animal and on the length of time a mature oocyte remains in the ovary without becoming atretic. In any case, these results suggest that the ribosomal RNA may under some conditions turn over in stage 6 oocytes. To investigate this possibility, Leonard and LaMarca (1975) injected ^{3}H-guanosine into female *Xenopus* and measured the specific activity of ribosomal RNA in the stage 6 oocytes at various times thereafter. Turnover was indeed detected, and half-lives ranging from 9 to 31 days were claimed. The turnover rate could vary greatly according to the dynamics of oocyte flow through the stages of oogenesis. In extremely rapidly growing oocytes such as those studied by Scheer (1973) the rate of ribosomal decay may be low or nonexistent, while in stored stage 6 oocytes waiting to be ovulated, it may be accelerated as the oocytes enter a steady state "waiting" condition. From equation (5.1) we may calculate the half-life expected for ribosomal RNA in a stage 6 oocyte which is maintaining itself in such a steady state. The oocyte contains about 3600 ng of ribosomal RNA (Chapter 4), and if we assume it is synthesizing this RNA at a rate of 12 pg min^{-1} (Table 8.2) the expected half-life is of the order of 145 days. This value greatly exceeds the half-lives measured in stage 6 oocytes by Leonard and LaMarca (1975). The latter therefore seem impossible to reconcile with the synthesis rate measurements of Anderson and Smith (1976), and we conclude that if there is a turnover of ribosomal RNA in stage 6 oocytes, it is very slow.

An interesting feature of ribosomal RNA synthesis in *Xenopus* oocytes is the transport of the ribosomes through the nuclear membrane into the cytoplasm. During the period of oocyte growth about 25% of the surface area of the nuclear membrane is occupied by "pore complexes" (Scheer, 1973). An electron micrograph of this remarkable structure is shown in Fig. 8.16. Data of Scheer (1973) show that in rapidly growing *Xenopus* oocytes 1–2 molecules of ribosomal RNA are transported through each pore complex per minute. This value is obtained from the number of pore complexes, about 24×10^6 at this stage, and the rate of ribosomal RNA accumulation in the cytoplasm (Table 8.2). The pore complexes are no doubt utilized for the transport of other species of ribonucleoprotein as well.

In all the growing oocytes which have been investigated, the overwhelming majority of the newly synthesized stable RNA's are found to be ribosomal. Examples include the various insect and amphibian oocytes mentioned above, the oocytes of *Urechis* (Davis and Wilt, 1972), sea urchins (Gross *et al.*, 1965b; Sconzo *et al.*, 1972), and mouse (Bachvarova, 1974), among others. The only clear exceptions in the literature are very young previtellogenesis oocytes of amphibians and teleosts, which synthesize mainly transfer RNA and 5 S RNA, as discussed above. The large

Fig. 8.16. Pore complexes in the nuclear membrane of *Xenopus* oocytes. Freeze-etch aspect of fractured nuclear envelope of an intact lampbrush stage oocyte. The pore margins are clearly visible. Note the high pore frequency. ×47,970. From U. Scheer (1973). *Dev. Biol.* **30**, 13.

amount of ribosomal RNA synthesis obscures the observation of hetero-geneous RNA species in total RNA extracts. Heterogeneous RNA's with high decay constants have been particularly difficult to detect in *in vivo* labeling experiments because of the large size of the oocyte precursor pools, and the necessity of labeling for relatively long periods in order to achieve measurable levels of incorporation. Nonetheless, heterogeneous RNA's have been noticed in several investigations. RNA species of various sizes, some very large, were reported by Sconzo *et al.* (1972) to be synthe-sized in sea urchin oocytes. Some heterogeneous RNA's are also reported to be transcribed in growing mouse oocytes, according to Bachvarova (1974) and Jahn *et al.* (1976). Davidson *et al.* (1964) described a labeled RNA displaying an unusually high uridylic acid content in Dumont stage 3 oocytes of *Xenopus*, after removal of ribosomal RNA in an initial extrac-tion. Mairy and Denis (1971) extracted RNA's from *Xenopus* oocytes of various stages which sedimented heterogeneously and were labeled within 4 hours of exposure to ^{3}H-guanosine. In the nuclei of previtellogenic (Dumont stage 1) oocytes of *Xenopus*, Thomas (1974) also found a sig-nificant amount of heterogeneous labeled RNA > 40 S in size. About 50% of the radioactivity was present in such RNA species 24 hours after label-ing began (Thomas, 1974). The heterogeneous fractions indicated by these observations could represent heterogeneous nuclear RNA's, mes-senger RNA's, or both, and we must turn to more specific studies to distinguish these various RNA classes.

Lampbrush Chromosomes and the Synthesis of Heterogeneous Nuclear RNA and Messenger RNA during Oogenesis

CHARACTER OF LAMPBRUSH CHROMOSOME RNA

In the first section of this chapter we reviewed data which led to the conclusion that the RNA of the lampbrush chromosome matrices may be of the heterogeneous nuclear type. There are several additional observa-tions on the newly synthesized RNA of lampbrush stage amphibian oo-cytes which support this view. Sommerville (1973) followed the fate of the ribonucleoprotein particles which appear to contain the newly synthe-sized RNA of the lampbrush chromosome loops. This material appears to be given off into the nuclear sap as particles composed of about 97% protein and 3% RNA. Both the protein and the RNA of the particles label rapidly, and the identity of the particles with the lampbrush loops is shown by the fact that they share the same proteins. Thus, as was illustrated in Fig. 8.4, fluorescein-labeled antibodies against the nuclear sap ribonucleoprotein particles react specifically with the lampbrush chromosome loops. Furthermore, treatment with actinomycin, which prevents reinitiation of transcription, results in the release of ribonucleoprotein from the loops and at the same time increases the amount of labeled RNA and protein in the nuclear sap particles. According to Malcolm and Sommerville (1974), the morphology of the ribonucleoprotein particles is identical with that of the loop matrix. Thus the nuclear sap particles are composed of the same 20 nm units as are seen in the loop matrices (Fig. 8.3f) and can be disaggregated into strands containing a beadlike array of these 20 nm units. It follows that the properties of the newly synthesized RNA in these particles can be considered characteristic of the RNA synthesized in the lampbrush chromosomes.

In many ways the oocyte nuclear sap particles resemble those in which the heterogeneous nuclear RNA of somatic cells is typically complexed (e.g., Samarina *et al.*, 1967; Moulé and Chauveau, 1968; Pederson, 1974; Kumar and Pederson, 1975). In *Triturus* oocyte nuclei (Sommerville, 1973), as in HeLa cells (Kumar and Pederson, 1975), these particles contain multiple species of proteins. The RNA extracted from the nuclear ribonucleoprotein particles in *Triturus* oocytes is very large, compared to cytoplasmic polysomal RNA (Sommerville, 1973). Sommerville and Malcolm (1976) isolated RNA's from these particles which are visualized in the electron microscope under denaturing conditions as linear forms 20 μm or more in length. Most of the nuclear RNA migrates in nondenaturing gradients (Sommerville, 1973) and gels (Sommerville and Malcolm, 1976) at 40–100 S and greater. It has a low GC base composition and displays a particularly high uridylic acid component (cf. Davidson *et al.*, 1964). In all these respects it is clearly of the heterogeneous nuclear RNA class. It is not yet possible to state what fraction if any of the newly synthesized lampbrush chromosome RNA turns over rapidly, i.e., at the rate of 20–30 minutes ($t_{1/2}$) observed for other heterogeneous RNA populations.

HETEROGENEOUS NUCLEAR RNA'S SYNTHESIZED IN MATURE OOCYTE NUCLEI

In Dumont stage 6 *Xenopus* oocytes rapidly decaying heterogeneous nuclear RNA's are clearly a major synthesis product. These have been studied by Anderson and Smith (1976). The rapidly decaying RNA's are confined to the oocyte nuclei, and their molecular weights are much higher than that of ribosomal precursor RNA. The labeling and turnover kinetics of these RNA's were derived by Anderson and Smith (1976) in experiments in which the precursor pool specific activity was measured. A molar incorporation curve for the >40 S RNA is shown in Fig. 8.17a [see equation (5.2)]. From these data a half-life of 30 minutes was calculated for a large portion of the nuclear RNA. In Fig. 8.17b the molar accumulation curves for the ribosomal RNA and the heterogeneous 4 S to 40 S RNA in whole stage 6 oocytes are shown for comparison. The 4 S to 40 S RNA clearly contains a component which decays rapidly, but in addition includes a component whose half-life is measured in hours. The latter may include messenger RNA, since at least a portion of this RNA is cytoplasmic in location. Anderson and Smith (1976) also found that the >40 S nuclear RNA of stage 6 oocytes displays a low GC base composition (42% GC) and an elevated uridylic acid component (34.3%).

The rate of synthesis of the rapidly decaying >40 S nuclear RNA in stage 6 oocytes is unusually high, compared to that in *Xenopus* embryo nuclei. From the data of Anderson and Smith (1976) k_s for the rapidly decaying nuclear RNA is about 11 pg min^{-1} (Table 8.2). This is significantly higher than the estimates for chromosomal RNA synthesis rate at the lampbrush stage shown in Table 8.2 and much greater than the approximate rate of heterogeneous nuclear RNA synthesis in *Xenopus* embryo cell nuclei (about 0.01–0.02 pg min^{-1} per nucleus). The rate of synthesis of heterogeneous nuclear RNA in the stage 6 oocyte nucleus appears to be about three orders of magnitude higher than in the somatic cell nuclei of the embryo, while a similar turnover rate probably prevails. Since it is probable that no more than about a tenfold increase in synthesis rate could result (compared to embryo nuclei) even if 100% of the genome were being transcribed, most of the high nuclear RNA synthesis rate in stage 6 oocytes must be due to dense packing of polymerases in the transcribed regions.

COMPLEXITY OF OOCYTE NUCLEAR RNA's

Unfortunately, little information on the single copy complexity of oocyte nuclear RNA's exists, other than the indirect inferences based on lampbrush chromosome structure reviewed in the first part of this chapter. It is known that the RNA's synthesized on Dumont stage 3 lampbrush chromosomes of *Xenopus* oocytes are highly complex, and they include significant single copy transcripts. Thus, Davidson and Hough (1969b) showed that the labeled RNA extracted from these oocytes hybridizes with excess single copy DNA. As will be recalled at least 5–10% of the genome seems to be included in the transcription units of the lampbrush chromosome loops (see first section of this chapter). Similarly in stage 6

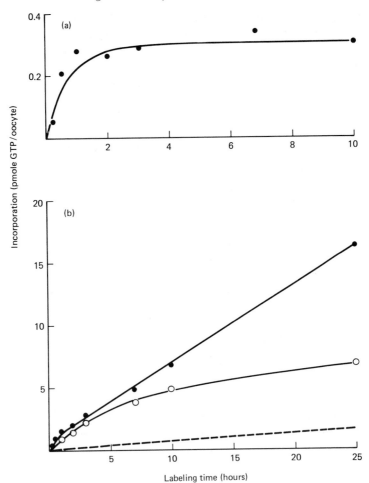

Fig. 8.17. Molar incorporation kinetics of the individual classes of RNA in *Xenopus* oocytes. Synthesis rates (k_s) for various RNA classes derived from these and other similar experiments are listed in Table 8.2 and in the footnotes to that table. (a) Incorporation kinetics for the RNA larger than the 40 S ribosomal precursor. This RNA turns over rapidly and was shown to be confined to the nucleus in experiments in which the RNA of isolated germinal vesicles was studied. The form of the incorporation kinetics shown is given by equation (5.2). k_d is 2.3×10^{-2} min^{-1}. (b) Incorporation kinetics for both ribosomal RNA (closed circles), and heterogeneous 4 S to 40 S RNA (open circles). The dashed line shows the accumulation of the stable component contained within the 4 S to 40 S fraction. The break in the ribosomal RNA curve at about 1 hour is probably due to the processing kinetics of the 40 S ribosomal RNA precursor (k_d is 1.15×10^{-2} min^{-1}). The molar accumulations were calculated from the measured specific activity of the GTP precursor pool and the fraction of the total incorporated radioactivity in each RNA class. D. M. Anderson and L. D. Smith (1976). *Dev. Biol.*, in press.

oocytes, if we consider maximum closest packing of polymerases to be about 100 nucleotides, a synthesis rate of 11 pg min^{-1} (Table 8.2) means that about 15% of the 4C oocyte genome could be included in transcription units. Calculations show that the sequence concentration of these putative complex nuclear RNA's would be so low that they could never have been detected in the hybridization experiments with total oocyte RNA reviewed in Chapter 6 (Davidson and Hough, 1971; Rosbash *et al.*, 1974). A direct measurement of heterogeneous nuclear RNA complexity in lambrush stage *Triturus* oocyte nuclei has been reported by Sommerville and Malcolm (1976). From both cDNA and saturation hybridization experiments with labeled single copy DNA they concluded that the complexity is at least 6×10^8 to 12×10^8 nucleotides (2–4% of the *Triturus* genome). These values agree reasonably well with the cytological estimates cited above, the lower limit of which was about 5% of the genome transcribed in lambrush stage oocytes. Sommerville and Malcolm (1976) also showed that as expected the RNA produced in the lambrush chromosomes contains both repetitive and nonrepetitive sequence transcripts.

HETEROGENEOUS REPETITIVE SEQUENCE TRANSCRIPTS ACCUMULATED IN LAMPBRUSH STAGE OOCYTES

The only other relevant information regarding the complexity of heterogeneous oocyte RNA's concerns the repetitive sequences transcribed in lambrush chromosomes. In Chapter 6 we reviewed low C_0t hybridization experiments of Davidson *et al.* (1966) and Crippa *et al.* (1967) which showed that the repetitive sequence transcripts present in mature oocytes persist beyond fertilization and are inherited by the embryo. As noted there, this transcript population could have consisted either of repetitive sequence elements transcribed in interspersed heterogeneous nuclear RNA's or of messenger RNA's transcribed from repetitive structural genes. In any case competition experiments of Davidson *et al.* (1966) showed that much the same set of repetitive sequence transcripts is present in the mature oocyte as is synthesized during the lambrush stage of oogenesis. An important fact is that these transcripts apparently accumulate in lambrush stage oocytes. The evidence for this statement is as follows: Hough and Davidson (1972) measured the fraction of the DNA to which this transcript population is homologous by the RNA excess hybridization method, using whole stage 6 oocyte RNA and an isolated repetitive sequence tracer (see Chapter 6). The results were closely comparable with those obtained by Davidson *et al.* (1966) and Crippa *et al.* (1967) who measured the amount of DNA represented in the repetitive sequence transcripts by saturating the DNA with *labeled* lampbrush stage oocyte RNA, and with *in vitro* labeled stage 6 oocyte RNA. To make this calculation it was assumed that the specific activity of the RNA labeled in the lampbrush stage oocytes was the same as that of the ribosomal RNA. As we have seen, the latter accumulates during the lampbrush stage and turns over slowly if at all. This specific activity assumption was justified by Davidson and Hough (1969b), who reextracted the lampbrush stage RNA from the RNA–DNA hybrids and compared the specific activity of the hybridizing RNA with that of the bulk (ribosomal) RNA of the starting preparation. These specific activities are found to be about the same. Hough and Davidson (1972) showed that $\geq 4\%$ of the repetitive tracer was hybridized by stage 6 oocyte RNA. Similarly Crippa *et al.* (1967) found that about 1.6% of the total DNA is hybridized by ^{32}P-RNA synthesized in Dumont stage 3 maximum lampbrush oocytes, when the mass of the hybridized ^{32}P-RNA at saturation was calculated from the specific activity of the ribosomal RNA. Since about 25% of the DNA of *Xenopus* is in repetitive sequence (cf. Chapter 1), these values are only about 1.5-fold apart. It follows from these measurements that the labeled repetitive sequence transcripts accumulate along with ribosomal RNA in lampbrush stage oocytes. This explains the ability of the mature oocyte RNA to compete effffectively in the hybridization of repetitive sequences transcribed at the lampbrush stage, since the heterogeneous repetitive transcripts remain about the same concentration relative to ribosomal RNA between Dumont stages 3 and 6. These transcripts are apparently retained throughout oogenesis and are passed on to the embryo (Davidson *et al.*, 1966; Crippa *et al.*, 1967).

SYNTHESIS AND ACCUMULATION OF MESSENGER RNA'S DURING OOGENESIS

The mature oocyte contains a complex set of maternal messenger RNA's according to information reviewed in Chapters 4, 5, and 6. In Table 6.1 we calculated that the mature *Xenopus* oocyte contains about 1.8×10^6 molecules of each complex RNA sequence. This may be an overestimate, since as noted in Chapter 6 it assumes that all the RNA's are represented equally and does not take into account the fraction of messenger RNA's transcribed from repetitive sequences (Rosbash *et al.*, 1974). However, it is instructive to calculate the length of time required to accumulate this many copies of each transcript. Many, if not all are maternal messenger RNA's (Chapter 6). Assuming the synthesis rate of 15 nucleotides sec^{-1} per polymerase (D. M. Anderson and L. D. Smith, personal communication) and assuming minimum polymerase packing intervals of only 100 nucleotides, it would require 35 days to synthesize this many copies, on the basis that the structural genes are single copy sequences and the chromosomal genome is 4C. This period is comparable to the minimum length of the lampbrush phase in *Xenopus* (Table 8.1). In this organism, therefore, little turnover of maternal messenger RNA could occur during the lampbrush phase of oogenesis if that were the period when the maternal message is synthesized. This assumes optimal physiological conditions when the oocytes are growing as rapidly as possible. The same arguments, however, suggest that in the relatively small marine eggs of invertebrates, such as sea urchin and *Urechis*, the length of time required to accumulate the complex RNA's of the oocyte could be far less. In these organisms, in which only about 10^3 copies of each transcript are present (see Table 6.1), the maternal messenger RNA could be synthesized very slowly, or during a short period of oogenesis, or could be synthesized and turned over many times. This results in an apparent difficulty for the conventional view that the lampbrush chromosome is the site of synthesis of stored maternal message. It seems unlikely that in echinoderms, for example, the lampbrush chromosome stage can be interpreted primarily as a phase in which maternal messenger RNA is synthesized and gradually *accumulated*. As indicated above the lampbrush chromosome stage probably lasts for some months in sea urchins compared to the day or two which would be required for maternal message synthesis assuming maximal rates. On the other hand the close packing of lampbrush transcripts in the echinoderm loop matrices must be equivalent to that in amphibian loops since their structure is so similar. That is, whatever RNA species are being synthesized on these loops are being produced at close to the maximum rates.

G. J. Dolecki and L. D. Smith (personal communication) have measured absolute synthesis rates for poly(A)-containing RNA during oogenesis in *Xenopus*. They found rates ranging from about 1.3 to 2.8 pg min^{-1} for Dumont stage 3 and 2.3 to 4.6 pg min^{-1} for stage 6 oocytes (Table 8.2). We can compare these rates with the rate of synthesis of the complex maternal messenger RNA of Table 6.1 which could be expected according to the optimal rate calculation above. The best estimate was that this message could be synthesized in no less than about 35 days, assuming no turnover. If turnover occurs, the required time would of course be greater. Thus, taking 40–70 ng as the quantity of poly(A) messenger RNA in the oocyte (Tables 4.2 and 6.1), the rate of synthesis of this message could be no greater than 0.8–1.4 pg min^{-1} per oocyte (Table 8.2). This rate may be lower than the rate of total poly(A) synthesis in lampbrush stage oocytes according to G. J.

Dolecki and L. D. Smith (personal communication) (Table 8.2), but it is not very different. If we suppose that the stored maternal message of the mature oocyte derives from poly(A)RNA synthesized in the lampbrush chromosomes, this poly(A)RNA could represent a slightly greater length of sequence than does the amount of maternal messenger poly(A)RNA stored in the mature oocyte. That is, a fraction of the poly(A)RNA nucleotides synthesized in Dumont stage 3 oocytes may not appear in maternal message per se, and could represent discarded portions of message precursor.

Anderson and Smith (1976) detected in stage 6 oocytes the synthesis of a species of RNA which probably is messenger RNA. This fraction sediments heterogeneously between 4 S and 40 S and appears to be kinetically stable, at least over a period of several hours. Its incorporation kinetics are shown in Fig. 8.17b. This RNA fraction could be cytoplasmic in location, and its rate of synthesis is about 1 pg min^{-1}. Thus it is synthesized so rapidly that most of it must eventually turn over in the cytoplasm of the mature oocyte, perhaps as ribosomal RNA does, since heterogeneous RNA's do not accumulate in stage 6 oocytes.

We found above that the synthesis rates of total chromosomal RNA or poly(A)RNA in lampbrush chromosomes of *Xenopus* oocytes are not inconsistent with the view that the transcripts are maternal message precursors, given the length of time the lampbrush chromosomes persist. However, there exists another item of evidence which suggests that at least in its simplest form this view is incorrect. Rosbash and Ford (1974) found that the amount of poly(A)RNA stored in the mature *Xenopus* oocyte is actually accumulated very early in oogenesis, *before* the lampbrush chromosome stage. In Fig. 8.18 data of Rosbash and Ford (1974) are shown which make this point clearly. The size of the oocytes studied by these authors and their data on RNA content show that only in oocytes which are still in Dumont stage 1 is there an amount of poly(A)RNA significantly less than that stored in the mature oocyte. The stage 2 oocytes already contain almost the final quantity of poly(A)RNA (Fig. 8.18a). After this, the poly(A)RNA simply gets diluted by ribosomal RNA as the oocyte grows (Fig. 8.18b). The messenger-like size distribution of the poly(A)RNA (Rosbash and Ford, 1974; Darnbrough and Ford, 1976) and its ability to form ribosomal initiation complexes (Darnbrough and Ford, 1976) certify its identification as messenger RNA. An additional item of evidence comes from studies of Ruderman and Pardue (1976), who identified histone messenger RNA's in *Xenopus* oocytes of various stages. Maternal messenger RNA's for histones are stored in the mature oocyte and are utilized in embryogenesis (see Chapter 4). RNA's extracted from Dumont stages 1, 2, 4, and 6 oocytes all could be translated in the

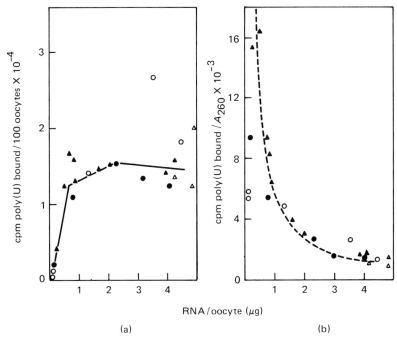

Fig. 8.18. Accumulation of poly(A)RNA during oogenesis in *Xenopus*. RNA was prepared from a fixed number of oocytes of each size class and a portion was hybridized with poly(U) in poly(U) excess. (a) The counts per minute of poly(U) rendered ribonuclease-resistant per 100 oocytes is plotted as a function of micrograms of RNA recovered per oocyte. Oocytes containing 1 μg RNA are in stage 2; 1.5 μg in stage 3; 4 μg in stage 6. (b) The counts per minute of poly(U) rendered ribonuclease-resistant per A_{260} RNA is plotted as a function of micrograms of RNA per oocyte. (O) Expt. 1; (●) Expt. 2; (▲) Expt. 3; (△) three preparations of laid eggs. (--) Theoretical dilution curve. From M. Rosbash and P. J. Ford (1974). *J. Mol. Biol.* **85**, 87.

cell-free wheat germ system to yield histones, and at all of these stages of oogenesis the histone messengers seem to be an abundant component. Darnbrough and Ford (1976) also investigated the translation products of poly(A)RNA's extracted from immature *Xenopus* oocytes. They demonstrated that the same set of about 40 prevalent proteins appears to be translated *in vitro* in the wheat germ system from RNA's of all stages of oocyte including stage 1, according to gel electrophoresis analysis. Yet Dumont (1972) showed that the lampbrush chromosomes are just in the process of expansion in stage 2 oocytes (see stage 3 in Duryee's classification, shown in Fig. 8.8) and do not reach full extension until stage 3. Therefore the data in Table 8.1 regarding the length of the lampbrush phase refer to a period *later* than the time when the poly(A)RNA stored in the oocyte is first synthesized, and when histone message and some other prevalent messages first appear. Little evidence exists on the duration of the stage 1 to stage 2 period. The important conclusion from Fig. 8.18 is that at least a large fraction of the putative maternal messenger RNA of the oocyte is made *before* the lampbrush phase is really begun. Any subsequent synthesis of the poly(A)RNA messengers, i.e., during the lampbrush stage, must be balanced by turnover. Of course, it remains possible that from the beginning the maternal message turns over and the final quantity represents a steady state level. This would require that the stage 1–stage 2 period be significantly longer than 35 days (see above), since about 3 months would be needed to attain the steady state level. More problematically, it would mean that the synthesis of maternal message goes on continuously, at more or less similar rates when the lampbrush chromosomes are present and when they are not, leaving their special function undefined.

EXAMPLES OF STRUCTURAL GENE PRODUCTS TRANSLATED DURING OOGENESIS AND REQUIRED IN EMBRYOGENESIS

Messenger RNA is translated during oogenesis as well as stored. Sommerville (1974) isolated polyribosomes from *Triturus* oocytes and studied the ribonucleoprotein particles released from them by EDTA treatment. The RNA in these particles labels within 4 hours and is active in cell-free protein synthesis systems. Though the oocytes contain relatively small quantities of polysomes, it seems clear that at least some of these are loaded with newly synthesized (and therefore turning over) messenger RNA. One conclusion which might be drawn from these observations is that the RNA synthesized in the maximum lampbrush chromosomes is unlikely to exist solely for the purpose of serving as precursor for the accumulated poly(A)RNA in the oocyte, since the latter is already present before the lampbrush chromosomes are fully active. Furthermore, according to Darnbrough and Ford (1976) the proteins synthesized *in vivo* by previtellogenic oocytes are different from those synthesized in cell-free translation systems from the RNA extracted from these oocytes. This suggests that the accumulated poly(A)RNA message and the message being utilized on the oocyte polysomes are distinct populations. Unfortunately, the amount of messenger RNA required to load the polysomes in lampbrush stage oocytes cannot be determined from the extant data.

Several examples exist in which the functional nature of some of the messenger RNA's translated in the oocyte polysomes can be specified. In *Drosophila* a number of mutations have been reported with the phenotype of a homozygous mutant female producing eggs unable to develop beyond very early stages (e.g., Zalokar *et al.*, 1975). This class of mutants includes *deep orange* (Garen and Gehring, 1972), *rudimentary* (e.g., Okada *et al.*, 1974b), and *cinnamon* (*cin*) (Baker, 1973). *Deep orange* and *cin* appear to affect pteridine metabolism. A similar pteridine metabolism deficiency which also behaves as a maternal effect mutant has been described in an annelid worm by Fischer (1974). Garen and Gehring (1972) have shown that injection of normal egg cytoplasm into *Drosophila* eggs produced by homozygous *deep orange* females rescues them from the otherwise lethal effect of the mutation. In the case of *cin*, the expression of the normal allele during embryogenesis in heterozygous females assures their survival. Injection of normal egg cytoplasm also rescues mutant eggs from the effects of the *rudimentary* mutation. This mutation is known to result in a critical defect in pyrimidine metabolism, and Okada *et al.* (1974b) found that the eggs of *rudimentary* mothers would survive if injected with pyrimidine nucleosides. These examples illustrate the requirement for enzymes which must be synthesized during oogenesis and transmitted to the embryo. Maternal developmental mutants are also known in the nematode *Caenorhabditis elegans*. Three temperature sensitive mutants of this type, designated *zyg-1*, *zyg-2*, and *zyg-3*, were reported by Vanderslice and Hirsh (1976). When adults bearing these mutations are exposed to the nonpermissive temperature, the embryos present in their reproductive tracts block development at stages specific to each mutant. These stages are first cleavage, 20- to 30-cell stage, and hatching, respectively. The period when the function of these genes is required and normally occurs was found to be during oogenesis in all three cases. Since the effect is genetically maternal, mutants exposed to high temperature must lack some necessary component of egg cytoplasm. Interestingly, these three genes are required for gonadogenesis as well as for embryogenesis. The temperature sensitive period for *zyg-3* also extends into cleavage, suggesting that its gene product, which is normally required for hatching, is synthesized both in oogenesis and in early embryogenesis. This is reminiscent of the sea urchin structural genes represented in maternal messenger RNA which also may be transcribed in the early embryo (see Chapter 6).

Another well-known example involving a substance synthesized in oogenesis which is needed during embryogenesis is the case of the *o* mutation in axolotl. This maternally acting mutation was discovered by Humphrey (1966) and was investigated by Briggs and Cassens (1966). Females homozygous for *o* produce eggs which are unable to complete gastrulation. However, Briggs and Cassens (1966) demonstrated that eggs from homozygous o females will gastrulate and develop if cytoplasm from mature o^+/o or o^+/o^+ eggs is injected into them. It is interesting that nucleoplasm obtained from stage 6 ovarian oocytes contains the active material(s), and these are much more concentrted in the stage 6 nucleoplasm than in the cytoplasm. Figure 8.19 (Briggs, 1972) illustrates the rescue phenomenon dramatically. Briggs and Justus (1968) showed that the o substance is a protein, and Briggs (1972) found that it is not species specific. Thus nuclear sap from normal oocytes of a variety of amphibian species is effective in correcting the o defect. Tadpoles bearing the o^+ allele which have survived early embryogenesis by virtue of an injection of normal oocyte nuclear sap continue to develop, while o homozygotes arrest (Fig. 8.19). This shows that the o gene normally begins to function in the period of organogenesis as do many other embryo genes (Chapter 2). Another interesting mutant known in the axolotl is cl (Carroll and Van Deusen, 1973). Mutant cl homozygous females again lay nonviable eggs, and in this case the defect appears to reside in the vegetal cortex of the egg. The early cleavage furrows in this region are defective, and a phenocopy of the cl effect can be produced by treatment with cytochalasin B. Animal pole cells from cl homozygous embryos can develop normally and give rise to neural vesicles, if grafted into the blastocoel roof of normal recipients. This example recalls the many morphogenetic agents which must be synthesized during oogenesis, are localized topographically in the eggs, and are required during early development (cf. Chapter 7).

These maternal effect mutations demonstrate that some of the messenger RNA's which are transcribed and translated during the long growth phase of oogenesis produce proteins which are essential in development. As shown in Chapter 6, the messenger RNA's present in ovarian sea urchin polysomes represent a very complex set of structural genes (see Figs. 6.10 and 6.11). If the polysomal messenger RNA in *Xenopus* oocytes is like that in sea urchin oocytes, a large fraction of the sequences in the stored maternal messenger RNA stockpile is also being translated during oogenesis. This might suggest continuing synthesis and turnover of the polysomal species.

IS THE MAIN FUNCTION OF LAMPBRUSH CHROMOSOMES THE SYNTHESIS OF MATERNAL MESSENGER RNA PRECURSORS?

It is now necessary to attempt to summarize and interpret the array of measurements reviewed here concerning nuclear and messenger RNA

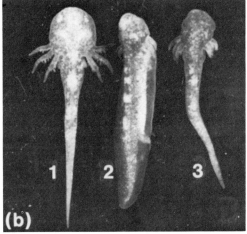

Fig. 8.19. Experimental rescue of *o* mutants in the axolotl. (a) Embryos derived from a spawning of an *o/o* female mated with an *o⁺/o* male. Eggs were *not* injected with *o⁺* substance. All expressed the maternal effect of *o*, and stopped developing at the stage shown in the photograph. ×3. 6. (b) Effect of injection of normal oocyte nuclear sap into eggs of the same spawning shown in (a). The maternal effect of *o* is corrected. All embryos develop to about stage 37. Half arrest at this stage (embryos 2 and 3). These are presumably the *o/o* zygotes. The remaining half, presumably the *o⁺/o*, continue to larval stages (embryo 1). ×5. From R. Briggs (1972). *J. Exp. Zool.* **181**, 271.

synthesis in oocytes. The messenger RNA falls into two classes, that stored for use in embryogenesis (maternal message) and that translated during oogenesis. In *Xenopus* the maternal message (and almost certainly the oocyte polysomal message as well) is single copy transcript. Given its restricted complexity (27×10^6 to 40×10^6 nucleotides) (Table 6.1) and considering the maximum synthesis rates possible, this message is accumulated in amounts which require a period of at least several weeks to synthesize (Table 8.2). However, the final quantity of maternal poly(A) messenger RNA, about 40–70 ng (Table 4.2) is already present in the oocytes by the *beginning* of the lampbrush phase. It is important to note that during the same previtellogenesis stages of oogenesis when the poly(A) messenger RNA is being accumulated, so are the relatively huge quantities of transfer RNA and 5 S ribosomal RNA which are eventually transferred to the embryo. Furthermore histone messenger RNA is already accumulated in the oocytes at these early stages. Therefore, the *pre*-lampbrush and perhaps the earliest lampbrush phases of oogenesis can be regarded as the period when various gene products are being accumulated and "packaged" for long-term storage and ultimate use in the embryo. Of course other classes of messenger RNA not containing long poly(A) tracts may be synthesized in the oocyte as well, and these could accumulate at different stages. The following comments refer only to that class of poly(A) maternal message so far identified, the accumulation of which is measured as indicated in Fig. 8.18.

In amphibian oocytes poly(A)RNA continues to be synthesized throughout the lampbrush stage and in stage 6 oocytes. Considering the chromosomal and poly(A)RNA synthesis rates summarized in Table 8.2, the possibility cannot be excluded that the poly(A)RNA made in the lampbrush stage consists of messenger RNA precursors perhaps a few times as large as the message, most of which is discarded during processing. However, any contribution of this RNA to the stored maternal message pool must be almost exactly equaled by message turnover, since continued increase of poly(A) messenger RNA is not observed during the lampbrush phase. This would be a rather puzzling result if the lampbrush chromosomes are considered the source of the maternal message, since it implies that in previtellogenesis prelampbrush stages the message is synthesized and *accumulated*, while in the following lampbrush stage the previously accumulated message is destroyed and replaced. An even less satisfactory correlation exists for echinoderms where it would take only a few days or less to accumulate the maternal message, while the lampbrush chromosome stage may persist for months. The stored maternal message aside, there are also newly synthesized messenger RNA's being translated in the oocyte, and these must turn over at some modest rate since they label rapidly.

A variety of evidence shows that lampbrush chromosomes synthesize RNA of the heterogeneous nuclear type. The RNA made in the lampbrush transcription units is immensely long; it displays a low GC and high U base composition; it probably has an interspersed sequence organization; and it is complexed with specific proteins. Observations of Sommerville and Malcolm (1976), and calculations based on the length of lampbrush transcription units and the number of chromosome loops show that this RNA also has a high complexity, many times greater than that of the stored maternal message. This discrepancy is probably even more extreme in *Triturus* oocytes than in *Xenopus*, since in the former more DNA is extruded in the lampbrush loops and is presumably being transcribed, while the complexities of the stored message are similar in the oocytes of the two species [*Triturus* DNA has much more single copy DNA sequence than does *Xenopus*, as well as more repetitive sequence (Rosbash *et al.*, 1974]. The rate of synthesis of lampbrush chromosomal RNA calculated from the close packing of polymerases observed in the lampbrush transcription units, about 3 pg min^{-1}, is about the same as the rate of synthesis measured for the poly(A)RNA made during the lampbrush stage, 1.3–2.8 pg min^{-1} (Table 8.2). Thus, like many nuclear RNA's, a large fraction of the lampbrush chromosome RNA's appears to be polyadenylated. In stage 6 oocytes heterogeneous nuclear RNA with a 30-minute half-life is found, but such observations have not yet been made in lampbrush stage oocytes. However, if the lampbrush chromosome RNA and the poly(A)RNA made in lampbrush stage oocytes are identical or overlapping populations, it seems unlikely that all of this RNA could turn over with anything like a 30-minute half-life. Such turnover kinetics would probably have been observed by G. J. Dolecki and L. D. Smith (personal communication) in their studies of stage 3 poly(A)-RNA, but were not. However, neither does lampbrush chromosome RNA accumulate. We know this because its complexity must be many times that of the *stored* RNA present in stage 6 oocytes and because much of it is polyadenylated but no further accumulation of poly(A)RNA takes place during the lampbrush stage. Therefore, the lampbrush chromosome RNA must turn over, but perhaps at a relatively moderate average rate, or a set of widely distributed rates. Suppose we assume that there is a steady state level of heterogeneous poly(A)RNA synthesized in the lampbrush chromosomes, which is only 5% of the level of stored poly(A) messenger RNA. Application of equation (5.1) shows that its half-life would be about 8–14 hours, assuming the synthesis rates for poly(A)RNA in stage 3 oocytes shown in Table 8.2. This of course refers to the major fraction of the RNA, and sequences representing only a few percent of the total incorporated nucleotides could be preserved if the remainder turns over at the rate indicated.

The functional significance of the lampbrush loops is strongly suggested by their structure. They appear to serve as marshaling sites on which the

RNA's being synthesized in the large transcription units are complexed with specific proteins. These ribonucleoprotein structures are then exported to the nuclear sap as particulate aggregates. The loop is polarized because transcription units are polarized, and when its matrix is released it collapses, as in actinomycin treatment or in stage 5 and 6 oocytes.

We have reached the tentative conclusion that the conventional proposal for lampbrush chromosomes as synthesis sites for maternal messages is in some ways paradoxical. RNA synthesis in lampbrush chromosomes is clearly occurring at the highest possible rates in the transcribed sequences. However, in amphibians the peak synthetic activity and extension of lampbrush chromosomes occur after the maternal message stockpile seems to be largely accumulated. In many organisms the lampbrush phase may last far longer than is needed for maternal message accumulation, and in others, it could be too short. Furthermore, there may be little relation between the complexity of the lampbrush transcript and that of the maternal message as different organisms of the same taxon are compared. The kinetic and complexity data on which many of the above arguments are based are still very scanty, however. It is not improbable that when further information is available a role in which lampbrush chromosome RNA serves as precursor for the maternal messenger RNA will seem more in harmony with experimental observations. Nonetheless, alternative speculations should be considered. The fact that the lampbrush chromosome stage always requires a long time, while the transcription units are maximally packed with polymerases, argues that some accumulation function is involved. This is also the direct implication of our comparative examination of meroistic oogenesis. The fact that most of the lampbrush RNA turns over, albeit slowly, means that what is accumulated must represent only a small fraction of each primary transcript. The only heterogeneous RNA class so far known to accumulate during the lampbrush chromosome phase is the repetitive sequence transcripts stored in the mature oocyte. Perhaps the function of the RNA transcribed on the lampbrush chromosome is to accumulate regulatory sequences in the oocyte of either an RNA or a protein nature (Britten and Davidson, 1969, 1971; Davidson and Britten, 1971; Davidson et al., 1976a). It could be a unique function of germ-line cells (i.e., either oocytes or nurse cells) to synthesize and store these sequences in sufficient concentration so that they can find their targets in the comparatively giant embryo. This speculation (Davidson and Britten, 1971) relates lampbrush chromosome function and the localization phenomenon, two of the most general features of animal development.

Bibliography

Adamson, E. D., and Woodland, H. R. (1974). Histone synthesis in early amphibian development: Histone and DNA synthesis are not coordinated. *J. Mol. Biol.* **88**, 263.

Adamson, E. D., and Woodland, H. R. (1976). Changes in the rate of histone synthesis during oocyte naturation and very early development of *Xenopus laevis*. *Dev. Biol.*, in press.

Aitkhozhin, M. A., Belitsina, N. V., and Spirin, A. S. (1964). Nucleic acids during early development of fish embryos (*Misgurnus fossilis*). *Biochemistry (Engl. Transl.)* **29**, 145.

Alfert, M. (1950). A cytochemical study of oogenesis and cleavage in the mouse. *J. Cell. Comp. Physiol.* **36**, 381.

Amma, K. (1911). Uber die Differenzierung der Keimbahnzellen bei den Copepoden. *Arch. Zellforsch.* **6**, 497.

Anderson, D. M., and Smith, L. D. (1976). Synthesis of heterogeneous nuclear RNA in full grown oocytes of *Xenopus laevis*. *Cell*, in press.

Anderson, D. M., Galau, G. A., Britten, R. J., and Davidson, E. H. (1976). Sequence complexity of the RNA accumulated in oocytes of *Arbacia punctulata*. *Dev. Biol.* **51**, 138.

Angelier, N., and Lacroix, J. C. (1975). Complexes de transcription d'origines nucléolaire et chromosomique d'ovocytes de *Pleurodeles waltlii* et *P. poireti* (Amphibiens, Urodèles). *Chromosoma* **51**, 323.

Angerer, R. C., Davidson, E. H., and Britten, R. J. (1975). DNA sequence organization in the mollusc *Aplysia californica*. *Cell* **6**, 29.

Arceci, R. J., Senger, D. R., and Gross, P. R. (1976). The programmed switch in lysine-rich histone synthesis at gastrulation. *Cell* **9**, 171.

Arnold, J. M. (1968). The role of the egg cortex in cephalopod development. *Dev. Biol.* **18**, 180.

Aronson, A. I. (1972). Degradation products and a unique endonuclease in heterogeneous nuclear RNA in sea urchin embryos. *Nature (London), New Biol.* **235**, 40.

Aronson, A. I., and Wilt, F. H. (1969). Properties of nuclear RNA in sea urchin embryos. *Proc. Natl. Acad. Sci. U.S.A.* **62**, 186.

Aronson, A. I., Wilt, F. H., and Wartiovaara, J. (1972). Characterization of pulse-labeled nuclear RNA in sea urchin embryos. *Exp. Cell Res.* **72**, 309.

383

Atkinson, J. W. (1971). Organogenesis in normal and lobeless embryos of the marine proso-
branch gastropod *Ilyanassa obsoleta. J. Morphol.* **133**, 339.

Auerbach, L. (1874). "Organologische Studien. Zur Charakteristik und Lebensgeschichte
der Zellkerne," p. 177. Morgenstern, Breslau.

Axel, R., Cedar, H., and Felsenfeld, G. (1973). Synthesis of globin ribonucleic acid from
duck-reticulocyte chromatin *in vitro. Proc. Natl. Acad. Sci. U.S.A.* **70**, 2029.

Axel, R., Feigelson, P., and Schutz, G. (1976). Analysis of the complexity and diversity
of mRNA from chicken liver and oviduct. *Cell* **7**, 247.

Bachvarova, R. (1974). Incorporation of tritiated adenosine into mouse ovum RNA. *Dev.
Biol.* **40**, 52.

Bachvarova, R., and Davidson, E. H. (1966). Nuclear activation at the onset of amphibian
gastrulation. *J. Exp. Zool.* **163**, 285.

Bachvarova, R., Davidson, E. H., Allfrey, V. G., and Mirsky, A. E. (1966). Activation of
RNA synthesis associated with gastrulation. *Proc. Natl. Acad. Sci. U.S.A.* **55**, 358.

Badman, W. S., and Brookbank, J. W. (1970). Serological studies of two hybrid sea urchins.
Dev. Biol. **21**, 243.

Bailey, G. S., and Dixon, G. H. (1973). Histone IIb1 from rainbow trout. Comparison in
amino acid sequence with calf thymus IIb1. *J. Biol. Chem.* **248**, 5463.

Baker, B. S. (1973). The maternal and zygotic control of development by *cinnamon*, a new
mutant in *Drosophila melanogaster. Dev. Biol.* **33**, 429.

Baker, T. G. (1963). A quantitative and cytological study of germ cells in human ovaries.
Proc. R. Soc. London, Ser. B **158**, 417.

Baker, T. G., and Franchi, L. L. (1966). Fine structure of the nucleus in the primordial
oocytes of primates. *J. Anat.* **100**, 697.

Baker, T. G., and Franchi, L. L. (1967). The fine structure of oogonia and oocytes in human
ovaries. *J. Cell Sci.* **2**, 213.

Baker, T. G., Beaumont, H. M., and Franchi, L. L. (1969). The uptake of tritiated uridine
and phenylalanine by the ovaries of rats and monkeys. *J. Cell Sci.* **4**, 655.

Baldwin, W. M. (1915). The action of ultraviolet rays upon the frog's eggs. I. The artificial
production of spina bifida. *Anat. Rec.* **9**, 365.

Ballard, W. W. (1973a). Morphogenetic movements in *Salmo gairdneri* Richardson. *J. Exp.
Zool.* **184**, 27.

Ballard, W. W. (1973b). A new fate map for *Salmo gairdneri. J. Exp. Zool.* **184**, 49.

Baltzer, F. (1910). Uber die Beziehung zwischen dem Chromatin und der Entwickelung und
Vererbungsrichtung bei Echinodermenbastarden. *Arch. Zellforsch.* **5**, 497.

Bantle, J. A., and Hahn, W. E. (1976). Complexity and characterization of polyadenylated
RNA in the mouse brain. *Cell* **8**, 139.

Bantock, C. R. (1970). Experiments on chromosome elimination in the gall midge,
Mayetiola destructor. J. Embryol. Exp. Morphol. **24**, 257.

Barrett, D., and Angelo, G. M. (1969). Maternal characteristics of hatching enzymes in
hybrid sea urchin embryos. *Exp. Cell Res.* **57**, 159.

Barros, C., Hand, G. S., Jr., and Monroy, A. (1966). Control of gastrulation in the starfish,
Asterias forbesii. Exp. Cell Res. **43**, 167.

Belitsina, N. V., Aitkhozhin, M. A., Gavrilova, L. P., and Spirin, A. S. (1964). Messenger
RNA of differentiating animal cells. *Biochemistry (Engl. Transl.)* **29**, 315.

Bell, E., and Reeder, R. (1967). The effect of fertilization on protein synthesis in the egg of
the surf clam, *Spisula solidissima. Biochim. Biophys. Acta* **142**, 500.

Benbow, R. M., and Ford, C. C. (1975). Cytoplasmic control of nuclear DNA synthesis
during early development of *Xenopus laevis:* A cell-free assay. *Proc. Natl. Acad. Sci.
U.S.A.* **72**, 2437.

Benbow, R. M., Pestell, R. Q. W., and Ford, C. C. (1975). Appearance of DNA polymerase activities during early development of *Xenopus laevis*. *Dev. Biol.* **43**, 159.

Bennett, M. V. L. (1973). Function of electrotonic junctions in embryonic and adult tissues. *Fed. Proc., Fed. Am. Soc. Exp. Biol.* **32**, 65.

Bennett, M. V. L., and Trinkaus, J. P. (1970). Electrical coupling between embryonic cells by way of extracellular space and specialized junctions. *J. Cell Biol.* **44**, 592.

Benoit, J. (1930). Contribution à l'étude de la lignée germinale chez le poulet. Destruction précoce des genocytes primaires par les rayons ultra-violet. *C. R. Seances Soc. Biol. Ses Fil.* **104**, 1329.

Benson, S. C., and Triplett, E. L. (1974a). The synthesis and activity of tyrosinase during development of the frog *Rana pipiens*. *Dev. Biol.* **40**, 270.

Benson, S. C., and Triplett, E. L. (1974b). Transcriptionally dependent accumulation of tyrosinase polyribosomes during frog embryonic development and its relationship to neural plate induction. *Dev. Biol.* **40**, 283.

Benttinen, L. C., and Comb, D. G. (1971). Early and late histones during sea urchin development. *J. Mol. Biol.* **57**, 355.

Berg, W. E. (1968). Kinetics of uptake and incorporation of valine in the sea urchin embryo. *Exp. Cell Res.* **49**, 379.

Berg, W. E. (1970). Further studies on the kinetics of incorporation of valine in the sea urchin embryo. *Exp. Cell Res.* **60**, 210.

Berg, W. E., and Mertes, D. H. (1970). Rates of synthesis and degradation of protein in the sea urchin embryo. *Exp. Cell Res.* **60**, 218.

Bergink, E. W., and Wallace, R. A. (1974). Precursor-product relationship between amphibian vitellogenin and the yolk proteins, lipovitellin and phosvitin. *J. Biol. Chem.* **249**, 2897.

Bernstein, R. M., and Mukherjee, B. B. (1972). Control of nuclear RNA synthesis in 2-cell and 4-cell mouse embryos. *Nature (London)* **238**, 457.

Berry, R. O. (1941). Chromosome behavior in the germ cells and development of the gonads in *Sciara ocellaris*. *J. Morphol.* **68**, 547.

Bibring, T., and Baxandall, J. (1971). Selective extraction of isolated mitotic apparatus. Evidence that typical microtubule protein is extracted by organic mercurial. *J. Cell Biol.* **48**, 324.

Bier, K. (1963). Synthese, interzellulärer Transport, und Abbua von Ribonukleinsäure in Ovar der Stubenfliege *Musca domestica*. *J. Cell Biol.* **16**, 436.

Bier, K. (1965). Zur Funktion der Nährzellen im meroistischen Insektenovar unter besonderer Berücksichtigung der Oogenese adephager Coleopteren. *Zool. Jahrb., Abt. Allg. Zool. Physiol. Tiere* **71**, 371.

Bier, K. (1967). Oogenese, das Wachstum von Riesenzellen. *Naturwissenschaften* **54**, 189.

Bier, K. Kunz, W., and Ribbert. D. (1969). Insect oogenesis with and without lampbrush chromosomes. *Chromosomes Today* **2**, 107.

Biggers, J. D., and Stern, S. (1973). Metabolism of the preimplantation mammalian embryo. *Adv. Reprod. Physiol.* **6**, 1.

Bird, A. P., Rochaix, J.-D., and Bakken, A. H. (1973). The mechanism of gene amplification in *Xenopus laevis* oöcytes. *In* "Molecular Cytogenetics" (B. A. Hamkalo and J. Papaconstantinou, eds.), p. 49. Plenum, New York.

Birnie, G. D., MacPhail, E., Young, B. D., Getz, M. J., and Paul, J. (1974). The diversity of the messenger RNA population in growing Friend cells. *Cell Differ.* **3**, 221.

Birnstiel, M., Telford, J., Weinberg, E., and Stafford, D. (1974). Isolation and some properties of the genes coding for histone proteins. *Proc. Natl. Acad. Sci. U.S.A.* **71**, 2900.

Bishop, J. O., and Freeman, K. B. (1974). DNA sequences neighboring the duck hemoglo-
bin genes. *Cold Spring Harbor Symp. Quant. Biol.* **38**, 707.
Bishop, J. O., Pemberton, R., and Baglioni, C. (1972). Reiteration frequency of haemoglo-
bin genes in the duck. *Nature (London), New Biol.* **235**, 231.
Bishop, J. O., Morton, J. G., Rosbash, M., and Richardson, M. (1974). Three abundance
classes in HeLa cell messenger RNA. *Nature (London)* **250**, 199.
Blackler, A. W. (1970). The integrity of the reproductive cell line in the amphibian. *Curr.
Top. Dev. Biol.* **5**, 71.
Blobel, G. (1971). Release, identification, and isolation of messenger RNA from mammalian
ribosomes. *Proc. Natl. Acad. Sci. U.S.A.* **68**, 832.
Blumenthal, A. B., Kriegstein, H. J., and Hogness, D. S. (1974). The units of DNA replica-
tion in *Drosophila melanogaster* chromsomes. *Cold Spring Harbor Symp. Quant. Biol.*
38, 205.
Boetdker, H. (1968). Dependence of the sedimentation coefficient on molecular weight of
RNA after reaction with formaldehyde. *J. Mol. Biol.* **35**, 61.
Boivin, A., Vendrely, R., and Vendrely, C. (1948). Biochimie de l'heredite.—L'acide de-
soxyribonucleique du noyan cellulaire, dépositaire des caractères heréditaires; argu-
ments d'ordre analytique. *C. R. Hebd. Seances Acad Sci.* **226**, 1061.
Bonner, J., Chalkley, G. R., Dahmus, M., Fambrough, D., Fujimura, F., Huang, R. C.,
Huberman, J., Jensen, R., Marushige, K., Ohlenbusch, H., Olivera, B., and Widholm,
J. (1968). Isolation and characterization of chromosomal nucleoproteins. *In* "Methods
in Enzymology" (L. Grossman and K. Moldave, eds.), Vol. 12, Part B, p. 3. Academic
Press, New York.
Bonner, T. I., Brenner, D. J., Neufeld, B. R., and Britten, R. J. (1973). Reduction in the rate
of DNA reassociation by sequence divergence. *J. Mol. Biol.* **81**, 123.
Bonnet, C. (1762). "Considérations sur les Corps Organisés." Chez Marc-Michel Rey,
Amsterdam.
Borum, K. (1967). Oogenesis in the mouse: A study of the origin of the mature ova. *Exp. Cell
Res.* **45**, 39.
Bottke, W. (1973). Lampenbürstenchromosomen und Amphinukleolen in Oocyten-kernen
der Schnecke *Bithynia tentaculata* L. *Chromosoma* **42**, 175.
Bounoure, L. (1937). Les suites de l'irradiation du déterminant germinal, chez la grenouille
rousse. Par les rayons ultra-violets: Résultats histologiques. *C. R. Seances Soc. Biol.
Ses. Fil.* **125**, 898.
Bounoure, L., Aubry, R., and Huck, M. L. (1954). Nouvelles recherches experimentales sur
les origines de la lignée reproductice chez la grenonille rousse. *J. Embryol. Exp. Mor-
phol.* **2**, 245.
Bourne, G. C. (1894). Epigenesis or evolution. *Sci. Prog. (Oxford)* **1**, 8.
Boveri, T. (1893). An organism produced sexually without characteristics of the mother.
Translated by T. H. Morgan. *Am. Nat.* **27**, 222.
Boveri, T. (1899). "Die Entwicklung von *Ascaris megalocephala* mit besonderer Rücksicht
auf die Kernverhältnisse," p. 383. F. C. von Kupffer, Jena.
Boveri, T. (1901). Die Polarität von Ovocyte, Ei und Larve des *Strongylocentrotus lividus.
Zool. Jahrb., Abt. Anat. Ontog. Tiere* **14**, 630.
Boveri, T. (1903). Uber den Einfluss der Sämenzalle auf die Larvencharaktere der Echini-
den. *Arch. Entwicklungsmech. Org.* **16**, 340.
Boveri, T. (1905). Zellenstudien. V. Uber die Abhängigkeit der Kerngrösse und Zellenzahl
der Seeigel-larven von der Chromosomenzahl der Ausgangszellen. *Jena. Z. Naturwiss.*
39, 1.
Boveri, T. (1918). Zwei Fehlerquellen bei Merogonieversuchen und die Entwicklungsfähig-
keit merogonischer partiellmerogonischer Seeigelbastarde. *Arch. Entwicklungsmech. Org.* **44**, 417.

Boyd, M. M. M. (1941). The structure of the ovary and the formation of the corpus luteum in *Hoplodactylus maculatus*, Gray. *Q. J. Microsc. Sci.* [N.S.] **82**, 337.

Boyer, B. C. (1971). Regulative development in a spiralian embryo as shown by cell deletion experiments on the acoel, *Childia. J. Exp. Zool.* **176**, 97.

Brachet, J. (1940). La localisation de l'acide thymonucléique pendant l'oogénèse et la maturation chez les Amphibiens. *Arch. Biol.* **51**, 151.

Brachet, J., and Denis, H. (1963). Effects of actinomycin D on morphogenesis. *Nature (London)* **198**, 205.

Brachet, J., Decroly, M., Ficq, A., and Quertier, J. (1963a). Ribonucleic acid metabolism in unfertilized and fertilized sea-urchin eggs. *Biochim. Biophys. Acta* **72**, 660.

Brachet, J., Ficq, A., and Tencer, R. (1963b). Amino acid incorporation into proteins of nucleate and anucleate fragments of sea urchin eggs: Effect of parthenogenetic activation. *Exp. Cell Res.* **32**, 168.

Brachet, J., Denis, H., and de Vitry, F. (1964). The effects of actinomycin D and puromycin on morphogenesis in amphibian eggs and *Acetabularia mediterranea*. *Dev. Biol.* **9**, 398.

Brandhorst, B. P. (1976). Two-dimensional gel patterns of protein synthesis before and after fertilization of sea urchin eggs. *Dev. Biol.*, in press.

Brandhorst, B. P., and Humphreys, T. (1971). Synthesis and decay rates of major classes of deoxyribonucleic acid-like ribonucleic acid in sea urchin embryos. *Biochemistry* **10**, 877.

Brandhorst, B. P., and Humphreys, T. (1972). Stabilities of nuclear and messenger RNA molecules in sea urchin embryos. *J. Cell Biol.* **53**, 474.

Brandhorst, B. P., and McConkey, E. H. (1974). Stability of nuclear RNA in mammalian cells. *J. Mol. Biol.* **85**, 451.

Bresslau, E. (1909). Die Entwicklungen der Acoelen. *Verh. Dtsch. Zool. Ges.* **19**, 314.

Breuer, M. E., and Pavan, C. (1955). Behavior of polytene chromosomes of *Rhynchosciara angelae* at different stages of larval development. *Chromosoma* **7**, 371.

Bridges, C. B. (1936). The *Bar* "gene": A duplication. *Science* **83**, 210.

Briggs, R. (1972). Further studies on the maternal effect of the *o* gene in the Mexican axolotl. *J. Exp. Zool.* **181**, 271.

Briggs, R., and Cassens, G. (1966). Accumulation in the oocyte nucleus of a gene product essential for embryonic development beyond gastrulation. *Proc. Natl. Acad. Sci. U.S.A.* **55**, 1103.

Briggs, R., and Justus, J. T. (1968). Partial characterization of the component from normal eggs which corrects the maternal effect of gene *o* in the Mexican axolotl (*Ambystoma mexicanum*). *J. Exp. Zool.* **167**, 105.

Briggs, R., and King, T. J. (1959). Nucleocytoplasmic interaction in eggs and embryos. *In* "The Cell" (J. Brachet and A. E. Mirsky, eds.), Vol. 1, p. 537. Academic Press, New York.

Briggs, R., Green, E. U., and King, T. J. (1951). An investigation of the capacity for cleavage and differentiation in *Rana pipiens* eggs lacking "functional" chromosomes. *J. Exp. Zool.* **116**, 455.

Brinster, R. L. (1971). Uptake and incorporation of amino acids by the preimplantation mouse embryo. *J. Reprod. Fertil.* **27**, 329.

Brinster, R. L. (1973). Protein synthesis and enzyme constitution of the preimplantation mammalian embryo. *In* "The Regulation of Mammalian Reproduction" (S. Segal *et al.*, eds.), p. 302. Thomas, Springfield, Illinois.

Bristow, D. A., and Deuchar, E. M. (1964). Changes in nucleic acid concentration during the development of *Xenopus laevis* embryos. *Exp. Cell Res.* **35**, 580.

Britten, R. J. (1969). The arithmetic of nucleic acid reassociation. *Carnegie Inst. Washington, Yearb.* **67**, 332.

Britten, R. J. (1972). DNA sequence interspersion and a speculation about evolution. *In*

"Evolution of Genetic Systems" (H. H. Smith, ed.), p. 80. Gordon & Breach, New York.

Britten, R. J., and Davidson, E. H. (1969). Gene regulation for higher cells: A theory. *Science* 165, 349.

Britten, R. J., and Davidson, E. H. (1971). Repetitive and non-repetitive DNA sequences and a speculation on the origins of evolutionary novelty. *Q. Rev. Biol.* 46, 111.

Britten, R. J., and Davidson, E. H. (1976). Studies on nucleic acid reassociation kinetics: Empirical equations describing DNA reassociation. *Proc. Natl. Acad. Sci. U.S.A.* 73, 415.

Britten, R. J., and Kohne, D. E. (1967). Nucleotide sequence repetition in DNA. *Carnegie Inst. Washington, Yearb.* 65, 78.

Britten, R. J., and Kohne, D. E. (1968a). Repeated sequences in DNA. *Science* 161, 529.

Britten, R. J., and Kohne, D. (1968b). Repeated nucleotide sequences. *Carnegie Inst. Washington, Yearb.* 66, 73.

Britten, R. J., Graham, D. E., and Neufeld, B. R. (1974). Analysis of repeating DNA sequences by reassociation. *In* "Methods in Enzymology" (L. Grossman and K. Moldave, eds.), Vol. 29, Part E, p. 363. Academic Press, New York.

Britten, R. J., Graham, D. E., Eden, F. C., Painchaud, D. M., and Davidson, E. H. (1976). Evolutionary divergence and length of repetitive sequences in sea urchin DNA. *J. Molec. Evol.*, in press.

Brookbank, J. W., and Cummins, J. E. (1972). Microspectrophotometry of nuclear DNA during the early development of a sea urchin, a sand dollar, and their interordinal hybrids. *Dev. Biol.* 29, 234.

Brown, D. D. (1966). The nucleolus and synthesis of ribosomal RNA during oogenesis and embryogenesis of *Xenopus laevis*. *Natl. Cancer Inst., Monogr.* 23, 297.

Brown, D. D., and Blackler, A. W. (1972). Gene amplification proceeds by a chromosome copy mechanism. *J. Mol. Biol.* 63, 75.

Brown, D. D., and Dawid, I. B. (1968). Specific gene amplification in oocytes. *Science* 160, 272.

Brown, D. D., and Gurdon, J. B. (1964). Absence of ribosomal RNA synthesis in the anucleolate mutant of *Xenopus laevis*. *Proc. Natl. Acad. Sci. U.S.A.* 51, 139.

Brown, D. D., and Gurdon, J. B. (1966). Size distribution and stability of DNA-like RNA synthesized during development of anucleolate embryos of *Xenopus laevis*. *J. Mol. Biol.* 19, 399.

Brown, D. D., and Littna, E. (1964). RNA synthesis during the development of *Xenopus laevis*, the South African clawed toad. *J. Mol. Biol.* 8, 669.

Brown, D. D., and Littna, E. (1966a). Synthesis and accumulation of DNA-like RNA during embryogenesis of *Xenopus laevis*. *J. Mol. Biol.* 20, 81.

Brown, D. D., and Littna, E. (1966b). Synthesis and accumulation of low molecular weight RNA during embryogenesis of *Xenopus laevis*. *J. Mol. Biol.* 20, 95.

Brown, D. D., and Weber, C. S. (1968). Gene linkage by RNA-DNA hybridization. I. Unique DNA sequences homologous to 4 S RNA, 5 S RNA and ribosomal RNA. *J. Mol. Biol.* 34, 661.

Brown, E. H., and King, R. C. (1964). Studies on the events resulting in the formation of an egg chamber in *Drosophila melanogaster*. *Growth* 28, 41.

Brown, I. R., and Church, R. B. (1971). RNA transcription from nonrepetitive DNA in the mouse. *Biochem. Biophys. Res. Commun.* 42, 850.

Brown, I. R., and Church, R. B. (1972). Transcription of nonrepeated DNA during mouse and rabbit development. *Dev. Biol.* 29, 73.

Brownlee, G. G., Cartwright, E. M., and Brown, D. D. (1974). Sequence studies on the 5 S DNA of *Xenopus laevis*. *J. Mol. Biol.* 89, 703.

Brutlag, D. L., and Peacock, W. J. (1975). Sequences of highly repeated DNA in *Drosophila melanogaster*. *In* "The Eukaryote Chromosome" (W. J. Peacock and R. D. Brock, eds.), p. 35. Aust. Nat. Univ. Press, Canberra.

Bull, A. L. (1966). Bicaudal, a genetic factor which affects the polarity of the embryo in *Drosophila melanogaster*. *J. Exp. Zool.* **161**, 221.

Burnett, A. L., Lowell, R., and Cyrlin, M. N. (1973). Regeneration of a complete *Hydra* from a single, differentiated somatic cell type. *In* "Biology of Hydra" (A. L. Burnett, ed.), p. 255. Academic Press, New York.

Burnside, B., Kozak, C., and Kafatos, F. C. (1973). Tubulin determination by an isotope dilution-vinblastine precipitation method. *J. Cell Biol.* **59**, 755.

Bütschli, O. (1875). Vorlaüfige Mittheilung über Untersuchungen betreffend die ersten Entwickelungsvorgänge im befruchteten Ei von Nematoden und Schnecken. *Z. Wiss. Zool.* **25**, 201.

Byrd, E. W., Jr., and Kasinsky, H. E. (1973). Nuclear accumulation of newly synthesized histones in early *Xenopus* development. *Biochim. Biophys. Acta* **331**, 430.

Calarco, P. G., and Brown, E. H. (1969). An ultrastructural and cytological study of preimplantation development of the mouse. *J. Exp. Zool.* **171**, 253.

Calarco, P. G., and Epstein, C. J. (1973). Cell surface changes during preimplantation development in the mouse. *Dev. Biol.* **32**, 208.

Callan, H. G. (1955). Recent work on the structure of cell nuclei. *Union Int. Sci. Biol., Ser. B* **21**, 89.

Callan, H. G. (1957). The lampbrush chromosomes of *Sepia officianalis* L., *Anilocra physodes* L., and *Scyllium catulus* Cuv. and their structural relationship to the lampbrush chromosomes of amphibia. *Pubbl. Stn. Zool. Napoli* **29**, 329.

Callan, H. G. (1963). The nature of lampbrush chromosomes. *Int. Rev. Cytol.* **15**, 1.

Callan, H. G. (1967). The organization of genetic units in chromosomes. *J. Cell Sci.* **2**, 1.

Callan, H. G. (1972). Replication of DNA in the chromosomes of eukaryotes. *Proc. R. Soc. London, Ser. B* **181**, 19.

Callan, H. G., and Lloyd, L. (1960). Lampbrush chromosomes of crested newts *Triturus cristatus* (Laurenti). *Philos. Trans. R. Soc. London, Ser. B* **243**, 135.

Camey, T., and Geilenkirchen, W. L. M. (1970). Cleavage delay and abnormal morphogenesis in *Lymnaea* eggs after pulse treatment with azide of successive stages in a cleavage cycle. *J. Embryol. Exp. Morphol.* **23**, 385.

Campo, M. S., and Bishop, J. O. (1974). Two classes of messenger RNA in cultured rat cells: Repetitive sequence transcripts and unique sequence transcripts. *J. Mol. Biol.* **90**, 649.

Cape, M., and Decroly, M. (1969). Mesure de la capacite "template" des acides ribonucleiques des oeufs de *Xenopus laevis* au cours du developpement. *Biochim. Biophys. Acta* **174**, 99.

Carroll, C. R., and Van Deusen, E. B. (1973). Experimental studies on a mutant gene (*cl*) in the Mexican axolotl which affects cell membrane formation in embryos from *cl/cl* females. *Dev. Biol.* **32**, 155.

Carroll, E. J., and Epel, D. (1975). Isolation and biological activity of the proteases released by sea urchin eggs following fertilization. *Dev. Biol.* **44**, 22.

Cassidy, J. D., and King, R. C. (1972). Ovarian development in *Habrobracon juglandis* (Ashmead) (Hymenoptera: Braconidae). I. The origin and differentation of the oocyte-nurse cell complex. *Biol. Bull.* **143**, 483.

Caston, J. D. (1962). Appearance of catecholamines during development of *Rana pipiens*. *Dev. Biol.* **5**, 468.

Cather, J. N. (1967). Cellular interactions in the development of the shell gland of the gastropod, *Ilyanassa*. *J. Exp. Zool.* **166**, 205.

Cather, J. N. (1971). Cellular interactions in the regulation of development in annelids and molluscs. *Adv. Morphog.* **9**, 67.

Cather, J. N. (1973). Regulation of apical cilia development by the polar lobe of *Ilyanassa* (Gastropoda: Nassariidae). *Malacologia* **12**, 213.

Cather, J. N., and Verdonk, N. H. (1974). The development of *Bithynia tentaculata* (Prosobranchia, Gastropoda) after removal of the polar lobe. *J. Embryol. Exp. Morphol.* **31**, 415.

Cave, M. D. (1973). Synthesis and characterization of amplified DNA in oocytes of the house cricket, *Acheta domesticus* (Orthoptera: Gryllidae). *Chromosoma* **42**, 1.

Cave, M. D. (1975). Absence of ribosomal DNA amplification in the meroistic (telotrophic) ovary of the large milkweed bug *Oncopeltus fasciatus* (Dallas) (Hemiptera: Lygaeidae). *J. Cell Biol.* **66**, 461.

Ceccarini, C., and Maggio, R. (1969). A study of aminoacyl transfer RNA synthetases by methylated albumin Kieselguhr column chromatography in *Paracentrotus lividus*. *Biochim. Biophys. Acta* **190**, 556.

Chaffee, R. R., and Mazia, D. (1963). Echinochrome synthesis in hybrid sea urchin embryos. *Dev. Biol.* **7**, 502.

Chamberlain, J. P. (1970). RNA synthesis in anucleate egg fragments and normal embryos of the sea urchin, *Arbacia punctulata*. *Biochim. Biophys. Acta* **213**, 183.

Chamberlain, J. P., and Metz, C. B. (1972). Mitochondrial RNA synthesis in sea urchin embryos. *J. Mol. Biol.* **64**, 593.

Chamberlin, M. E., Britten, R. J., and Davidson, E. H. (1975). Sequence organization in *Xenopus* DNA studied by the electron microscope. *J. Mol. Biol.* **96**, 317.

Chan, L., Means, A. R., and O'Malley, B. W. (1973). Rates of induction of specific translatable messenger RNAs for ovalbumin and avidin by steroid hormones. *Proc. Natl. Acad. Sci. U.S.A.* **70**, 1870.

Chan, L.-N., and Gehring, W. (1971). Determination of blastoderm cells in *Drosophila melanogaster*. *Proc. Natl. Acad. Sci. U.S.A.* **68**, 2217.

Chapman, V. M., Whitten, W. K., and Ruddle, F. H. (1971). Expression of paternal glucose phosphate isomerase-1 (Gpi-1) in preimplantation stages of mouse embryos. *Dev. Biol.* **26**, 153.

Chase, J. W. (1970). Formation of mitochondria during embryogenesis of *Xenopus laevis*. *Carnegie Inst. Washington, Yearb.* **68**, 517.

Chase, J. W., and Dawid, I. B. (1972). Biogenesis of mitochondria during *Xenopus laevis* development. *Dev. Biol.* **27**, 504.

Chen, P. S. (1960). The rate of oxygen consumption in the lethal hybrid between *Triton* ♀ and *Salamandra* ♂. *Exp. Cell Res.* **19**, 621.

Chen, P. S. (1967). Biochemistry of nucleo-cytoplasmic interactions in morphogenesis. *In* "The Biochemistry of Animal Development" (R. Weber, ed.), Vol. 2, p. 115. Academic Press, New York.

Chetsanga, C. J., Poccia, D. L., Hill, R. J., and Doty, P. (1970). Stage-specific RNA transcription in developing sea urchins and their chromatins. *Cold Spring Harbor Symp. Quant. Biol.* **35**, 629.

Child, C. M. (1900). The early development of *Arenicola* and *Sternapsis*. *Arch. Entwicklungsmech. Org.* **9**, 587.

Child, C. M. (1940). Lithium and echinoderm exogastrulation: With a review of the physiological-gradient concept. *Physiol. Zool.* **13**, 4.

Chun, C. (1880). "Die Ctenophoren des Golfes von Neapel. Fauna and Flora des Golfes von Neapel," Monogr. I. Engelmann, Leipzig.

Chung, H.-M., and Malacinski, G. M. (1975). Repair of ultraviolet irradiation damage to a cytoplasmic component required for neural induction in the amphibian egg. *Proc. Natl. Acad. Sci. U.S.A.* **72**, 1235.

Church, R. B., and Brown, I. R. (1972). Tissue specificity of genetic transcription. *In* "Results and Problems in Cell Differentiation" (H. Ursprung, ed.), Vol. 3, p. 11. Springer-Verlag, Berlin and New York.

Church, R. B., and McCarthy, B. J. (1967). Ribonucleic acid synthesis in regenerating and embryonic liver. II. The synthesis of RNA during embryonic liver development and its relationship to regenerating liver. *J. Mol. Biol.* **23**, 477.

Church, R. B., and McCarthy, B. J. (1970). Unstable nuclear RNA synthesis following estrogen stimulation. *Biochim. Biophys. Acta* **199**, 103.

Clandinin, M. T., and Schultz, G. A. (1975). Levels and modification of methionyl-transfer RNA in preimplantation rabbit embryos. *J. Mol. Biol.* **93**, 517.

Clarkson, S. G., Birnstiel, M. L., and Serra, V. (1973a). Reiterated transfer RNA genes of *Xenopus laevis*. *J. Mol. Biol.* **79**, 391.

Clarkson, S. G., Birnstiel, M. L., and Purdom, I. F. (1973b). Clustering of transfer RNA genes of *Xenopus laevis*. *J. Mol. Biol.* **79**, 411.

Clegg, J. S. (1967). Metabolic studies of cryptobiosis in encysted embryos of *Artemia salina*. *Comp. Biochem. Physiol.* **20**, 801.

Clegg, J. S., and Golub, A. L. (1969). Protein synthesis in *Artemia salina* embryos. II. Resumption of RNA and protein synthesis upon cessation of dormancy in the encysted gastrula. *Dev. Biol.* **19**, 178.

Clegg, K. B., and Denny, P. C. (1974). Synthesis of rabbit globin in a cell-free protein synthesis system utilizing sea urchin egg and zygote ribosomes. *Dev. Biol.* **37**, 263.

Clement, A. C. (1952). Experimental studies on germinal localization in *Ilyanassa*. I. The role of the polar lobe in determination of the cleavage pattern and its influence in later development. *J. Exp. Zool.* **121**, 593.

Clement, A. C. (1956). Experimental studies on germinal localization in *Ilyanassa*. II. The development of isolated blastomeres. *J. Exp. Zool.* **132**, 427.

Clement, A. C. (1962). Development of *Ilyanassa* following removal of the D macromere at successive cleavage stages. *J. Exp. Zool.* **149**, 193.

Clement, A. C. (1963). Effects of micromere deletion on development in *Ilyanassa*. *Biol. Bull.* **125**, 375.

Clement, A. C. (1967). The embryonic value of the micromeres in *Ilyanassa obsoleta*, as determined by deletion experiments. I. The first quartet cells. *J. Exp. Zool.* **166**, 77.

Clement, A. C. (1968). Development of the vegetal half of the *Ilyanassa* egg after removal of most of the yolk by centrifugal force, compared with the development of animal halves of similar visible composition. *Dev. Biol.* **17**, 165.

Clement, A. C., and Tyler, A. (1967). Protein-synthesizing activity of the anucleate polar lobe of the mud snail *Ilyanassa obsoleta*. *Science* **158**, 1457.

Coggins, L. W. (1973). An ultrastructural and radioautographic study of early oogenesis in the toad *Xenopus laevis*. *J. Cell Sci.* **12**, 71.

Coggins, L. W., and Gall, J. G. (1972). The timing of meiosis and DNA synthesis during early oogenesis in the toad, *Xenopus laevis*. *J. Cell Biol.* **52**, 569.

Cohen, G. H., and Iverson, R. M. (1967). High-resolution density-gradient analysis of sea urchin polysomes. *Biochem. Biophys. Res. Commun.* **29**, 349.

Cohen, L. H., Mahowald, A. P., Chalkley, R., and Zweidler, A. (1973). Histones of early embryos. *Fed. Proc., Fed. Am. Soc. Exp. Biol.* **32**, 588.

Cohen, L. H., Newrock, K. M., and Zweidler, A. (1975). Stage-specific switches in histone synthesis during embryogenesis of the sea urchin. *Science* **190**, 994.

Cohen, S. (1954). The metabolism of $C^{14}O_2$ during amphibian development. *J. Biol. Chem.* **211**, 337.

Cohen, W. O., and Rebhun, L. I. (1970). An estimate of the amount of microtubule protein in the isolated mitotic apparatus. *J. Cell Sci.* **6**, 159.

Collier, J. R. (1960). The localization of ribonucleic acid in the egg of *Ilyanassa obsoleta*. *Exp. Cell Res.* **21**, 126.

Collier, J. R. (1961). Nucleic acid and protein metabolism of the *Ilyanassa* embryo. *Exp. Cell Res.* **24**, 320.

Collier, J. R. (1965a). Morphogenetic significance of biochemical patterns in mosaic embryos. *In* "Biochemistry of Animal Development" (R. Weber, ed.), Vol. 1, p. 203. Academic Press, New York.

Collier, J. R. (1965b). Ribonucleic acids of the *Ilyanassa* embryo. *Science* **147**, 150.

Collier, J. R. (1966). The transcription of genetic information in the spiralian embryo. *Curr. Top. Dev. Biol.* **1**, 39.

Collier, J. R. (1975a). Polyadenylation of nascent RNA during the embryogenesis of *Ilyanassa obsoleta*. *Exp. Cell Res.* **95**, 263.

Collier, J. R. (1975b). Nucleic acid synthesis in the normal and lobeless embryo of *Ilyanassa obsoleta*. *Exp. Cell Res.* **95**, 254.

Collier, J. R., and Schwartz, R. (1969). Protein synthesis during *Ilyanassa* embryogenesis. *Exp. Cell Res.* **54**, 403.

Collier, J. R., and Yuyama, S. (1969). Characterization of DNA-like RNA in the *Ilyanassa* embryo. *Exp. Cell Res.* **56**, 281.

Colman, A. (1974). Synthesis of RNA in oocytes of *Xenopus laevis* during culture *in vitro*. *J. Embryol. Exp. Morphol.* **32**, 515.

Comb, D. G., and Brown, R. (1964). Preliminary studies on the degradation and synthesis of RNA components during sea urchin development. *Exp. Cell Res.* **34**, 360.

Comb, D. G., Katz, S., Branda, R., and Pinzino, C. J. (1965). Characterization of RNA species synthesized during early development of sea urchins. *J. Mol. Biol.* **14**, 195.

Conklin, E. G. (1897). The embryology of *Crepidula*. *J. Morphol.* **13**, 1.

Conklin, E. G. (1905). The organization and cell lineage of the ascidian egg. *J. Acad. Natl. Sci. Philadelphia* **13**, 1.

Cordeiro-Stone, M., and Lee, C. S. (1976). Studies on the satellite DNAs of *Drosophila nasutoides*: Their buoyant densities, melting temperatures, reassociation rates, and localizations in polytene chromosomes. *J. Mol. Biol.* **104**, 1.

Costello, D. P. (1945). Experimental studies of germinal localization in *Nereis*. I. The development of isolated blastomeres. *J. Exp. Zool.* **100**, 19.

Costello, D. P. (1956). Cleavage, blastulation and gastrulation. *In* "Analysis of Development" (B. H. Willier, P. A. Weiss, and V. Hamburger, eds.), p. 213. Saunders, Philadelphia, Pennsylvania.

Counce, S. J. (1973). The causal analysis of insect embryogenesis. *In* "Developmental Systems: Insects" (S. J. Counce and C. H. Waddington, eds.), Vol. 2, p. 1. Academic Press, New York.

Craig, S. P. (1970). Synthesis of RNA in non-nucleate fragments of sea urchin eggs. *J. Mol. Biol.* **47**, 615.

Craig, S. P., and Piatigorsky, J. (1971). Protein synthesis and development in the absence of cytoplasmic RNA synthesis in nonnucleate egg fragments and embryos of sea urchins: Effect of ethidium bromide. *Dev. Biol.* **24**, 214.

Crain, W. R., Eden, F. C., Pearson, W. R., Davidson, E. H., and Britten, R. J. (1976a). Absence of short period interspersion of repetitive and nonrepetitive sequences in the DNA of *Drosophila melanogaster*. *Chromosoma* **56**, 309.

Crain, W. R., Davidson, E. H., and Britten, R. J. (1976b). Contrasting patterns of DNA sequence arrangement in *Apis mellifera* (honeybee) and in *Musca domestica* (housefly). *Chromosoma*, in press.

Crampton, H. E. (1896). Experimental studies on gasteropod development. *Arch. Entwicklungsmech. Org.* **3**, 1.

Crawford, R. B., and Wilde, C. E., Jr. (1966). Cellular differentiation in the anamniota. IV. Relationship between RNA synthesis and aerobic metabolism in *Fundulus heteroclitus* embryos. *Exp. Cell Res.* **44**, 489.

Crick, F. (1971). General model for the chromosomes of higher organisms. *Nature (London)* **234**, 25.

Crick, F. H., and Lawrence, P. A. (1975). Compartments and polyclones in insect development. *Science* **189**, 340.

Crippa, M., Davidson, E. H., and Mirsky, A. E. (1967). Persistence in early amphibian embryos of informational RNA's from the lampbrush chromosome stage of oögenesis. *Proc. Natl. Acad. Sci. U.S.A.* **57**, 885.

Crouse, G. F., Fodor, E. J. B., and Doty, P. (1976). *In vitro* transcription of chromatin in the presence of a mercurated nucleotide. *Proc. Natl. Acad. Sci. U.S.A.* **73**, 1564.

Curtis, A. S. G. (1962). Morphogenetic interactions before gastrulation in the amphibian *Xenopus laevis*—the cortical field. *J. Embryol. Exp. Morphol.* **10**, 410.

Czihak, G. (1965). Entwicklungsphysiologische Untersuchungen an Echiniden. Ribonucleinsäure-synthese in den Mikromeren und Entodermdifferenzierung ein Beitrag zum Problem der Induktion. *Wilhelm Roux' Arch. Entwicklungmech. Org.* **156**, 504.

Czihak, G., and Hörstadius, S. (1970). Transplantation of RNA-labeled micromeres into animal halves of sea urchin embryos. A contribution to the problem of embryonic induction. *Dev. Biol.* **22**, 15.

Daentl, D. L., and Epstein, C. J. (1971). Developmental interrelationships of uridine uptake, nucleotide formation and incorporation into RNA by early mammalian embryos. *Dev. Biol.* **24**, 428.

Daentl, D. L., and Epstein, C. J. (1973). Uridine transport by mouse blastocysts. *Dev. Biol.* **31**, 316.

Daneholt, B., and Hosick, H. (1974). The transcription unit in Balbiani ring 2 of *Chironomus tentans*. *Cold Spring Harbor Symp. Quant. Biol.* **38**, 629.

Daniel, J. C., and Flickinger, R. A. (1971). Nuclear DNA-like RNA in developing frog embryos. *Exp. Cell Res.* **64**, 285.

Daniel, J. C., Jr. (1964). Early growth of rabbit trophoblast. *Am. Nat.* **98**, 85.

Dapples, C. C., and King, R. C. (1970). The development of the nucleolus of the ovarian nurse cell of *Drosophila melanogaster*. *Z. Zellforsch. Mikrosk. Anat.* **103**, 34.

Darnbrough, C., and Ford, P. J. (1976). Cell-free translation of messenger RNA from oocytes of *Xenopus laevis*. *Dev. Biol.* **50**, 285.

Darnell, J. E., and Balint, R. (1970). The distribution of rapidly hybridizing RNA sequences in heterogeneous nuclear RNA and mRNA from HeLa cells. *J. Cell. Physiol.* **76**, 349.

Darnell, J. E., Jelinek, W. R., and Molloy, G. R. (1973). Biogenesis of mRNA: Genetic regulation in mammalian cells. *Science* **181**, 1215.

Davidson, E. H. (1968). "Gene Activity in Early Development." Academic Press, New York.

Davidson, E. H., and Britten, R. J. (1971). Note on the control of gene expression during development. *J. Theor. Biol.* **32**, 123.

Davidson, E. H., and Britten, R. J. (1973). Organization, transcription, and regulation in the animal genome. *Q. Rev. Biol.* **48**, 565.

Davidson, E. H., and Britten, R. J. (1974). Molecular aspects of gene regulation in animal cells. *Cancer Res.* **34**, 2034.

Davidson, E. H., and Hough, B. R. (1969a). Synchronous oogenesis in *Engystomops pus-*

tulosus, a neotropic anuran suitable for laboratory studies; localization in the embryo of RNA synthesized at the lampbrush stage. *J. Exp. Zool.* **172**, 25.

Davidson, E. H., and Hough, B. R. (1969b). High sequence diversity in the RNA synthesized at the lampbrush stage of oögenesis. *Proc. Natl. Acad. Sci. U.S.A.* **63**, 342.

Davidson, E. H., and Hough, B. R. (1971). Genetic information in oocyte RNA. *J. Mol. Biol.* **56**, 491.

Davidson, E. H., Allfrey, V. G., and Mirsky, A. E. (1964). On the RNA synthesized during the lampbrush phase of amphibian oogenesis. *Proc. Natl. Acad. Sci. U.S.A.* **52**, 501.

Davidson, E. H., Haslett, G. W., Finney, R. J., Allfrey, V. G., and Mirsky, A. E. (1965). Evidence for prelocalization of cytoplasmic factors affecting gene activation in early embryogenesis. *Proc. Natl. Acad. Sci. U.S.A.* **54**, 696.

Davidson, E. H., Crippa, M., Kramer, F. R., and Mirsky, A. E. (1966). Genomic function during the lampbrush chromosome stage of amphibian oogenesis. *Proc. Natl. Acad. Sci. U.S.A.* **56**, 856.

Davidson, E. H., Crippa, M., and Mirsky, A. E. (1968). Evidence for the appearance of novel gene products during amphibian development. *Proc. Natl. Acad. Sci. U.S.A.* **60**, 152.

Davidson, E. H., Hough, B. R., Amenson, C. S., and Britten, R. J. (1973). General interspersion of repetitive with non-repetitive sequence elements in the DNA of *Xenopus*. *J. Mol. Biol.* **77**, 1.

Davidson, E. H., Graham, D. E., Neufeld, B. R., Chamberlin, M. E., Amenson, C. S., Hough, B. R., and Britten, R. J. (1974). Arrangement and charcterization of repetitive sequence elements in animal DNA's. *Cold Spring Harbor Symp. Quant. Biol.* **38**, 295.

Davidson, E. H., Galau, G. A., Angerer, R. C., and Britten, R. J. (1975a). Comparative aspects of DNA organization in metazoa. *Chromosoma* **51**, 253.

Davidson, E. H., Hough, B. R., Klein, W. H., and Britten, R. J. (1975b). Structural genes adjacent to interspersed repetitive DNA sequences. *Cell* **4**, 217.

Davidson, E. H., Klein, W. H., and Britten, R. J. (1976a). Sequence organization in animal DNA and a speculation on hnRNA as a coordinate regulatory transcript. *Dev. Biol.*, in press.

Davidson, E. H., Klein, W. H., Hough-Evans, B. R., Smith, M. J., Galau, G. A., Crain, W. R., Angerer, R. C., Eden, F. C., Wold, B. J., Davis, M. M., and Britten, R. J. (1976b). The organization of functional DNA sequences in animal genomes. *In* "Organization and Expression of the Eukaryote Genome," Tehran Symposium in Molecular Biology (K. Javaherian and E. M. Bradbury, eds.). Academic Press, New York. In press.

Davis, F. C., Jr. (1975). Unique sequence DNA transcripts present in mature oocytes of *Urechis caupo*. *Biochim. Biophys. Acta* **390**, 33.

Davis, F. C., Jr., and Wilt, F. H. (1972). RNA synthesis during oogenesis in the echiuroid worm *Urechis caupo*. *Dev. Biol.* **27**, 1.

Davis, R. W., Simon, M., and Davidson, N. (1971). Electron microscope heteroduplex methods for mapping regions of base sequence homology in nucleic acids. *In* "Methods in Enzymology" (L. Grossman and K. Moldave, eds.), Vol. 21, Part D, p. 413. Academic Press, New York.

Dawid, I. B. (1965). Deoxyribonucleic acid in amphibian eggs. *J. Mol. Biol.* **12**, 581.

Dawid, I. B. (1972). Cytoplasmic DNA. *In* "Oogenesis" (J. D. Biggers and A. W. Schuetz, eds.), p. 215. Univ. Park Press, Baltimore, Maryland.

Dawid, I. B., and Brown, D. D. (1970). The mitochondrial and ribosomal DNA components of oocytes of *Urechis caupo*. *Dev. Biol.* **22**, 1.

de Angelis, E., and Runnström, J. (1970). The effect of temporary treatment of animal half embryos with lithium and the modification of this effect by simultaneous exposure to actinomycin D. *Wilhelm Roux' Arch. Entwicklungsmech. Org.* **164**, 236.

Decroly, M., Cape, M., and Brachet, J. (1964). Studies on the synthesis of ribonucleic acids in embryonic stages of *Xenopus laevis. Biochim. Biophys. Acta* 87, 34.

de Lange, R. J., Fambrough, D. M., Smith, E. L., and Bonner, J. (1968). Calf and pea histone IV. I. Amino acid compositions and the identical COOH-terminal 19-residue sequence. *J. Biol. Chem.* 243, 5906.

de Lange, R. J., Fambrough, D. M., Smith, E. L., and Bonner, J. (1969a). Calf and pea histone IV. II. The complete amino acid sequence of calf thymus histone IV; presence of ϵ-N-acetyllysine. *J. Biol. Chem.* 244, 319.

de Lange, R. J., Fambrough, D. M., Smith, E. L., and Bonner, J. (1969b). Calf and pea histone IV. III. Complete amino acid sequence of pea seedling histone IV; comparison with the homologous calf thymus histone. *J. Biol. Chem.* 244, 5669.

de Lange, R. J., Hooper, J. A., and Smith, E. L. (1973). Histone III. III. Sequence studies on the cyanogen bromide peptides; complete amino acid sequence of calf thymus histone III. *J. Biol. Chem.* 248, 3261.

Delobel, N. (1971). Etude descriptive des chromosomes en écouvillon chez *Echinaster sepositus* (Échinoderme, Astéride). *Ann. Embryol. Morphog.* 4, 383.

Demerec, M., Kaufmann, B. P., Fano, U., Sutton, E., and Sansome, E. R. (1942). The gene. *Carnegie Inst. Washington, Yearb.* 41, 190.

Denis, H. (1966). Gene expression in amphibian development. II. Release of the genetic information in growing embryos. *J. Mol. Biol.* 22, 285.

Denis, H., and Brachet, J. (1969). Gene expression in interspecific hybrids, I. DNA synthesis in the lethal cross *Arbacia lixula* ♂ × *Paracentrotus lividus* ♀. *Proc. Natl. Acad. Sci. U.S.A.* 62, 194.

Denis, H., and Mairy, M. (1972). Recherches biochimiques sur l'oogénèse. 2. Distribution intracellulaire du RNA dans les petits oocytes de *Xenopus laevis. Eur. J. Biochem.* 25, 524.

Denis, H., and Wegnez, M. (1973). Recherches biochimiques sur l'oogénèse. 7. Synthèse et maturation du RNA 5 S dans les petits oocytes de *Xenopus laevis. Biochimie* 55, 1137.

Denis, H., Wegnez, M., and Williem, R. (1972). Recherches biochimiques sur l'oogénèse. V. Comparaison entre le RNA 5 S somatique et le RNA 5 S des oocytes de *Xenopus laevis. Biochimie* 54, 1189.

Denis, H., Mazabraud, A., and Wegnez, M. (1975). Biochemical research on oogenesis. Comparison between transfer RNAs from somatic cells and from oocytes in *Xenopus laevis. Eur. J. Biochem.* 58, 43.

Denny, P. C., and Reback, P. (1970). Active polysomes in sea urchin eggs and zygotes: Evidence for an increase in translatable messenger RNA after fertilization. *J. Exp. Zool.* 175, 133.

Denny, P. C., and Tyler, A. (1964). Activation of protein biosynthesis in non-nucleate fragments of sea urchin eggs. *Biochem. Biophys. Res. Commun.* 14, 245.

De Petrocellis, B., and Monroy, A. (1974). Regulatory processes of DNA synthesis in the embryo. *Endeavour* 33, 92.

De Petrocellis, B., and Vittorelli, M. L. (1975). Role of cell interactions in development and differentiation of the sea urchin *Paracentrotus lividus*. Changes in the activity of some enzymes of DNA biosynthesis after cell dissociation. *Exp. Cell Res.* 94, 392.

Destree, O. H. J., d'Adelhart Toorop, H. A., and Charles, R. (1973). Analysis of histones from different tissues and embryos of *Xenopus laevis* (Daudin). II. Qualitative and quantitative aspects of nuclear histones during early stages of development. *Cell Differ.* 2, 229.

De Vincentiis, M., and Lancieri, M. (1970). Observations on the development of the sea urchin embryos in the presence of actinomycin. *Exp. Cell Res.* 59, 479.

Devlin, R. (1976). Mitochondrial poly(A)RNA synthesis during early sea urchin development. *Dev. Biol.* 50, 443.

D'Hollander, F. (1904). Recherches sur l'oogénèse et sur la structure et la signification du noyau vitellin de Balbiani chez les oizeaux. *Arch. Anat. Microsc. Morphol. Exp.* **7**, 117.

Dicaprio, R. A., French, A. S., and Sanders, E. J. (1975). Intercellular connectivity in the eight-cell *Xenopus* embryo. Correlation of electrical and morphological investigations. *Biophys. J.* **15**, 373.

Dohmen, M. R., and Lok, D. (1975). The ultrastructure of the polar lobe of *Crepidula fornicata* (Gasteropoda, Prosobranchia). *J. Embryol. Exp. Morphol.* **34**, 419.

Dohmen, M. R., and Verdonk, N. H. (1974). The structure of a morphogenetic cytoplasm, present in the polar lobe of *Bithynia tentaculata* (Gastropoda, Prosobranchia). *J. Embryol. Exp. Morphol.* **31**, 423.

Donohoo, P., and Kafatos, F. C. (1973). Differences in the proteins synthesized by the progeny of the first two blastomeres of *Ilyanassa*, a "mosaic" embryo. *Dev. Biol.* **32**, 224.

Driesch, H. (1891). Entwicklungsmechanische Studien. I. Der Werth der beiden ersten Furchungszellen in der Echinodermentwicklung. Experimentelle Erzeugung von Theil-und Doppelbildungen. II. Über die Beziehungen des Lichtes zur ersten Etappe der thierischen Formbildung. *Z. Wiss. Zool.* **53**, 160.

Driesch, H. (1892). Entwickelungsmechanisches. *Anat. Anz.* **7**, 584.

Driesch, H. (1898). Über rein-mütterliche Charaktere an Bastardlarven von Echiniden. *Arch. Entwicklungsmech. Org.* **7**, 65.

Driesch, H. (1900). Die isolirten Blastomeren des Echinidenkeimes. Eine Nachprüfung und Erweiterung früherer Untersuchungen. *Arch. Entwicklungsmech. Org.* **10**, 361.

Driesch, H., and Morgan, T. H. (1896). Zur Analysis der ersten Entwickelungsstadien des Ctenophoreneies. I. Von der Entwickelung einzelner Ctenophorenblastomeren. *Arch. Entwicklungsmech. Org.* **2**, 204.

Dubroff, L. M., and Nemer, M. (1975). Molecular classes of heterogeneous nuclear RNA in sea urchin embryos. *J. Mol. Biol.* **95**, 455.

Ducibella, T., and Anderson, E. (1975). Cell shape and membrane changes in the eight-cell mouse embryo: Prerequisites for morphogenesis of the blastocyst. *Dev. Biol.* **47**, 45.

Dumont, J. N. (1972). Oogenesis in *Xenopus laevis* (Daudin). 1. Stages of oocyte development in laboratory maintained animals. *J. Morphol.* **136**, 153.

Dunn, L. C., and Gleucksohn-Waelsch, S. (1953). Genetic analysis of seven newly discovered mutant alleles at locus T in the house mouse. *Genetics* **38**, 261.

Duryee, W. (1950). Chromosomal physiology in relation to nuclear structure. *Ann. N.Y. Acad. Sci.* **50**, 920.

Easton, D., and Chalkley, R. (1972). High resolution electrophoretic analysis of the histones from embryos and sperm of *Arbacia punctulata*. *Exp. Cell Res.* **72**, 502.

Easton, D. P., Chamberlain, J. P., Whiteley, A. H., and Whiteley, H. R. (1974). Histone gene expression in interspecies hybrid echinoid embryos. *Biochem. Biophys. Res. Commun.* **57**, 513.

Ecker, R. E. (1972). The regulation of protein synthesis in anucleate frog embryos. *In* "Biology and Radiobiology of Anucleate Systems" (S. Bonotto *et al.*, eds.) Vol. I, p. 165. Academic Press, New York.

Ecker, R. E., and Smith, L. D. (1966). The kinetics of protein synthesis in early amphibian development. *Biochim. Biophys. Acta* **129**, 186.

Ecker, R. E., and Smith, L. D. (1968). Protein synthesis in amphibian oocytes and early embryos. *Dev. Biol.* **18**, 232.

Ecker, R. E., and Smith, L. D. (1971). The nature and fate of *Rana pipiens* proteins synthesized during maturation and early cleavage. *Dev. Biol.* **24**, 559.

Ecker, R. E., Smith, L. D., and Subtelny, S. (1968). Kinetics of protein synthesis in enucleate frog oocytes. *Science* **160**, 1115.

Eddy, E. M. (1975). Germ plasm and the differentiation of the germ cell line. *Int. Rev. Cytol.* **43**, 229.

Edström, J. E., and Gall, J. G. (1963). The base composition of ribonucleic acid in lampbrush chromosomes, nucleoli, nuclear sap, and cytoplasm of *Triturus* oocytes. *J. Cell Biol.* **19**, 279.

Efstratiadis, A., Crain, W. R., Britten, R. J., Davidson, E. H., and Kafatos, F. C. (1976). DNA sequence organization in the lepidopteran *Antheraea pernyi*. *Proc. Natl. Acad. Sci. U.S.A.* **73**, 2289.

Ellem, K. A. O., and Gwatkin, R. B. L. (1968). Patterns of nucleic acid synthesis in the early mouse embryo. *Dev. Biol.* **18**, 311.

Elsdale, T. R., Fischberg, M., and Smith, S. (1958). A mutation that reduces nucleolar number in *Xenopus laevis*. *Exp. Cell Res.* **14**, 642.

Emerson, C. P., Jr., and Humphreys, T. (1970). Regulation of DNA-like RNA and the apparent activation of ribosomal RNA synthesis in sea urchin embryos: Quantitative measurements of newly synthesized RNA. *Dev. Biol.* **23**, 86.

Emerson, C. P., Jr., and Humphreys, T. (1971). Ribosomal RNA synthesis and the multiple, atypical nucleoli in cleaving embryos. *Science* **171**, 898.

Epel, D. (1967). Protein synthesis in sea urchin eggs: A "late" response to fertilization. *Proc. Natl. Acad. Sci. U.S.A.* **57**, 899.

Epel, D. (1975). The program of and mechanisms of fertilization in the echinoderm egg. *Am. Zool.* **15**, 507.

Epstein, C. J., and Daentl, D. L. (1971). Precursor pools and RNA synthesis in preimplantation mouse embryos. *Dev. Biol.* **26**, 517.

Epstein, C. J., and Daentl, D. L. (1972). Molecular events in preimplantation mammalian development. *In* "Cell Differentiation" (R. Harris, P. Allin, and D. Viza, eds.), p. 69. Munksgaard, Copenhagen.

Epstein, C. J., and Smith, S. A. (1973). Amino acid uptake and protein synthesis in preimplantation mouse embryos. *Dev. Biol.* **33**, 171.

Epstein, C. J., and Smith, S. A. (1974). Electrophoretic analysis of proteins synthesized by preimplantation mouse embryos. *Dev. Biol.* **40**, 233.

Erickson, R. P., Betlach, C. J., and Epstein, C. J. (1974). Ribonucleic acid and protein metabolism of t^{12}/t^{12} embryos and T/t^{12} spermatozoa. *Differentiation* **2**, 203.

Evans, L. E., and Ozaki, H. (1973). Nuclear histones of unfertilized sea urchin eggs. *Exp. Cell Res.* **79**, 228.

Fankhauser, G. (1934). Cytological studies on egg fragments of the salamander *Triton*. IV. The cleavage of egg fragments without the egg nucleus. *J. Exp. Zool.* **67**, 349.

Frankhauser, G. (1956). The role of nucleus and cytoplasm. *In* "Analysis of Development" (B. H. Willier, P. A. Weiss, and V. Hamburger, eds.), p. 126. Saunders, Philadelphia, Pennsylvania.

Fansler, B., and Loeb, L. A. (1969). Sea urchin nuclear DNA polymerase. II. Changing localization during early development. *Exp. Cell Res.* **57**, 305.

Fansler, B., and Loeb, L. A. (1972). Sea urchin nuclear DNA polymerase. IV. Reversible association of DNA polymerase with nuclei during the cell cycle. *Exp. Cell Res.* **75**, 433.

Farfaglio, G. (1963a). Experiments on the formation of combs in the ctenophores. *Experientia* **19**, 303.

Farfaglio, G. (1963b). Experiments on the formation of the ciliated plates in ctenophores. *Acta Embryol. Morphol. Exp.* **6**, 191.

Farquhar, M. N., and McCarthy, B. J. (1973). Histone mRNA in eggs and embryos of *Strongylocentrotus purpuratus*. *Biochem. Biophys. Res. Commun.* **53**, 515.

Fausto-Sterling, A., Zheutlin, L. M., and Brown, P. R. (1974). Rates of RNA synthesis during early embryogenesis in *Drosophila melanogaster*. *Dev. Biol.* **40**, 78.

Fawcett, D. W., Ito, S., and Slautterback, D. B. (1959). The occurrence of intercellular bridges in groups of cells exhibiting synchronous differentiation. *J. Biophys. Biochem. Cytol.* **5**, 453.

Fedecka-Bruner, B., Anderson, M., and Epel, D. (1971). Control of enzyme synthesis in early sea urchin development: Aryl sulfatase activity in normal and hybrid embryos. *Dev. Biol.* **25**, 655.

Feigenbaum, L., and Goldberg, E. (1965). Effect of actinomycin D on morphogenesis in *Ilyanassa*. *Am. Zool.* **5**, 198.

Firtel, R. A. (1972). Changes in the expression of single-copy DNA during development of the cellular slime mold *Dictyostelium discoideum*. *J. Mol. Biol.* **66**, 363.

Firtel, R. A., and Monroy, A. (1970). Polysomes and RNA synthesis during early development of the surf clam *Spisula solidissima*. *Dev. Biol.* **21**, 87.

Fischel, A. (1898). Experimentelle Untersuchungen am Ctenophorene. II. Von der Künstlichen Erzeugung (halber) Doppel-und Missbildungen. III. Über Regulation der Entwickelung. IV. Über den Entwickelungsgang und die Organisations-Stufe des Ctenophoreneies. *Arch. Entwicklungsmech. Org.* **7**, 557.

Fischel, A. (1903). Entwickelung und Organ-differenzirung. *Arch. Entwicklungsmech. Org.* **15**, 670.

Fischer, A. (1974). Activity of a gene in transplanted oocytes in the annelid, *Platynereis*. *Wilhelm Roux' Arch. Entwicklungsmech. Org.* **174**, 250.

Flamm, W. G., Walker, P. M. B., and McCallum, M. J. (1969). Renaturation and isolation of single strands from the nuclear DNA of the guinea pig. *J. Mol. Biol.* **42**, 441.

Flickinger, R. A., Jr. (1954). Utilization of $C^{14}O_2$ by developing amphibian embryos with special reference to regional incorporation into individual embryos. *Exp. Cell Res.* **6**, 172.

Foe, V. E., Wilkinson, L. E., and Laird, C. D. (1976). Comparative organization of active transcription units in *Oncopeltus fasciatus*. *Cell* **9**, 131.

Fol, H. (1877). Sur les phénomènes intimes de la fécondation. *C. R. Hebd. Seances Acad. Sci.* **84**, 268.

Fol, H. (1878). Recherches sur la fécondation et le commencement de l'héogénie chez divers animaux. *Mem. Soc. Phys. Geneve* **26**, 12.

Ford, C. C., and Woodland, H. R. (1975). DNA synthesis in oocytes and eggs of *Xenopus laevis* injected with DNA. *Dev. Biol.* **43**, 189.

Ford, C. C., Pestell, R. Q. W., and Benbow, R. M. (1975). Template preferences of DNA polymerase activities appearing during early development of *Xenopus laevis*. *Dev. Biol.* **43**, 175.

Ford, P. J. (1971). Non-coordinated accumulation and synthesis of 5 S ribonucleic acid by ovaries of *Xenopus laevis*. *Nature (London)* **233**, 561.

Ford, P. J., and Southern, E. M. (1973). Different sequences for 5 S RNA in kidney cells and ovaries of *Xenopus laevis*. *Nature (London), New Biol.* **241**, 7.

Franchi, L. L., and Mandl, A. M. (1963). The ultrastructure of oogonia and oocytes in the foetal and neonatal rat. *Proc. R. Soc. London, Ser. B* **157**, 99.

Franchi, L. L., Mandl, A. M., and Zuckerman, S. (1962). The development of the ovary and the process of oogenesis. In "The Ovary" (S. Zuckerman, ed.), Vol. 1, p. 1. Academic Press, New York.

Frederiksen, S., Hellung-Larsen, P., and Engberg, J. (1973). Small molecular weight RNA components in sea urchin embryos. *Exp. Cell Res.* **78**, 287.

Freeman, G. (1976). The role of cleavage in the localization of developmental potential in the ctenophore *Mnemiopsis leidyi*. *Dev. Biol.* **49**, 143.

Freeman, G., and Reynolds, G. T. (1973). The development of bioluminescence in the ctenophore *Mnemiopsis leidyi*. *Dev. Biol.* **31**, 61.

Freeman, S. B. (1971). A comparison of certain isozyme patterns in lobeless and normal embryos of the snail, *Ilyanassa obsoleta*. *J. Embryol. Exp. Morphol.* **26**, 339.

Fromson, D., and Duchastel, A. (1975). Poly(A)-containing polyribosomal RNA in sea urchin embryos: Changes in proportion during development. *Biochim. Biophys. Acta* **378**, 394.

Fromson, D., and Verma, D. P. S. (1976). Translation of nonpolyadenylated messenger RNA of the sea urchin embryos. *Proc. Natl. Acad. Sci. U.S.A.* **73**, 148.

Fry, B. J., and Gross, P. R. (1970a). Patterns and rates of protein synthesis in sea urchin embryos. I. Uptake and incorporation of amino acids during the first cleavage cycle. *Dev. Biol.* **21**, 125.

Fry, B. J., and Gross, P. R. (1970b). Patterns and rates of protein synthesis in sea urchin embryos. II. The calculation of absolute rates. *Dev. Biol.* **21**, 125.

Furshpan, E. J., and Potter, D. D. (1968). Low-resistance junctions between cells in embryos and tissue culture. *Curr. Top. Dev. Biol.* **3**, 95.

Gabrielli, F., and Baglioni, C. (1975). Maternal messenger RNA and histone synthesis in embryos of the surf clam *Spisula solidissima*. *Dev. Biol.* **43**, 254.

Galau, G. A., Britten, R. J., and Davidson, E. H. (1974). A measurement of the sequence complexity of polysomal messenger RNA in sea urchin embryos. *Cell* **2**, 9.

Galau, G. A., Lipson, E. D., Britten, R. J., and Davidson, E. H. (1976a). Synthesis and turnover of polysomal mRNAs in sea urchin embryos. *Cell*, in press.

Galau, G. A., Britten, R. J., and Davidson, E. H. (1976b). Studies on nucleic acid reassociation kinetics. III. Rate of excess RNA-DNA hybridization compared to the rate of DNA renaturation. *Proc. Natl. Acad. Sci. U.S.A.*, in press.

Galau, G. A., Klein, W. H., Davis, M. M., Wold, B. J., Britten, R. J., and Davidson, E. H. (1976c). Structural gene sets active in embryos and adult tissues of the sea urchin. *Cell* **7**, 487.

Gall, J. G. (1954). Lampbrush chromosomes from oocyte nuclei of the newt. *J. Morphol.* **94**, 283.

Gall, J. G. (1955). On the sub-microscopic structure of chromosomes. *Brookhaven Symp. Biol.* **8**, 17.

Gall, J. (1958). Chromosomal differentiation. *In* "The Chemical Basis of Development" (W. D. McElroy and B. Glass, eds.), p. 103. Johns Hopkins Press, Baltimore, Maryland.

Gall, J. (1963). Kinetics of deoxyribonuclease action on chromosomes. *Nature (London)* **198**, 36.

Gall, J. G. (1966). Techniques for the study of lampbrush chromosomes. *Methods Cell Physiol.* **2**, 37.

Gall, J. G. (1968). Differential synthesis of the genes for ribosomal RNA during amphibian oogenesis. *Proc. Natl. Acad. Sci. U.S.A.* **60**, 553.

Gall, J. G. (1969). The genes for ribosomal RNA during oogenesis. *Genetics* **61**, Suppl., 1.

Gall, J. G., and Callan, H. G. (1962). ^{3}H-uridine incorporation in lampbrush chromosomes. *Proc. Natl. Acad. Sci. U.S.A.* **49**, 544.

Gall, J. G., and Rochaix, J.-D. (1974). The amplified ribosomal DNA of dytiscid beetles. *Proc. Natl. Acad. Sci. U.S.A.* **71**, 1819.

Gallien, C.-L., Aimar, C., and Guillet, F. (1973). Nucleocytoplasmic interactions during ontogenesis in individuals obtained by intra- and interspecific nuclear transplantation in the genus *Pleurodeles* (Urodele Amphibian). Morphology, analysis of two enzymatic systems (LDH and MDH) and immunity reactions. *Dev. Biol.* **33**, 154.

Gardner, R. L. (1968). Mouse chimaeras obtained by the injection of cells into the blastocyst. *Nature (London)* **220**, 596.

Gardner, R. L. (1971). Manipulations on the blastocyst. *Adv. Biosci.* **6**, 279.

Gardner, R. L. (1972). An investigation of inner cell mass and trophoblast tissues following their isolation from the mouse blastocyst. With an appendix by M. H. Johnson. *J. Embryol. Exp. Morphol.* **28**, 279.

Gardner, R. L., and Papaioannou, V. E. (1975). Differentiation in the trophectoderm and inner cell mass. *In* "The Early Development of Mammals" (M. Balls and A. E. Wild, ed.), p. 107. Cambridge Univ. Press, London and New York.

Garen, A., and Gehring, W. (1972). Repair of the lethal developmental defect in *deep orange* embryos of *Drosophila* by injection of normal egg cytoplasm. *Proc. Natl. Acad. Sci. U.S.A.* **69**, 2982.

Geigy, R. (1931). Action de l'ultra-violet sur le pole germinal dans l'oeuf de *Drosophila melanogaster* (castration et mutabilité). *Rev. Suisse Zool.* **38**, 187.

Geilenkirchen, W. L. M. (1966). Cell division and morphogenesis of *Limnaea* eggs after treatment with heat pulses at successive stages in early division cycles. *J. Embryol. Exp. Morphol.* **17**, 367.

Geilenkirchen, W. L. M. (1967). Programming of gastrulation during the second cleavage cycle in *Limnaea stagnalis:* A study with lithium chloride and actinomycin D. *J. Embryol. Exp. Morphol.* **16**, 321.

Geilenkirchen, W. L. M., Verdonk, N. H., and Timmermans, L. P. M. (1970). Experimental studies on morphogenetic factors localized in the first and the second polar lobe of *Dentalium* eggs. *J. Embryol. Exp. Morphol.* **23**, 237.

Gelderman, A. H., Rake, A. V., and Britten, R. J. (1971). Transcription of nonrepeated DNA in neonatal and fetal mice. *Proc. Natl. Acad. Sci. U.S.A.* **68**, 172.

Georgiev, G. P. (1972). The structure of transcriptional units in eukaryotic cells. *Curr. Top. Dev. Biol.* **7**, 1.

Geuskens, M. (1969). Mise en évidence au microscope électronique de polysomes actifs dans des lobes polaires isolés d'*Ilyanassa*. *Exp. Cell Res.* **54**, 263.

Geyer-Duszynska, I. (1966). Genetic factors in oogenesis and spermatogenesis in *Cecidomyidae*. *Chromosomes Today* **1**, 174.

Gibbins, J. R., Tilney, L. G., and Porter, K. R. (1969). Microtubules in the formation and development of the primary mesenchyme in *Arbacia punctulata*. *J. Cell Biol.* **41**, 201.

Gilchrist, F. G. (1933). The time relations of determination in early amphibian development. *J. Exp. Zool.* **66**, 15.

Gilmour, R. S., and Paul, J. (1973). Tissue-specific transcription of the globin gene in isolated chromatin. *Proc. Natl. Acad. Sci. U.S.A.* **70**, 3440.

Gilmour, R. S., Harrison, P. R., Windass, J. D., Affara, N. A., and Paul, J. (1974). Globin messenger RNA synthesis and processing during haemoglobin induction in Friend cells. I. Evidence for transcriptional control in clone M2. *Cell Differ.* **3**, 9.

Giorgi, F., and Galleni, L. (1972). The lampbrush chromosomes of *Rana esculenta* L. (Amphibia-Anura). *Caryologia* **25**, 107.

Giudice, G. (1973). "Developmental Biology of the Sea Urchin Embryo." Academic Press, New York.

Giudice, G., and Mutolo, V. (1967). Synthesis of ribosomal RNA during sea-urchin development. *Biochim. Biophys. Acta* **138**, 276.

Giudice, G., Mutolo, V., and Donatuti, G. (1968). Gene expression in sea urchin development. *Wilhelm Roux' Arch. Entwicklungsmech. Org.* **161**, 118.

Giudice, G., Sconzo, G., Ramirez, F., and Albanese, I. (1972). Giant RNA is also found in the cytoplasm in sea urchin embryos. *Biochim. Biophys. Acta* **262**, 401.

Giudice, G., Sconzo, G., Albanese, I., Ortolani, G., and Cammarata, M. (1974). Cytoplasmic giant RNA in sea urchin embryos. I. Proof that it is not derived from artifactual nuclear leakage. *Cell Differ.* **3**, 287.

Glišin, V. R., and Glišin, M. V. (1964). Ribonucleic acid metabolism following fertilization in sea urchin eggs. *Proc. Natl. Acad. Sci. U.S.A.* **52**, 1548.

Glišin, V. R., Glišin, M. V., and Doty, P. (1966). The nature of messenger RNA in the early stages of sea urchin development. *Proc. Natl. Acad. Sci. U.S.A.* **56**, 285.

Golbus, M. S., Calarco, P. G., and Epstein, C. J. (1973). The effects of inhibitors of RNA synthesis (α-amanitin and actinomycin D) on preimplantation mouse embryogenesis. *J. Exp. Zool.* **186**, 207.

Goldberg, R. B., Galau, G. A., Britten, R. J., and Davidson, E. H. (1973). Nonrepetitive DNA sequence representation in sea urchin embryo messenger RNA. *Proc. Natl. Acad. Sci. U.S.A.* **70**, 3516.

Goldberg, R. B., Crain, W. R., Ruderman, J. V., Moore, G. P., Barnett, T. R., Higgins, R. C., Gelfand, R. A., Galau, G. A., Britten, R. J., and Davidson, E. H. (1975). DNA sequence organization in the genomes of five marine invertebrates. *Chromosoma* **51**, 225.

Goldstein, E. S., and Penman, S. (1973). Regulation of protein synthesis in mammalian cells. V. Further studies on the effect of actinomycin D on translation control in HeLa cells. *J. Mol. Biol.* **80**, 243.

Golub, A., and Clegg, J. S. (1968). Protein synthesis in *Artemia salina* embryos. I. Studies on polyribosomes. *Dev. Biol.* **17**, 644.

Gottesfeld, J. M., Garrard, W. T., Bagi, G., Wilson, R. F., and Bonner, J. (1974). Partial purification of the template-active fraction of chromatin: A preliminary report. *Proc. Natl. Acad. Sci. U.S.A.* **71**, 2193.

Gould, M. C. (1969). A comparison of RNA and protein synthesis in fertilized and unfertilized eggs of *Urechis caupo*. *Dev. Biol.* **19**, 482.

Govaert, J. (1957). Etude quantitative de la teneur en acide désoxyribonucléique des noyaux des cellules somatiques et germinatives chez *Fasciola hepatica*. *Arch. Biol.* **68**, 165.

Graham, D. E., Neufeld, B. R., Davidson, E. H., and Britten, R. J. (1974). Interspersion of repetitive and non-repetitive DNA sequences in the sea urchin genome. *Cell* **1**, 127.

Grainger, R. M., and Wilt, F. H. (1976). The incorporation of $^{13}C–^{15}N$ nucleosides and measurements of RNA synthesis and turnover in sea urchin embryos. *J. Mol. Biol.* **104**, 589.

Grant, P. (1958). The incorporation of P^{32} and glycine-2-C^{14} into nucleic acids during early embryonic development of *Rana pipiens*. *J. Cell. Comp. Physiol.* **52**, 249.

Grant, P. (1969). Nucleo-cortical interactions during amphibian development. *In* "Biology of Amphibian Tumors" (M. Mizell, ed.), p. 43. Springer-Verlag, Berlin and New York.

Grant, P., and Wacaster, J. F. (1972). The amphibian gray crescent region—a site of developmental information? *Dev. Biol.* **28**, 454.

Green, H., Goldberg, B., Schwartz, M., and Brown, D. D. (1968). The synthesis of collagen during the development of *Xenopus laevis*. *Dev. Biol.* **18**, 391.

Greenberg, J. R., and Perry, R. P. (1971). Hybridization properties of DNA sequences directing the synthesis of messenger RNA and heterogeneous nuclear RNA. *J. Cell Biol.* **50**, 774.

Greene, R. F., and Flickinger, R. A., Jr. (1970). Qualitative changes in DNA-like RNA during development in *Rana pipiens*. *Biochim. Biophys. Acta* **217**, 447.

Greenhouse, G. A., Hynes, R. O., and Gross, P. R. (1971). Sea urchin embryos are permeable to actinomycin. *Science* **171**, 686.

Gregg, J. R., and Løvtrup, S. (1960). A reinvestigation of DNA synthesis in lethal amphibian hybrids. *Exp. Cell Res.* **19**, 621.

Grell, R. F., and Chandley, A. C. (1965). Evidence bearing on the coincidence of exchange and DNA replication in the oocyte of *Drosophila melanogaster*. *Proc. Natl. Acad. Sci. U.S.A.* **53**, 1340.

Griffiths, M. (1965). A study of the synthesis of naphthaquinone pigments by the larvae of two species of sea urchins and their reciprocal hybrids. *Dev. Biol.* **11**, 433.

Grippo, P., and Lo Scavo, A. (1972). DNA polymerase activity during maturation in *Xenopus laevis* oocytes. *Biochem. Biophys. Res. Commun.* **48**, 280.

Gross, K. W., Jacobs-Lorena, M., Baglioni, C., and Gross, P. R. (1973a). Cell-free translation of maternal messenger RNA from sea urchin eggs. *Proc. Natl. Acad. Sci. U.S.A.* 70, 2614.

Gross, K. W., Ruderman, J., Jacobs-Lorena, M., Baglioni, C., and Gross, P. R. (1973b). Cell-free synthesis of histones directed by messenger RNA from sea urchin embryos. *Nature (London), New Biol.* 241, 272.

Gross, K. W., Probst, E., Schaffner, W., and Birnstiel, M. (1976a). Molecular analysis of the histone gene cluster of *Psammechinus miliaris*. I. Fractionation and identification of five individual histone mRNAs. *Cell* 8, 455.

Gross, K. W., Schaffner, W., Telford, J., and Birnstiel, M. (1976b). Molecular analysis of the histone gene cluster of *Psammechinus miliaris*. II. Polarity and asymmetry of the histone-coding sequences. *Cell* 8, 479.

Gross, P. R. (1967). The control of protein synthesis in embryonic development and differentiation. *Curr. Top. Dev. Biol.* 2, 1.

Gross, P. R., and Cousineau, G. H. (1963a). Effects of actinomycin D on macromolecule synthesis and early development in sea urchin eggs. *Biochem. Biophys. Res. Commun.* 10, 321.

Gross, P. R., and Cousineau, G. H. (1963b). Synthesis of spindle-associated proteins in early cleavage. *J. Cell Biol.* 19, 260.

Gross, P. R., and Cousineau, G. H. (1964). Macromolecule synthesis and the influence of actinomycin on early development. *Exp. Cell Res.* 33, 368.

Gross, P. R., Malkin, L. I., and Moyer, W. A. (1964). Templates for the first proteins of embryonic development. *Proc. Natl. Acad. Sci. U.S.A.* 51, 407.

Gross, P. R., Kraemer, K., and Malkin, L. I. (1965a). Base composition of RNA synthesized during cleavage of the sea urchin embryo. *Biochem. Biophys. Res. Commun.* 18, 569.

Gross, P. R., Malkin, L. I., and Hubbard, M. (1965b). Synthesis of RNA during oogenesis in the sea urchin. *J. Mol. Biol.* 13, 463.

Grossbach, U. (1974). Chromosome puffs and gene expression in polytene cells. *Cold Spring Harbor Symp. Quant. Biol.* 38, 619.

Groudine, M., and Weintraub, H. (1975). Rous sarcoma virus activates embryonic globin genes in chicken fibroblasts. *Proc. Natl. Acad. Sci. U.S.A.* 72, 4464.

Grouse, L., Chilton, M.-D., and McCarthy, B. J. (1972). Hybridization of ribonucleic acid with unique sequences of mouse deoxyribonucleic acid. *Biochemistry* 11, 798.

Grunstein, M., and Schedl, P. (1976). Isolation and sequence analysis of sea urchin (*Lytechinus pictus*) histone H4 messenger RNA. *J. Mol. Biol.* 104, 323.

Grunstein, M., Levy, S., Schedl, P., and Kedes, L. (1974). Messenger RNAs for individual histone proteins: Fingerprint analysis and *in vitro* translation. *Cold Spring Harbor Symp. Quant. Biol.* 38, 717.

Grunstein, M., Schedl, P., and Kedes, L. (1976). Sequence analysis and evolution of sea urchin (*Lytechinus pictus* and *Strongylocentrotus purpuratus*) histone H4 messenger RNAs. *J. Mol. Biol.* 104, 351.

Guerrier, P. (1967). Les facteurs de polarisation dans les premiers stades du développement chez *Parascaris equorum*. *J. Embryol. Exp. Morphol.* 18, 121.

Guerrier, P. (1968). Origine et stabilité de la polarité animale-végétative chez quelques Spiralia. *Ann. Embryol. Morphog.* 1, 119.

Guerrier, P. (1970a). Les caractères de la segmentation et la détermination de la polarité dorsoventrale dans le développement de quelques Spiralia. I. Les formes à premier clivage ègal. *J. Embryol. Exp. Morphol.* 23, 611.

Guerrier, P. (1970b). Les caractères de la segmentation et la détermination de la polarité dorsoventrale dans le développement de quelques Spiralia. II. *Sabellaria alveolata* (Annélide polychète). *J. Embryol. Exp. Morphol.* 23, 639.

Guerrier, P. (1970c). Les caractères de la segmentation et la détermination de la polarité dorsoventrale dans le développement de quelques Spiralia. III. *Pholas dactylus* et *Spisula subtruncata* (Mollusques Lamellibranches). *J. Embryol. Exp. Morphol*. **23**, 667.

Guerrier, P. (1970d). Nouvelles données expérimentales sur la segmentation et l'organogénèse chez *Limax maximus* (Gasteropode Pulmone). *Ann. Embryol. Morphog*. **3**, 283.

Guerrier, P., and Freyssinet, G. (1974). Protein synthesis during embryogenesis of *Sabellaria alveolata* L. (polychaete annelid). *Exp. Cell Res*. **87**, 290.

Gurdon, J. B. (1962). Adult frogs derived from the nuclei of single somatic cells. *Dev. Biol*. **4**, 256.

Gurdon, J. B. (1963). Nuclear transplantation in amphibia and the importance of stable nuclear changes in promoting cellular differentiation. *Q. Rev. Biol*. **38**, 54.

Gurdon, J. B. (1967). Control of gene activity during the early development of *Xenopus laevis*. *In* "Heritage from Mendel" (A. Brink, ed.), p. 203. Univ. of Wisconsin Press, Wisconsin.

Gurdon, J. B. (1968). Nucleic acid synthesis in embryos and its bearing on cell differentiation. *Essays Biochem*. **4**, 26.

Gurdon, J. B., and Brown, D. D. (1965). Cytoplasmic regulation of RNA synthesis and nucleolus formation in developing embryos of *Xenopus laevis*. *J. Mol. Biol*. **12**, 27.

Gurdon, J. B., and Ford, P. J. (1967). Attachment of rapidly labelled RNA to polysomes in the absence of ribosomal RNA synthesis during normal cell differentiation. *Nature (London)* **216**, 666.

Gurdon, J. B., and Laskey, R. A. (1970). The transplantation of nuclei from single cultured cells into enucleate frogs' eggs. *J. Embryol. Exp. Morphol*. **24**, 227.

Gurdon, J. B., and Uehlinger, V. (1966). "Fertile" intestinal nuclei. *Nature (London)* **210**, 1240.

Gurdon, J. B., Birnstiel, M. L., and Speight, V. A. (1969). The replication of purified DNA introduced into living egg cytoplasm. *Biochim. Biophys. Acta* **174**, 614.

Gurdon, J. B., Lane, C. D., Woodland, H. R., and Marbaix, G. (1971). Use of frog eggs and oocytes for the study of messenger RNA and its translation in living cells. *Nature (London)* **233**, 177.

Gurdon, J. B., Lingrel, J. B., and Marbaix, G. (1973). Message stability in injected frog oocytes: Long life of mammalian α and β globin messages. *J. Mol. Biol*. **80**, 539.

Gurdon, J. B., Woodland, H. R., and Lingrel, J. B. (1974). The translation of mammalian globin mRNA injected into fertilized eggs of *Xenopus laevis*. I. Message stability in development. *Dev. Biol*. **39**, 125.

Gustafson, T., and Wolpert, L. (1963). The cellular basis of morphogenesis and sea urchin development. *Int. Rev. Cytol*. **15**, 139.

Hagström, B. E., and Lönning, S. (1969). Time-lapse and electron microscopic studies of sea urchin micromeres. *Protoplasma* **68**, 271.

Hahn, W. E., and Laird, C. D. (1971). Transcription of nonrepeated DNA in mouse brain. *Science* **173**, 158.

Hallberg, R. L., and Brown, D. D. (1969). Co-ordinated synthesis of some ribosomal proteins and ribosomal RNA in embryos of *Xenopus laevis*. *J. Mol. Biol*. **46**, 393.

Hardisty, M. W., and Cosh, J. (1966). Primordial germ cells and fecundity. *Nature (London)* **210**, 1370.

Harel, L., and Montagnier, L. (1971). Homology of double stranded RNA from rat liver cells with the cellular genome. *Nature (London), New Biol*. **229**, 106.

Harris, S. E., and Forrest, H. S. (1967). RNA and DNA synthesis in developing eggs of the milkweed bug, *Oncopeltus fasciatus* (Dallas). *Science* **156**, 1613.

Harris, S. E., Rosen, J. M., Means, A. R., and O'Malley, B. W. (1975). Use of a specific probe for ovalbumin messenger RNA to quantitate estrogen-induced gene transcripts. *Biochemistry* 14, 2072.

Harrison, P. R., Birnie, G. D., Hell, A., Humphries, S., Young, B. D., and Paul, J. (1974). Kinetic studies of gene frequency. I. Use of a DNA copy of reticulocyte 9S RNA to estimate globin gene dosage in mouse tissue. *J. Mol. Biol.* 84, 539.

Hartmann, J. F., and Comb, D. G. (1969). Transcription of nuclear and cytoplasmic genes during early development of sea urchin embryos. *J. Mol. Biol.* 41, 155.

Hartmann, J. F., Ziegler, M. M., and Comb, D. G. (1971). Sea urchin embryogenesis. I. RNA synthesis by cytoplasmic and nuclear genes during development. *Dev. Biol.* 25, 209.

Harvey, E. B. (1936). Parthenogenetic merogony or cleavage without nuclei in *Arbacia punctulata*. *Biol. Bull.* 71, 101.

Harvey, E. B. (1940). A comparison of the development of nucleate and nonnucleate eggs of *Arbacia punctulata*. *Biol. Bull.* 79, 166.

Hastie, N. D., and Bishop, J. O. (1976). The expression of three abundance classes of messenger RNA in mouse tissues. *Cell*, in press.

Hatt, P. (1931). La fusion expérimentale d'oeufs de *Sabellaria alveolata* L. et leur développement. *Arch. Biol.* 42, 303.

Hatt, P. (1932). Essais expérimentaux sur les localisations germinales dans l'oeuf d'un annelide (*Sabellaria alveolata* L.). *Arch. Anat. Microsc. Morphol. Exp.* 28, 81.

Hay, E. D. (1968). Dedifferentiation and metaplasia in vertebrate and invertebrate regeneration. *In* "The Stability of the Differentiated State" (H. Ursprung, ed.), p. 85. Springer-Verlag, Berlin and New York.

Hay, E. D., and Gurdon, J. B. (1967). Fine structure of the nucleolus in normal and mutant *Xenopus* embryos. *J. Cell Sci.* 2, 151.

Hebard, C. N., and Herold, R. C. (1967). The ultrastructure of the cortical cytoplasm in the unfertilized egg and first cleavage zygote of *Xenopus laevis*. *Exp. Cell Res.* 46, 553.

Hegner, R. W. (1911). Experiments with chrysomelid beetles. III. The effects of killing parts of the eggs of *Leptinotarsa decemlineata*. *Biol. Bull.* 20, 237.

Hegner, R. W. (1914). "The Germ-Cell Cycle in Animals." Macmillan, New York.

Hennen, S. (1963). Chromosomal and embryological analyses of nuclear changes occurring in embryos derived from transfers of nuclei between *Rana pipiens* and *Rana sylvatica*. *Dev. Biol.* 6, 133.

Hennen, S. (1970). Influence of spermine and reduced temperature on the ability of transplanted nuclei to promote normal development in eggs of *Rana pipiens*. *Proc. Natl. Acad. Sci. U.S.A.* 66, 630.

Hennig, W. (1967). Untersuchungen zur Struktur and Funktion des Lampenbürstein-Y-Chromosoms in der Spermatogenese von *Drosophila*. *Chromosoma* 22, 294.

Hennig, W. (1968). Ribonucleic acid synthesis of the Y chromosome of *Drosophila hydei*. *J. Mol. Biol.* 38, 227.

Hennig, W., Meyer, G. F., Hennig, I., and Leoncini, O. (1974). Structure and function of the Y chromosome of *Drosophila hydei*. *Cold Spring Harbor Symp. Quant. Biol.* 38, 673.

Herbst, C. (1892). Experimentelle Untersuchungen über den Einfluss der veränderten chemischen Zusammensetzung des umgebeden Médiums auf die Entwicklung der Tiere. *Z. Wiss. Zool.* 55, 446.

Hetwig, O. (1876). Beiträge zur Kenntniss der Bildung, Befruchtung und Teilung des tierischen Eies. *Morphol. Jahrb.* 1, 347.

Hertwig, O. (1885). The problem of fertilization and isotropy of the egg, a theory of inheritance. *In* "The Chromosome Theory of Inheritance: Classic Papers in Development and Heredity" (B. R. Voeller, ed.), p. 26. Appleton, New York, 1968 (transl.).

Hertwig, O. (1894). Präformation oder Epigenese. Grundziige einer Entwicklungstheorie der Organisme. *Z. Streitfragen Biol. (Jena).* 1, 143.

Hertwig, O., and Hertwig, R. (1887). Über den Befruchtungs- und Theilungs-vorgang des thierischen Eies unter dem Einfluss äusserer Agentien. *Jena. Z. Naturwiss.* 20, Part 1, 120.

Hess, O. (1965). The effect of X-rays on the functional structures of the Y chromosome in spermatocytes of *Drosophila hydei. J. Cell Biol.* 25, 169.

Hess, O. (1966). Structural modifications of the Y-chromosome in *Drosophila hydei* and their relation to gene activity. *Chromosomes Today* 1, 167.

Hess, O. (1970). Genetic function correlated with unfolding of lampbrush loops by the Y-chromosome in spermatocytes of *Drosophila hydei. Mol. Gen. Genet.* 106, 328.

Hess, O. (1971). Fresh water gastropoda. *In* "Experimental Embryology of Marine and Fresh-Water Invertebrates" (G. Reverberi, ed.), p. 215. North-Holland Publ., Amsterdam.

Hill, R. J., Poccia, D. L., and Doty, P. (1971). Towards a total macromolecular analysis of sea urchin embryo chromatin. *J. Mol. Biol.* 61, 445.

Hill, R. N., and McConkey, E. H. (1972). Coordination of ribosomal RNA synthesis in vertebrate cells. *J. Cell. Physiol.* 79, 15.

Hillman, N. (1972). Autoradiographic studies of t^{12}/t^{12} mouse embryos. *Am. J. Anat.* 134, 41.

Hillman, N. and Tasca, R. J. (1969). Ultrastructural and autoradiographic studies on mouse cleavage stages. *Am. J. Anat.* 126, 151.

Hillman, N., and Tasca, R. J. (1973). Synthesis of RNA in t^{12}/t^{12} mouse embryos. *J. Reprod. Fertil.* 33, 501.

Hillman, N., Hillman, R., and Wileman, G. (1970). Ultrastructural studies of cleavage stage t^{12}/t^{12} mouse embryos. *Am. J. Anat.* 128, 311.

Hillman, N., Sherman, M. I., and Graham, C. (1972). The effect of spatial arrangement on cell determination during mouse development. *J. Embryol. Exp. Morphol.* 28, 263.

Hinegardner, R. T. (1967). Echinoderms. *In* "Methods in Developmental Biology" (F. H. Wilt and N. K. Wessells, eds.), p. 139. Crowell-Collier, New York.

Hinegardner, R. T. (1969). Growth and development of the laboratory cultured sea urchin. *Biol. Bull.* 137, 465.

Hinegardner, R. T. (1974). Cellular DNA content of the Echinodermata. *Comp. Biochem. Physiol. B* 49, 219.

Hirama, M. N., and Mano, Y. (1974). Polysomes of the sea urchin embryo. Identification of tubulin-synthesizing polysomes. *Exp. Cell Res.* 86, 15.

His, W. (1874). "Unsere Körperform und das physiologische Problem ihrer Entstehung." Vogel, Leipzig.

Hoberman, H. D., Metz, C. B., and Graff, J. (1952). Uptake of deuterium into proteins of fertilized and unfertilized *Arbacia* eggs suspended in heavy water. *J. Gen. Physiol.* 35, 639.

Hogan, B., and Gross, P. R. (1972). Nuclear RNA synthesis in sea urchin embryos. *Exp. Cell Res.* 72, 101.

Hogue, M. J. (1910). Über die Wirkung der Centrifugalkraft auf die Eier von *Ascaris megalocephala. Arch. Entwicklungsmech. Org.* 29, 109.

Holmes, D. S., and Bonner, J. (1974a). Sequence composition of rat nuclear deoxyribonucleic acid and high molecular weight nuclear ribonucleic acid. *Biochemistry* 13, 841.

Holmes, D. S., and Bonner, J. (1974b). Interspersion of repetitive and single-copy sequences in nuclear ribonucleic acid of high molecular weight. *Proc. Natl. Acad. Sci. U.S.A.* **71**, 1108.

Holtfreter, J., and Hamburger, V. (1956). Embryogenesis: Progressive differentiation. Amphibians. *In* "Analysis of Development" (B. H. Willier, P. A. Weiss, and V. Hamburger, eds.), p. 230. Saunders, Philadelphia, Pennsylvania.

Honjo, T., and Reeder, R. H. (1973). Preferential transcription of *Xenopus laevis* ribosomal RNA in interspecies hybrids between *Xenopus laevis* and *Xenopus mulleri. J. Mol. Biol.* **80**, 217.

Hooper, J. A., Smith, E. L., Sommer, K. R., and Chalkley, R. (1973). Histone III. IV. Amino acid sequence of histone III of the testes of the carp, *Letiobus bubalus. J. Biol. Chem.* **248**, 3275.

Hörstadius, S. (1928). Über die Determination des Keimes bei Echinodermen. *Acta Zool.* (*Stockholm*) **9**, 1.

Hörstadius, S. (1936). Studien über heterosperme Seeigalmerogone nebst Bemerkungen über einige Keimblattchimären. *Mem. Mus. Hist. Nat. Belg.* [2] **3**, 801.

Hörstadius, S. (1937a). Investigations as to the localization of the micromere, the skeleton, and the entoderm-forming material in the unfertilized egg of *Arbacia punctulata. Biol. Bull.* **73**, 295.

Hörstadius, S. (1937b). Experiments on determination in the early development of *Cerebratulus lacteus. Biol. Bull.* **73**, 317.

Hörstadius, S. (1939). The mechanics of sea urchin development, studied by operative methods. *Biol. Rev. Cambridge Philos. Soc.* **14**, 132.

Hörstadius, S. (1973). "Experimental Embryology of Echinoderms." Oxford Univ. Press (Clarendon), London and New York.

Hough, B. R., and Davidson, E. H. (1972). Studies on the repetitive sequence transcripts of *Xenopus* oocytes. *J. Mol. Biol.* **70**, 491.

Hough, B. R., Yancey, P. H., and Davidson, E. H. (1973). Persistence of maternal RNA in *Engystomops* embryos. *J. Exp. Zool.* **185**, 357.

Hough, B. R., Smith, M. J., Britten, R. J., and Davidson, E. H. (1975). Sequence complexity of heterogeneous nuclear RNA in sea urchin embryos. *Cell* **5**, 291.

Hough-Evans, B. R., Wold, B. J., Ernst, S. G., Britten, R. J., and Davidson, E. H. (1976). Appearance, persistence, and disappearance of complex maternal RNA sequences in sea urchin development. *Dev. Biol.*, in press.

Hourcade, D., Dressler, D., and Wolfson, J. (1973). The amplification of ribosomal RNA genes involves a rolling circle intermediate. *Proc. Natl. Acad. Sci. U.S.A.* **70**, 2926.

Hourcade, D., Dressler, D., and Wolfson, J. (1974). The nucleolus and the rolling circle. *Cold Spring Harbor Symp. Quant. Biol.* **38**, 537.

Huberman, J. A., and Riggs, A. D. (1968). On the mechanism of DNA replication in mammalian chromosomes. *J. Mol. Biol.* **32**, 327.

Hughes, M., and Berry, S. J. (1970). The synthesis and secretion of ribosomes by nurse cells of *Antheraea polyphemus. Dev. Biol.* **23**, 651.

Hultin, T. (1952). Incorporation of N^{15}-labeled glycine and alanine into the proteins of developing sea urchin eggs. *Exp. Cell Res.* **3**, 494.

Hultin, T. (1961a). The effect of puromycin on protein metabolism and cell division in fertilized sea urchin eggs. *Experientia* **17**, 410.

Hultin, T. (1961b). Activation of ribosomes in sea urchin eggs in response to fertilization. *Exp. Cell Res.* **25**, 405.

Hultin, T., and Morris, J. E. (1968). The ribosomes of encysted embryos of *Artemia salina* during cryptobiosis and resumption of development. *Dev. Biol.* **17**, 143.

Humphrey, R. R. (1966). A recessive factor (o, for ova deficient) determining a complex of abnormalities in the Mexican axolotl (*Ambystoma mexicanum*). *Dev. Biol.* **13**, 57.

Humphreys, T. (1969). Efficiency of translation of messenger RNA before and after fertilization in sea urchins. *Dev. Biol.* **20**, 435.

Humphreys, T. (1969). Efficiency of translation of messenger RNA before and after fertilization of sea urchin eggs. *Dev. Biol.* **26**, 201.

Humphreys, T. (1973). RNA and protein synthesis during early animal embryogenesis. *In* "Developmental Regulation. Aspects of Cell Differentiation" (S. J. Coward, ed.), p. 1. Academic Press, New York.

Humphries, S., Windass, J., and Williamson, R. (1976). Mouse globin gene expression in erythroid and non-erythroid tissues. *Cell* 7, 267.

Hunt, J. A. (1974). Rate of synthesis and half-life of globin messenger ribonucleic acid. Rate of synthesis of globin messenger ribonucleic acid calculated from data of cell haemoglobin content. *Biochem. J.* **138**, 499.

Hutton, J. R., and Wetmur, J. G. (1973). Length dependence of the kinetic complexity of mouse satellite DNA. *Biochem. Biophys. Res. Commun.* **52**, 1148.

Huxley, T. H. (1878). Evolution in biology. "Encyclopedia Brittanica," 9th ed., p. 187. Scribner's, New York.

Hyman, L. H. (1940). "The Invertebrates: Protozoa through Ctenophora," Vol. I. McGraw-Hill, New York.

Hyman, L. H. (1951). "The Invertebrates: Platyhelminthes and Rhynchocoela. The Acoelomate Bilateria," Vol. II. McGraw-Hill, New York.

Hyman, L. H. (1955). "The Invertebrates: Echinodermata," Vol. IV. McGraw-Hill, New York.

Hynes, R. O., and Gross, P. R. (1972). Informational RNA sequences in early sea urchin embryos. *Biochim. Biophys. Acta* **259**, 104.

Hynes, R. O., Greenhouse, G. A., Minkoff, R., and Gross, P. R. (1972). Properties of the three cell types in sixteen-cell sea urchin embryos: RNA synthesis. *Dev. Biol.* **27**, 457.

Illmensee, K. (1972). Developmental potencies of nuclei from cleavage, preblastoderm, and syncytial blastoderm transplanted into unfertilized eggs of *Drosophila melanogaster*. *Wilhelm Roux' Arch. Entwicklungsmech. Org.* **170**, 267.

Illmensee, K., and Mahowald, A. P. (1974). Transplantation of posterior polar plasm in *Drosophila*. Induction of germ cells at the anterior pole of the egg. *Proc. Natl. Acad. Sci. U.S.A.* **71**, 1016.

Illmensee, K., and Mahowald, A. P. (1976). The autonomous function of germ plasm in a somatic region of the *Drosophila* egg. *Exp. Cell Res.* **97**, 127.

Illmensee, K., Mahowald, A. P., and Loomis, M. R. (1976). The ontogeny of germ plasm during oogenesis in *Drosophila*. *Dev. Biol.* **49**, 40.

Infante, A. A., and Nemer, M. (1967). Accumulation of newly synthesized RNA templates in a unique class of polyribosomes during embryogenesis. *Proc. Natl. Acad. Sci. U.S.A.* **58**, 681.

Iwai, K., Ishikawa, K., and Hayashi, H. (1970). Amino-acid sequence of slightly lysine-rich histone. *Nature (London)* **226**, 1056.

Izawa, M., Allfrey, V. G., and Mirsky, A. E. (1963). Composition of the nucleus and chromosomes in the lampbrush stage of the newt oocyte. *Proc. Natl. Acad. Sci. U.S.A.* **50**, 811.

Izquierdo, L., and Ortiz, M. E. (1975). Differentiation in the mouse morulae. *Wilhelm Roux' Arch. Entwicklungsmech. Org.* **177**, 67.

Jacob, E., Malacinski, G., and Birnstiel, M. L. (1976). Reiteration frequency of the histone genes in the genome of the amphibian, *Xenopus laevis. Eur. J. Biochem.*, in press.

Jahn, C. L., Baran, M. M., and Bachvarova, R. (1976). *In vivo* labeling of RNA of growing mouse oocytes and the retention of labeled RNA to ovulation. *J. Exp. Zool.* 197, 161.

Jelinek, W., and Darnell, J. E. (1972). Double-stranded regions in heterogeneous nuclear RNA from HeLa cells. *Proc. Natl. Acad. Sci. U.S.A.* 69, 2537.

Jelinek, W., Molloy, G., Fernandez-Munoz, R., Salditt, M., and Darnell, J. E. (1974). Secondary structure in heterogeneous nuclear RNA: Involvement of regions from repeated DNA sites. *J. Mol. Biol.* 82, 361.

Jenkins, N., Taylor, M. W., and Raff, R. A. (1973). *In vitro* translation of oogenetic messenger RNA of sea urchin eggs and picornavirus RNA with a cell-free system from sarcoma 180. *Proc. Natl. Acad. Sci. U.S.A.* 70, 3287.

Johnson, A. W., and Hnilica, L. S. (1971). Cytoplasmic and nuclear basic protein synthesis during early sea urchin development. *Biochim. Biophys. Acta* 246, 141.

Johnson, J. H., and King, R. C. (1972). Studies on *fes*, a mutation affecting cystocyte cytokinesis, in *Drosophila melanogaster*. *Biol. Bull.* 143, 525.

Johnson, K. E. (1969). Altered contact behavior of presumptive mesodermal cells from hybrid amphibian embryos arrested at gastrulation. *J. Exp. Zool.* 170, 325.

Johnson, K. E. (1970). The role of changes in cell contact behavior in amphibian gastrulation. *J. Exp. Zool.* 175, 391.

Johnson, K. E. (1971). A biochemical and cytological investigation of differentiation in the interspecific hybrid amphibian embryo *Rana pipiens* ♀ × *Rana sylvatica* ♂ . *J. Exp. Zool.* 177, 191.

Johnson, K. E., and Chapman, V. M. (1972). Expression of paternal genes during embryogenesis in the viable interspecific hybrid amphibian embryo *Rana pipiens* ♀ × *Rana palustris* ♂ . Five enzyme systems. *J. Exp. Zool.* 178, 313.

Jörgenssen, M. (1913). Die Ei-und Nährzellen von *Pisciola*. *Arch. Zellforsch.* 10, 127.

Kafatos, F. C. (1972). The cocoonase zymogen cells of silk moths: A model of terminal cell differentiation for specific protein synthesis. *Curr. Top. Dev. Biol.* 7, 125.

Kafatos, F. C., and Gelinas, R. (1974). mRNA stability and the control of specific protein synthesis in highly differentiated cells. *In* "MTP International Review of Science— Biochemistry of Differentiation and Development" (J. Paul, ed.), Vol. 9, p. 223. Medical and Technical Publ. Co., Oxford.

Kalt, M. R. (1971). The relationship between cleavage and blastocoel formation in *Xenopus laevis*. II. Electron microscopic observations. *J. Embryol. Exp. Morphol.* 26, 51.

Kalt, M. R., and Gall, J. G. (1974). Observations on early germ cell development and premeiotic ribosomal DNA amplification in *Xenopus laevis*. *J. Cell Biol.* 62, 460.

Kalthoff, K. (1971a). Photoreversion of UV induction of the malformation "double abdomen" in the egg of *Smittia* spec. (Diptera, Chironomidae). *Dev. Biol.* 25, 119.

Kalthoff, K. (1971b). Position of targets and period of competence for UV-induction of the malformation "double abdomen" in the egg of *Smittia* spec. (Diptera, Chironomidae). *Wilhelm Roux' Arch. Entwicklungsmech. Org.* 168, 63.

Kalthoff, K. (1973). Action spectra for UV induction and photoreversal of a switch in the developmental program of the egg of an insect (*Smittia*). *Photochem. Photobiol.* 18, 355.

Kalthoff, K., and Sander, K. (1968). Der Entwicklungsgang der Mißbildung "Doppelabdomen" im UV-bestrahlen Ei von *Smittia parthenogenica* (Dipt., Chironomidae). *Wilhelm Roux' Arch. Entwicklungsmech. Org.* 161, 129.

Kandler-Singer, I., and Kalthoff, K. (1976). RNase sensitivity of an anterior morphogenetic determinant in an insect egg (*Smittia* spec., Chironomidae, Diptera). *Proc. Natl. Acad. Sci. U.S.A.*, in press.

Karasaki, S. (1965). Electron microscopic examination of the sites of nuclear RNA synthesis during amphibian embryogenesis. *J. Cell Biol.* 26, 937.

Karasaki, S. (1968). The ultrastructure and RNA metabolism of nucleoli in early sea urchin embryos. *Exp. Cell Res.* **52**, 13.

Karnofsky, D. A., and Simmel, E. B. (1963). Effects of growth-inhibiting chemicals on the sand-dollar embryo, *Echinarachnius parma*. *Prog. Exp. Tumor Res.* **3**, 254.

Karp, G. C., and Whiteley, A. H. (1973). DNA-RNA hybridization studies of gene activity during the development of the gastropod, *Acmaea scutum*. *Exp. Cell Res.* **78**, 236.

Karp, G. C., Manes, C., and Hahn, W. E. (1973). RNA synthesis in the preimplantation rabbit embryo: Radioautographic analysis. *Dev. Biol.* **31**, 404.

Karp, G. C., Manes, C., and Hahn, W. E. (1974). Ribosome production and protein synthesis in the preimplantation rabbit embryo. *Differentiation* **2**, 65.

Kaulenas, M. S., and Fairbairn, D. (1968). RNA metabolism of fertilized *Ascaris lumbricoides* eggs during uterine development. *Exp. Cell Res.* **52**, 233.

Kaulenas, M. S., Foor, W. E., and Fairbairn, D. (1969). Ribosomal RNA synthesis during cleavage of *Ascaris lumbricoides* eggs. *Science* **163**, 1201.

Kedes, L. H., and Birnstiel, M. L. (1971). Reiteration and clustering of DNA sequences complementary to histone messenger RNA. *Nature (London), New Biol.* **230**, 165.

Kedes, L. H., and Gross, P. R. (1969). Synthesis and function of messenger RNA during early embryonic development. *J. Mol. Biol.* **42**, 559.

Kedes, L. H., and Stavy, L. (1969). Structural and functional identity of ribosomes from eggs and embryos of sea urchins. *J. Mol. Biol.* **43**, 337.

Kedes, L. H., Gross, P. R., Cognetti, G., and Hunter, A. L. (1969). Synthesis of nuclear and chromosomal proteins on light polyribosomes during cleavage in the sea urchin embryo. *J. Mol. Biol.* **45**, 337.

Kedes, L. H., Cohn, R. H., Lowry, J. C., Chang, A. C. Y., and Cohen, S. N. (1975a). The organization of sea urchin histone genes. *Cell* **6**, 359.

Kedes, L. H., Chang, A. C. Y., Housman, D., and Cohen, S. N. (1975b). Isolation of histone genes from unfractionated sea urchin DNA by subculture cloning in *E. coli*. *Nature (London)* **255**, 533.

Kelly, S. J. (1975). Studies of the potency of the early cleavage blastomeres of the mouse. *In* "The Early Development of Mammals" (M. Balls and A. E. Wild, eds.), p. 97. Cambridge Univ. Press, London and New York.

Kennelly, J. J., Foote, R. H., and Jones, R. C. (1970). Duration of premeiotic deoxyribonucleic acid synthesis and the stages of prophase I in rabbit ocytes. *J. Cell Biol.* **47**, 577.

Kerr, J. B., and Dixon, K. E. (1974). An ultrastructural study of germ plasm in spermatogenesis of *Xenopus laevis*. *J. Embryol. Exp. Morphol.* **32**, 573.

Kidder, G. M. (1972a). Gene transcription in mosaic embryos. I. Pattern of RNA synthesis in early development of the coot clam, *Mulinia lateralis*. *J. Exp. Zool.* **180**, 55.

Kidder, G. M. (1972b). Gene transcription in mosaic embryos. II. Polyribosomes and messenger RNA in early development of the coot clam, *Mulinia lateralis*. *J. Exp. Zool.* **180**, 75.

Kijima, S., and Wilt, F. H. (1969). Rate of nuclear ribonucleic acid turnover in sea urchin embryos. *J. Mol. Biol.* **40**, 235.

Kiknadze, I. I. (1963). On the existence of nucleoli at early stages of cleavage. *Tsitologiya* **5**, 319.

Kinderman, N. B., and King, R. C. (1973). Oogenesis in *Drosophila virilis*. I. Interactions between the ring canal rims and the nucleus of the oocyte. *Biol. Bull.* **144**, 331.

King, R. C. (1970). "Ovarian Development in *Drosophila melanogaster*." Academic Press, New York.

King, R. C., and Aggarwal, S. K. (1965). Oogenesis in *Hyalophora cecropia*. *Growth* **29**, 17.

Klein, W. H., Murphy, W., Attardi, G., Britten, R. J., and Davidson, E. H. (1974). Distribution of repetitive and nonrepetitive sequence transcripts in HeLa mRNA. *Proc. Natl. Acad. Sci. U. S. A.* **71**, 1785.

Klug, W. S., King, R. C., and Wattiaux, J. M. (1970). Oogenesis in the *suppressor*[2] of *hairy-wing* mutant of *Drosophila melanogaster*. II. Nucleolar morphology and *in vitro* studies of RNA protein synthesis. *J. Exp. Zool.* **174**, 125.

Knowland, J. S. (1970). Polyacrylamide gel electrophoresis of nucleic acids synthesized during the early development of *Xenopus laevis* Daudin. *Biochim. Biophys. Acta* **204**, 416.

Knowland, J. S., and Graham, C. (1972). RNA synthesis at the two-cell stage of mouse development. *J. Embryol. Exp. Morphol.* **27**, 167.

Knowland, J. S., and Miller, L. (1970). Reduction of ribosomal RNA synthesis and ribosomal RNA genes in a mutant of *Xenopus laevis* which organizes only a partial nucleolus. I. Ribosomal RNA synthesis in embryos of different nucleolar types. *J. Mol. Biol.* **53**, 321.

Kobel, H. R., Brun, R. B., and Fischberg, M. (1973). Nuclear transplantation with melanophores, ciliated epidermal cells, and the established cell line A-8 in *Xenopus laevis*. *J. Embryol. Exp. Morphol.* **29**, 539.

Koch, E. A., Smith, P. A., and King, R. C. (1967). The division and differentiation of *Drosophila* cystocytes. *J. Morphol.* **121**, 55.

Kohl, D. M., Norman, J., and Brooks, S. (1973). The relationship between species of RNA synthesized and levels of different DNA-dependent RNA polymerases in isolated nuclei of *Rana pipiens* embryos. *Cell Differ.* **2**, 21.

Kohne, D. E., and Byers, M. J. (1973). Amplification and evolution of deoxyribonucleic acid sequences expressed as ribonucleic acid. *Biochemistry* **12**, 2373.

Kölreuter, J. G. (1761). Preliminary report of experiments and observations concerning some aspects of the sexuality of plants. *In* "The Chromosome Theory of Inheritance: Classic Papers in Development and Heredity" (B. R. Voeller, ed.), p. 15. Appleton, New York, 1968 (transl.).

Kornberg, R. D. (1974). Chromatin structure: A repeating unit of histones and DNA. *Science* **184**, 868.

Kornberg, R. D., and Thomas, J. O. (1974). Chromatin structure: Oligomers of the histones. *Science* **184**, 865.

Koser, R. B., and Collier, J. R. (1972). Characterization of the ribosomal RNA precursor in *Ilyanassa*. *Exp. Cell Res.* **70**, 124.

Koser, R. B., and Collier, J. R. (1976). An electrophoretic analysis of RNA synthesis in normal and lobeless *Ilyanassa* embryo. *Differentiation* **6**, 47.

Kriegstein, H. J., and Hogness, D. S. (1974). Mechanism of DNA replication in *Drosophila* chromosomes: Structure of replication forks and evidence for bidirectionality. *Proc. Natl. Acad. Sci. U. S. A.* **71**, 135.

Krigsgaber, M. R., and Neyfakh, A. A. (1972). Investigation of the mode of nuclear control over protein synthesis in early development of loach and sea urchin. *J. Embryol. Exp. Morphol.* **28**, 491.

Kronenberg, L. H., and Humphreys, T. (1972). Double-stranded ribonucleic acid in sea urchin embryos. *Biochemistry* **11**, 2020.

Kumar, A., and Pederson, T. (1975). Comparison of proteins bound to heterogeneous nuclear RNA and messenger RNA in HeLa cells. *J. Mol. Biol.* **96**, 353.

Kung, C. S. (1974). On the size relationship between nuclear and cytoplasmic RNA in sea urchin embryos. *Dev. Biol.* **36**, 343.

Kunz, W. (1967a). Funktionsstrukturen im Oocytenkern von *Locusta migratoria*. *Chromosoma* **20**, 332.

Kunz, W. (1967b). Lampenbürstenchromosomen und multiple Nukleolen bei Orthopteren. *Chromsoma* **21**, 446.

Kunz, W., Trepte, H.-H., and Bier, K. (1970). On the function of the germ line chromosomes in the oogenesis of *Wachtliella persicariae* (Cecidomyiidae). *Chromosoma* **30**, 180.

Kutsky, P. B. (1950). Phosphate metabolism in the early development of *Rana pipiens. J. Exp. Zool.* **115**, 429.

Laird, C. D. (1971). Chromatid structure: Relationship between DNA content and nucleotide sequence diversity. *Chromosoma* **32**, 378.

LaMarca, M. J., Smith, L. D., and Strobel, M. C. (1973). Quantitative and qualitative analysis of RNA synthesis in stage 6 and stage 4 oocytes of *Xenopus laevis. Dev. Biol.* **34**, 106.

LaMarca, M. J., Fidler, M. C. S., Smith, L. D., and Keem, K. (1975). Hormonal effects on RNA synthesis by stage 6 oocytes of *Xenopus laevis. Dev. Biol.* **47**, 384.

Lambert, C. C. (1971). Genetic transcription during the development and metamorphosis of the tunicate, *Ascidia callosa. Exp. Cell Res.* **66**, 401.

Landesman, R., and Gross, P. R. (1968). Patterns of macromolecule synthesis during development of *Xenopus laevis*. I. Incorporation of radioactive precursors into dissociated embryos. *Dev. Biol.* **18**, 571.

Landesman, R., and Gross, P. R. (1969). Patterns of macromolecule synthesis during development of *Xenopus laevis*. II. Identification of the 40 S precursor to ribosomal RNA. *Dev. Biol.* **19**, 244.

Landström, U., and Løvtrup, S. (1975). On the determination of the dorsoventral polarity in *Xenopus laevis* embryos. *J. Embryol. Exp. Morphol.* **33**, 879.

Lane, C. D., Marbaix, G., and Gurdon, J. B. (1971). Rabbit haemoglobin synthesis in frog cells: The translation of reticulocyte 9 S RNA in frog oocytes. *J. Mol. Biol.* **61**, 73.

Lane, N. J. (1967). Spheroidal and ring nucleoli in amphibian oocytes. Patterns of uridine incorporation and fine structural features. *J. Cell Biol.* **35**, 421.

Lankester, E. R. (1877). Notes on the embryology and classification of the animal kingdom: Comprising a revision of speculations relative to the origin and significance of the germ-layers. 1. The planula theory. *Q. J. Microsc. Sci.* [N.S.] **17**, 399.

Laskey, R. A., and Gurdon, J. B. (1970). Genetic content of adult somatic cells tested by nuclear transplantation from cultured cells. *Nature (London)* **228**, 1332.

Laskey, R. A., Gerhart, J. C., and Knowland, J. S. (1973). Inhibition of ribosomal RNA synthesis in neurula cells by extracts from blastulae of *Xenopus laevis. Dev. Biol.* **33**, 241.

Lawrence, P. A. (1973). A clonal analysis of segment development in *Oncopeltus* (Hemiptera). *J. Embryol. Exp. Morphol.* **30**, 681.

Lee, C. S., and Pavan, C. (1974). Replicating DNA molecules from fertilized eggs of *Cochliomyia hominivorax* (Diptera). *Chromosoma* **47**, 429.

Lentz, T. L., and Trinkaus, J. P. (1971). Differentiation of the junctional complex of surface cells in the developing *Fundulus* blastoderm. *J. Cell Biol.* **48**, 455.

Leonard, D. A., and LaMarca, M. J. (1975). *In vivo* synthesis and turnover of cytoplasmic ribosomal RNA by stage 6 oocytes of *Xenopus laevis. Dev. Biol.* **45**, 199.

Levner, M. H. (1974). RNA transcription in mature sea urchin eggs. *Exp. Cell Res.* **85**, 296.

Levy, S., Wood, P., Grunstein, M., and Kedes, L. (1975). Individual histone messenger RNAs: identification by template activity. *Cell* **4**, 239.

Levy W., B. and McCarthy, B. J. (1975). Messenger RNA complexity in *Drosophila melanogaster. Biochemistry* **14**, 2440.

Lewis, J. C., and McMillan, D. B. (1965). The development of the ovary of the sea lamprey (*Petromyzon marinus*). *J. Morphol.* **117**, 425.

Liarakos, C. D., Rosen, J. M., and O'Malley, B. W. (1973). Effect of estrogen on gene expression in the chick oviduct. II. Transcription of chick tritiated unique deoxyribonucleic acid as measured by hybridization in ribonucleic acid excess. *Biochemistry* **12**, 2809.

Lifton, R. P., and Kedes, L. H. (1976). Size and sequence homology of masked maternal and embryonic histone messenger RNAs. *Dev. Biol.* **48**, 47.

Lillie, F. R. (1902). Differentiation without cleavage in the egg of the annelid *Chaetopterus pergamentaceus. Arch. Entwicklungsmech. Org.* **14**, 477.

Lin, T. P. (1969). Microsurgery of the inner cell mass of mouse blastocysts. *Nature (London)* **222**, 480.

Lingrel, J. B., and Woodland, H. R. (1974). Initiation does not limit the rate of globin synthesis in message-injected *Xenopus* oocytes. *Eur. J. Biochem.* **47**, 47.

Littau, V. C., Allfrey, V. G., Frenster, J. H., and Mirsky, A. E. (1964). Active and inactive regions of nuclear chromatin as revealed by electron microscopic autoradiography. *Proc. Natl. Acad. Sci. U.S.A.* **52**, 93.

Lockshin, R. A. (1966). Insect embryogenesis: Macromolecular synthesis during early development. *Science* **154**, 775.

Loeb, L. A. (1970). Molecular association of DNA polymerase with chromatin in sea urchin embryos. *Nature (London)* **226**, 448.

Loeb, L. A., and Fansler, B. (1970). Intracellular migration of DNA polymerase in early developing sea urchin embryos. *Biochim. Biophys. Acta* **217**, 50.

Loeb, L. A., Fansler, B., Williams, R., and Mazia, D. (1969). Sea urchin nuclear DNA polymerase. I. Localization in nuclei during rapid DNA synthesis. *Exp. Cell Res.* **57**, 298.

Lohs-Schardin, M., and Sander, K. (1976). A dicephalic monster embryo of *Drosophila melanogaster. Wilhelm Roux' Arch. Entwicklungsmech. Org.* **179**, 159.

Lowell, R., and Burnett, A. L. (1973). Regeneration from isolated epidermal explants. *In* "Biology of Hydra" (A. L. Burnett, ed.), p. 223. Academic Press, New York.

Loyez, M. (1905). Recherches sur le développement ovarien des oeufs méroblastiques. *Arch. Anat. Microsc. Morphol. Exp.* **8**, 69.

Lundblad, G. (1955). Proteolytic activity in sea urchin gametes. IV. Further investigation of the proteolytic enzymes of the egg. *Arch. Kemi* **7**, 127.

Lützeler, I. E., and Malacinski, G. M. (1974). Modulations in the electrophoretic spectrum of newly synthesized protein in early axolotl (*Ambystoma mexicanum*) development. *Differentiation* **2**, 287.

McCarthy, B. J., and Hoyer, B. H. (1964). Identity of DNA and diversity of messenger RNA molecules in normal mouse. *Proc. Natl. Acad. Sci. U. S. A.* **52**, 915.

McClay, D. R., and Hausman, R. E. (1975). Specificity of cell adhesion: Differences between normal and hybrid sea urchin cells. *Dev. Biol.* **47**, 454.

McClendon, J. F. (1910). The development of isolated blastomeres of the frog's egg. *Am. J. Anat.* **10**, 425.

McColl, R. S., and Aronson, A. I. (1974). Transcription from unique and redundant DNA sequences in sea urchin embryos. *Biochem. Biophys. Res. Commun.* **56**, 47.

MacGregor, H. C., and Kezer, J. (1970). Gene amplification in oocytes with 8 germinal vesicles from the tailed frog *Ascaphus truei* Stejneger. *Chromosoma* **29**, 189.

MacGregor, H. C., Mizuno, S., and Vlad, M. (1976). Chromosomes and DNA sequences in salamanders. *Chromosomes Today* **5**, 331.

MacKintosh, F. R., and Bell, E. (1969). Regulation of protein synthesis in sea urchin eggs. *J. Mol. Biol.* **41**, 365.

McKnight, S. L., and Miller, O. L. (1976). Ultrastructural patterns of RNA synthesis during early embryogenesis of *Drosophila melanogaster. Cell* **8**, 305.

McLean, K. W., and Whiteley, A. H. (1974). RNA synthesis during the early development of the Pacific oyster, *Crassostrea gigas. Exp. Cell Res.* **87**, 132.

Maggio, R., and Catalano, C. (1963). Activation of amino acids during sea urchin development. *Arch. Biochem. Biophys.* **103**, 164.

Maggio, R., Vittorelli, M. L., Rinaldi, A. M., and Monroy, A. (1964). *In vitro* incorporation of amino acids into proteins stimulated by RNA from unfertilized sea urchin eggs. *Biochem. Biophys. Res. Commun.* **15**, 436.

Mahowald, A. P. (1962). Fine structure of pole cells and polar granules in *Drosophila melanogaster*. *J. Exp. Zool.* **151**, 201.

Mahowald, A. P. (1968). Polar granules of *Drosophila*. II. Ultrastructural changes during early embryogenesis. *J. Exp. Zool.* **167**, 237.

Mahowald, A. P. (1971a). Polar granules of *Drosophila*. III. The continuity of polar granules during the life cycle of *Drosophila*. *J. Exp. Zool.* **176**, 329.

Mahowald, A. P. (1971b). Polar granules of *Drosophila*. IV. Cytochemical studies showing loss of RNA from polar granules during early stages of embryogenesis. *J. Exp. Zool.* **176**, 345.

Mahowald, A. P., and Hennen, S. (1971). Ultrastructure of the "germ plasm" in eggs and embryos of *Rana pipiens*. *Dev. Biol.* **24**, 37.

Mairy, M., and Denis, H. (1971). Recherches biochimiques sur l'oogénèse. I. Synthèse et accumulation du RNA pendant l'oogénèse du crapaud Sud-Africain *Xenopus laevis*. *Dev. Biol.* **24**, 143.

Mairy, M., and Denis, H. (1972). Recherches biochimiques sur l'oogénèse. 3. Assemblage des ribosomes pendant le grand accroissement des oocytes de *Xenopus laevis*. *Eur. J. Biochem.* **25**, 535.

Malacinski, G. M. (1972). Identification of a presumptive morphogenetic determinant from the amphibian oocyte germinal vesicle nucleus. *Cell Differ.* **1**, 253.

Malacinski, G. M. (1974). Biological properties of a presumptive morphogenetic determinant from the amphibian oocyte germinal vesicle nucleus. *Cell Differ.* **3**, 31.

Malacinski, G. M., Benford, H., and Chung, H.-M. (1974). Association of an ultraviolet irradiation sensitive cytoplasmic localization with the future dorsal side of the amphibian egg. *J. Exp. Zool.* **191**, 97.

Malcolm, D. B., and Sommerville, J. (1974). The structure of chromosome-derived ribonucleoprotein in oocytes of *Triturus cristatus carnifex* (Laurenti). *Chromosoma* **48**, 137.

Malkin, L. I., Gross, P. R., and Romanoff, P. (1964). Polyribosomal protein synthesis in fertilized sea urchin eggs: The effect of actinomycin treatment. *Dev. Biol.* **10**, 378.

Mancino, G., and Barsacchi, G. (1969). The maps of the lampbrush chromosomes of *Triturus* (Amphibia Urodela). III. *Triturus italicus*. *Ann. Embryol. Morphog.* **2**, 355.

Mancino, G., Barsacchi, G., and Nardi, I. (1969). The lampbrush chromosomes of *Salamandra salamandra* (L.) (Amphibia Urodela). *Chromosoma* **26**, 365.

Manes, C. (1969). Nucleic acid synthesis in preimplantation rabbit embryos. I. Quantitative aspects, relationship to early morphogenesis and protein synthesis. *J. Exp. Zool.* **172**, 303.

Manes, C. (1971). Nucleic acid synthesis in preimplantation rabbit embryos. II. Delayed synthesis of ribosomal RNA. *J. Exp. Zool.* **176**, 87.

Manes, C. (1973). The participation of the embryonic genome during early cleavage in the rabbit. *Dev. Biol.* **32**, 453.

Manes, C., and Daniel, J. C., Jr. (1969). Quantitative and qualitative aspects of protein synthesis in the preimplantation rabbit embryo. *Exp. Cell Res.* **55**, 261.

Manes, C., and Sharma, O. K. (1973). Hypermethylated tRNA in cleaving rabbit embryos. *Nature (London)* **244**, 283.

Mangold, O. (1920). Fragen der Regulation und Determination an ungeordneten Furchungsstadien und verschmolzenen Keimen von *Triton*. *Arch. Entwicklungsmech. Org.* **47**, 249.

Manning, J. E., Schmid, C. W., and Davidson, N. (1975). Interspersion of repetitive and non-repetitive DNA sequences in the *Drosophila melanogaster* genome. *Cell* **4**, 141.

Mano, Y. (1966). Role of a trypsin-like protease in "informosomes" in a trigger mechanism of activation of protein synthesis by fertilization in sea urchin eggs. *Biochem. Biophys. Res. Commun.* **25**, 216.

Mano, Y. (1970). Cytoplasmic regulation and cyclic variation in protein synthesis in the early cleavage stage of the sea urchin embryo. *Dev. Biol.* **22**, 433.

Mano, Y. (1971a). Mechanism of increase in the basal rate of protein synthesis in the early cleavage stage of the sea urchin. *J. Biochem. (Tokyo)* **69**, 1.

Mano, Y. (1971b). Cell-free cyclic variation of protein synthesis associated with the cell cycle of sea urchin embryos. *J. Biochem. (Tokyo)* **69**, 11.

Mano, Y. (1971c). Participation of the sulfhydryl groups of a protein in the cyclic variation in the rate of protein synthesis in a cell-free system from sea urchin cells. *Arch. Biochem. Biophys.* **146**, 237.

Mano, Y., and Nagano, H. (1970). Mechanism of release of maternal messenger RNA induced by fertilization in sea urchin eggs. *J. Biochem. (Tokyo)* **67**, 611.

Marbaix, G., and Lane, C. D. (1972). Rabbit haemoglobin synthesis in frog cells. II. Further characterization of the products of translation of reticulocyte 9 s RNA. *J. Mol. Biol.* **67**, 517.

Maréchal, J. (1907). Sur l'ovogénèse des selaciens et de quelques autres chordates. Premier memoire: Morphologie de l'élément chromosomique dans l'ovocyte. I. Chez les selaciens, les téléostéens, les tuniciers et l'amphioxus. *Cellule* **24**, 1.

Mariano, E. E., and Schram-Doumont, A. (1965). Rapidly labelled ribonucleic acid in *Xenopus laevis* embryonic cells. *Biochim. Biophys. Acta* **103**, 610.

Marmur, J., Rownd, R., and Schildkraut, C. L. (1963). Denaturation and renaturation of deoxyribonucleic acid. *Prog. Nucleic Acid Res.* **1**, 231.

Maximow, A. (1927). Cultures of blood leucocytes from lymphocyte and monocyte to connective tissue. *Arch. Exp. Zellforsch. Besonders Gewebezuecht* **5**, 169.

Mazabraud, A., Wegnez, M., and Denis, H. (1975). Biochemical research on oogenesis. RNA accumulation in the oocytes of teleosts. *Dev. Biol.* **44**, 326.

Mazia, D. (1966). Biochemical aspects of mitosis. *Proc. Int. Symp. Cell Nucl.-Metab. Radiosensitivity*, 1966, p. 15.

Meeker, G. L., and Iverson, R. M. (1971). Tubulin synthesis in fertilized sea urchin eggs. *Exp. Cell Res.* **64**, 129.

Melli, M., Whitfield, C., Rao, K. V., Richardson, M., and Biship, J. O. (1971). DNA-RNA hybridization in vast DNA excess. *Nature (London), New Biol.* **231**, 8.

Mel'nikova, N. L., Timofeeva, M. Ya., Rott, N. N., and Ignat'eva, G. M. (1972). Synthesis of ribosomal RNA in the early embryogenesis of the trout. *Ontogenez (Sov. J. Dev. Biol.) (Engl. Transl.)* **3**, 67.

Mescher, A., and Humphreys, T. (1974). Activation of maternal mRNA in the absence of poly(A) formation in fertilised sea urchin eggs. *Nature (London)* **249**, 138.

Metafora, S., Felicetti, L., and Gambino, R. (1971). The mechanism of protein synthesis activation after fertilization of sea urchin eggs. *Proc. Natl. Acad. Sci. U.S.A.* **68**, 600.

Meyer, G. F. (1968). Experimental studies on spermiogenesis in *Drosophila*. *Genetics* **61**, Suppl. 1, 79.

Miller, L. (1972). Initiation of the synthesis of ribosomal ribonucleic acid precursor in different regions of frog (*Rana pipiens*) gastrulae. *Biochem. J.* **127**, 733.

Miller, L. (1973). Control of 5 S RNA synthesis during early development of anucleolate and partial nucleolate mutants of *Xenopus laevis*. *J. Cell Biol.* **59**, 624.

Miller, L., and Knowland, J. (1970). Reduction of ribosomal RNA synthesis and ribosomal

RNA genes in a mutant of *Xenopus laevis* which organizes only a partial nucleolus. II. The number of ribosomal RNA genes in animals of different nucleolar types. *J. Mol. Biol.* **53**, 329.

Miller, O. L., Jr. (1965). Fine structure of lampbrush chromosomes. *Natl. Cancer Inst., Monogr.* **18**, 79.

Miller, O. L., Jr. (1966). Structure and composition of peripheral nucleoli of salamander oocytes. *Natl. Cancer Inst., Monogr.* **23**, 53.

Miller, O. L., Jr., and Bakken, A. H. (1972). Morphological studies of transcription. *Acta Endocrinol. (Copenhagen), Suppl.* **168**, 155.

Miller, O. L., Jr., and Beatty, B. R. (1969a). Portrait of a gene. *J. Cell Physiol.* **74**, Suppl. 1, 225.

Miller, O. L., Jr., and Beatty, B. R. (1969b). Extrachromosomal nucleolar genes in amphibian oocytes. *Genetics, Suppl.* **61**, 1.

Miller, O. L., Jr., and Beatty, B. R. (1969c). Visualization of nucleolar genes. *Science* **164**, 955.

Millonig, G. (1966). The morphological changes of the nucleolus during oogenesis and embryogenesis of echinoderms. *Electron Microsc., Proc. Int. Congr. 6th, 1966*, Vol. II, p. 345.

Minganti, A. (1951). Ricerche istochimiche sulla localizzazione del territorio presuntivo degli organi sensoriali nelle larve di ascidie. *Pubbl. Stn. Zool. Napoli* **23**, 52.

Minganti, A. (1959a). Androgenetic hybrids in ascidians. I. *Ascidia malaca* (♀) × *Phallusia mamillata* (♂). *Acta Embryol. Morphol. Exp.* **2**, 244.

Minganti, A. (1959b). Lo sviluppo embrionale e il comportamento dei cromosomi in ibridi tra 5 specie di ascidie. *Acta Embryol. Morphol. Exp.* **2**, 269.

Mintz, B. (1962). Formation of genotypically mosaic mouse embryos. *Am. Zool.* **2**, 432.

Mintz, B. (1964). Synthetic processes and early development in the mammalian egg. *J. Exp. Zool.* **157**, 85.

Mintz, B. (1965). Experimental genetic mosaicism in the mouse. *Preimplantation Stages Pregnancy, Ciba Found. Symp., 1965*, p. 194.

Mirkes, P. E. (1972). Polysomes and protein synthesis during development of *Ilyanassa obsoleta*. *Exp. Cell Res.* **74**, 503.

Mirsky, A. E. (1951). Some chemical aspects of the cell nucleus. *In* "Genetics in the Twentieth Century" (L. C. Dunn, ed.), p. 127. Macmillan, New York.

Mirsky, A. E. (1953). The chemistry of heredity. *Sci. Am.* **188**, 47.

Mirsky, A. E., and Ris, H. (1949). Variable and constant components of chromosomes. *Nature (London)* **163**, 666.

Mizuno, S., Lee, Y. R., Whiteley, A. H., and Whiteley, H. R. (1974). Cellular distribution of RNA populations in 16-cell stage embryos of the sand dollar, *Dendraster excentricus*. *Dev. Biol.* **37**, 18.

Moar, V. A., Gurdon, J. B., Lane, C. D., and Marbaix, G. (1971). Translational capacity of living frog eggs and oocytes, as judged by messenger RNA injection. *J. Mol. Biol.* **61**, 93.

Moav, B., and Nemer, M. (1971). Histone synthesis. Assignment to a special class of polyribosomes in sea urchin embryos. *Biochemistry* **10**, 881.

Molinaro, M., and Farace, M. G. (1972). Changes in codon recognition and chromatographic behavior of tRNA species during embryonic development of the sea urchin *Paracentrotus lividus*. *J. Exp. Zool.* **181**, 223.

Molloy, G. R., Jelinek, W., Salditt, M., and Darnell, J. E. (1974). Arrangement of specific oligonucleotides with poly(A) terminated hnRNA molecules. *Cell* **1**, 43.

Monesi, V., and Salfi, V. (1967). Macromolecular synthesis during early development in the mouse embryo. *Exp. Cell Res.* **46**, 632.

Monesi, V., Molinaro, M., Spalletta, E., and Davoli, C. (1970). Effect of metabolic inhibitors on macromolecular synthesis and early development in the mouse embryo. *Exp. Cell Res.* 59, 197.

Monroy, A. (1965). "Chemistry and Physiology of Fertilization." Holt, New York.

Monroy, A., and Tyler, A. (1963). Formation of active ribosomal aggregates (polysomes) upon fertilization and development of sea urchin eggs. *Arch. Biochem. Biophys.* 103, 431.

Monroy, A., Maggio, R., and Rinaldi, A. M. (1965). Experimentally induced activation of the ribosomes of the unfertilized sea urchin egg. *Proc. Natl. Acad. Sci. U.S.A.* 54, 107.

Moore, G. P. M. (1975). The RNA polymerase activity of the preimplantation mouse embryo. *J. Embryol. Exp. Morphol.* 34, 291.

Moore, J. A. (1941). Developmental rate of hybrid frogs. *J. Exp. Zool.* 86, 405.

Morata, G., and Lawrence, P. A. (1975). Control of compartment development by the *engrailed* gene in *Drosophila*. *Nature (London)* 255, 614.

Morgan, T. H., ed. (1927). "Experimental Embryology." p. 468. Columbia Univ. Press, New York.

Morgan, T. H. (1934). "Embryology and Genetics." Columbia Univ. Press, New York.

Moritz, K. B., and Roth, G. E. (1976). Complexity of germline and somatic DNA in *Ascaris*. *Nature (London)* 259, 55.

Morrill, G. A., and Kostellow, A. B. (1965). Phospholipid and nucleic acid gradients in the developing amphibian embryos. *J. Cell Biol.* 25, 21.

Morrill, J. B. (1973). Biochemical and electrophoretic analysis of acid and alkaline phosphatase activity in the developing embryo of *Physa fontinalis* (Gastropoda, Pulmonata). *Acta Embryol. Exp.* p. 61.

Morrill, J. B., and Norris, E. (1965). Electrophoretic analysis of hydrolytic enzymes in the *Ilyanassa* embryo. *Acta Embryol. Morphol. Exp.* 8, 232.

Morrow, J. F. (1974). Mapping the SV40 chromosome by use of restriction enzymes. Ph.D. Thesis, Standord University, Stanford, California.

Motomura, I. (1966). Secretion of a mucosubstance in the cleaving egg of the sea urchin. *Acta Embryol. Morphol. Exp.* 9, 56.

Moulé, Y., and Chauveau, J. (1968). Particules ribonucléoprotéiques 40s des noyaux de foie de rat. *J. Mol. Biol.* 33, 465.

Müller, W. P. (1974). The lampbrush chromosomes of *Xenopus laevis* (Daudin). *Chromosoma* 47, 283.

Mulnard, J. G. (1954). Etude morphologique et cytochimique de l'oogénèse chez *Acanthoscelides obtecteus* Say (Bruchide-Coléoptère). *Arch. Biol.* 65, 261.

Mulnard, J. G. (1965). Studies of regulation of mouse ova *in vitro*. *Preimplantation Stages Pregnancy*, *Ciba Found. Symp.*, 1965, p. 123.

Nadel, M., Carroll, A., and Kafatos, F. C. (1976). Changes in the pattern of protein synthesis after fertilization in *Spisula solidissima*. *Dev. Biol.*, in press.

Nadijcka, M., and Hillman, N. (1974). Ultrastructural studies of the mouse blastocyst substages. *J. Embryol. Exp. Morphol.* 32, 675.

Nägeli, C. (1884). "Mechanisch-physiologische theorie der abstammungslehre." R. Oldenbourg, Munich & Leipzig.

Nakano, E., and Monroy, A. (1958). Incorporation of S^{35} methionine in the cell fractions of sea urchin eggs and embryos. *Exp. Cell Res.* 14, 236.

Nakatsuji, N. (1974). Studies on the gastrulation of amphibian embryos: Pseudopodia in the gastrula of *Bufo bufo japonicus* and their significance to gastrulation. *J. Embryol. Exp. Morphol.* 32, 795.

Namenwirth, M. (1974). The inheritance of cell differentiation during limb regeneration in the axolotl. *Dev. Biol.* 41, 42.

Nemer, M. (1962). Interrelation of messenger polyribonucleotides and ribosomes in the sea urchin egg during embryonic development. *Biochem. Biophys. Res. Commun.* **8**, 511.

Nemer, M. (1963). Old and new RNA in the embryogenesis of the purple sea urchin. *Proc. Natl. Acad. Sci. U.S.A.* **50**, 217.

Nemer, M. (1975). Developmental changes in the synthesis of sea urchin embryo messenger RNA containing and lacking polyadenylic acid. *Cell* **6**, 559.

Nemer, M., and Bard, S. G. (1963). Polypeptide synthesis in sea urchin embryogenesis: An examination with synthetic polyribonucleotides. *Science* **140**, 664.

Nemer, M., and Infante, A. A. (1965). Messenger RNA in early sea urchin embryos: Size classes. *Science* **150**, 217.

Nemer, M., and Infante, A. A. (1967a). Ribosomal ribonucleic acid of the sea urchin egg and its fate during embryogenesis. *J. Mol. Biol.* **27**, 73.

Nemer, M., and Infante, A. A. (1967b). Early control of gene expression. *In* "The Control of Nuclear Activity" (L. Goldstein, ed.), p. 101. Prentice-Hall, Englewood Cliffs, New Jersey.

Nemer, M., and Lindsay, D. T. (1969). Evidence that the s-polysomes of early sea urchin embryos may be responsible for the synthesis of chromosomal histones. *Biochem. Biophys. Res. Commun.* **35**, 156.

Nemer, M., Graham, M., and Dubroff, L. M. (1974). Co-existence of non-histone messenger RNA species lacking and containing polyadenylic acid in sea urchin embryos. *J. Mol. Biol.* **89**, 435.

Nemer, M., Dubroff, L. M., and Graham, M. (1975). Properties of sea urchin embryo messenger RNA containing and lacking poly(A). *Cell* **6**, 171.

Newman, H. H. (1914). Modes of inheritance in teleost hybrids. *J. Exp. Zool.* **16**, 447.

Newman, H. H. (1915). Development and heredity in heterogenic teleost hybrids. *J. Exp. Zool.* **18**, 511.

Newrock, K. M., and Raff, R. A. (1975). Polar lobe specific regulation of translation in embryos of *Ilyanassa obsoleta*. *Dev. Biol.* **42**, 242.

Neyfakh, A. A. (1964). Radiation investigation of nucleo-cytoplasmic interrelations in morphogenesis and biochemical differentiation. *Nature (London)* **201**, 880.

Neyfakh, A. A., Kostomarova, A. A., and Rubakova, T. A. (1974). Study of RNA transfer from nucleus to cytoplasm in early loach (*Misgurnus fossilis* L.) and hybrid loach ♀ × goldfish (*Carassius auratus auratus*) ♂ embryos. *Ontogenez (Sov. J. Dev. Biol.) (Engl. Transl.)* **4**, 307.

Nicholas, J. S., and Hall, B. V. (1942). Experiments on developing rats. II. The development of isolated blastomeres and fused eggs. *J. Exp. Zool.* **90**, 441.

Nieuwkoop, P. D. (1969). The formation of the mesoderm in urodelean amphibians. II. The origin of the dorso-ventral polarity of the mesoderm. *Wilhelm Roux' Arch. Entwicklungsmech. Org.* **163**, 298.

Nieuwkoop, P. D. (1973). The "organization center" of the amphibian embryo: Its origin, spatial organization, and morphogenetic action. *Adv. Morphog.* **10**, 1.

Nieuwkoop, P. D., and Faber, J. (1956). "Normal Table of *Xenopus laevis* (Daudin)." North-Holland Publ., Amsterdam.

Nilsson, M. O., and Hultin, T. (1974). Characteristics and intracellular distribution of messengerlike RNA in encysted embryos of *Artemia salina*. *Dev. Biol.* **38**, 138.

Noll, M. (1974). Subunit structure of chromatin. *Nature (London)* **251**, 249.

Novikoff, A. B. (1938). Embryonic determination in the annelid *Sabellaria vulgaris*. II. Transplantation of polar lobes and blastomeres as a test of their inducing capacities. *Biol. Bull.* **74**, 211.

Okada, M., Kleinman, I. A., and Schneiderman, H. A. (1974a). Restoration of fertility in sterilized *Drosophila* eggs by transplantation of polar cytoplasm. *Dev. Biol.* **37**, 43.

Okada, M., Kleinman, I. A., and Schneiderman, H. A. (1974b). Repair of a genetically-caused defect in oogenesis in Drosophila melanogaster by transplantation of cytoplasm from wild-type eggs and by injection of pyrimidine nucleosides. Dev. Biol. 37, 55.

Okada, M., Kleinman, I. A., and Schneiderman, H. A. (1974c). Chimeric Drosophila adults produced by transplantation of nuclei into specific regions of fertilized eggs. Dev. Biol. 39, 286.

Okazaki, K. (1960). Skeleton formation of sea urchin larvae. II. Organic matrix of the spicule. Embryologia 5, 283.

Okazaki, K. (1965). Skeleton formation of sea urchin larvae. V. Continuous observations of the process of matrix formation. Exp. Cell Res. 40, 585.

Okkelberg, P. (1921). The early history of the germ cells in the brook lamprey, Entosphenus wilderi (Gage), up to and including the period of sex differentiation. J. Morphol. 35, 1.

Olds, P. J., Stern, S., and Biggers, J. D. (1973). Chemical estimates of the RNA and DNA contents of the early mouse embryo. J. Exp. Zool. 186, 39.

Olins, A. L., and Olins, D. E. (1974). Spheroid chromatin units (ν bodies). Science 183, 330.

O'Melia, A. F., and Villee, C. A. (1972). De novo synthesis of transfer and 5S RNA$^{1\prime}$ in cleaving sea urchin embryos. Nature (London) New Biol. 239, 51.

Ortolani, G. (1964). Origine dell'organo apicale e di derivati mesodermice nello sviluppo embrionale di ctenophori. Acta Embryol. Morphol. Exp. 7, 191.

Oudet, P., Gross-Bellard, M., and Chambon, P. (1975). Electron microscopic and biochemical evidence that chromatin structure is a repeating unit. Cell 4, 281.

Ozaki, H. (1971). Developmental studies of sea urchin chromatin. Chromatin isolated from spermatozoa of the sea urchin Strongylocentrotus purpuratus. Dev. Biol. 26, 209.

Ozaki, H. (1974). Localization and multiple forms of acetylcholinesterase in sea urchin embryos. Dev., Growth & Differ. 16, 267.

Ozaki, H. (1975). Regulation of isozymes in interspecies sea urchin hybrid embryos. In "Isozymes" (C. L. Markert, ed.), Vol. III, p. 543. Academic Press, New York.

Ozaki, H., and Whiteley, A. H. (1970). L-Malate dehydrogenase in the development of the sea urchin Strongylocentrotus purpuratus. Dev. Biol. 21, 196.

Packman, S., Aviv, H., Ross, J., and Leder, P. (1972). A comparison of globin genes in duck reticulocytes and liver cells. Biochem. Biophys. Res. Commun. 49, 813.

Pagoulatos, G. N., and Darnell, J. E. (1970). A comparison of the heterogeneous nuclear RNA of HeLa cells in different periods of the cell growth cycle. J. Cell Biol. 44, 476.

Palmiter, R. D. (1973). Rate of ovalbumin messenger ribonucleic acid synthesis in the oviduct of estrogen-primed chicks. J. Biol. Chem. 248, 8260.

Pardue, M. L., and Gall, J. G. (1969). Molecular hybridization of radioactive DNA to the DNA of cytological preparations. Proc. Natl. Acad. Sci. U.S.A. 64, 600.

Pardue, M. L., Brown, D. D., and Birnstiel, M. L. (1973). Location of the genes for 5S ribosomal RNA in Xenopus laevis. Chromosoma 42, 191.

Patterson, J. B., and Stafford, D. W. (1971). Characterization of sea urchin ribosomal satellite deoxyribonucleic acid. Biochemistry 10, 2775.

Paul, M., Goldsmith, M. R., Hunsley, J. R., and Kafatos, F. C. (1972). Specific protein synthesis in cellular differentiation. J. Cell Biol. 55, 653.

Peacock, W. J. (1965). Chromosome replication. Natl. Cancer Inst., Monogr. 18, 101.

Pederson, T. (1974). Proteins associated with heterogeneous nuclear RNA in eukaryotic cells. J. Mol. Biol. 83, 163.

Peltz, R. (1973). The integrity of "giant" nuclear RNA. Biochim. Biophys. Acta 308, 148.

Penners, A. (1922). Die Furchung von Tubifex rivulorum Lam. Zool. Jahrb. Part I., Abt. Anat. Ontog. Tiere 43, 323.

Penners, A. (1926). Experimentelle Untersuchungen zum Determinationsproblem am Keim von *Tubifex rivulorum* Lam. II. Die Entwicklungs teilweise abgetöteter Keime. *Z. Wiss. Zool.* **127**, 1.

Perkowska, E., MacGregor, H. C., and Birnstiel, M. L. (1968). Gene amplification in the oocyte nucleus of mutant and wild-type *Xenopus laevis*. *Nature (London)* **217**, 649.

Pestell, R. Q. W. (1975). Microtubule protein synthesis during oogenesis and early embryogenesis in *Xenopus laevis*. *Biochem. J.* **145**, 527.

Petrakis, N. L., Davis, M., and Lucia, S. P. (1961). The *in vivo* differentiation of human leukocytes into histiocytes, fibroblasts and fat cells in subcutaneous diffusion chambers. *Blood* **17**, 109.

Piatigorsky, J. (1968). Ribonuclease and trypsin treatment of ribosomes and polyribosomes from sea urchin eggs. *Biochim. Biophys. Acta* **166**, 142.

Piatigorsky, J., and Tyler, A. (1970). Changes upon fertilization in the distribution of RNA-containing particles in sea urchin eggs. *Dev. Biol.* **21**, 13.

Pietruschka, F., and Bier, K. (1972). Autoradiographische Untersuchungen zur RNS- und Protein-Synthesis in der fruehen Embryogenese von *Musca domestica*. *Wilhelm Roux' Arch. Entwicklungsmech. Org.* **169**, 56.

Pikó, L. (1970). Synthesis of macromolecules in early mouse embryos cultured *in vitro*: RNA, DNA, and a polysaccharide component. *Dev. Biol.* **21**, 257.

Poccia, D. L., and Hinegardner, R. T. (1975). Developmental changes in chromatin proteins of the sea urchin from blastula to mature larva. *Dev. Biol.* **45**, 81.

Pollack, S. B., and Telfer, W. H. (1969). RNA in *Cecropia* moth ovaries: Sites of synthesis, transport, and storage. *J. Exp. Zool.* **170**, 1.

Pucci-Minafra, I., Minafra, S., and Collier, J. R. (1969). Distribution of ribosomes in the egg of *Ilyanassa obsoleta*. *Exp. Cell Res.* **57**, 167.

Pukkila, P. J. (1975). Identification of the lampbrush chromosome loops which transcribe 5 S ribosomal RNA in *Notophthalmus (Triturus) viridescens*. *Chromosoma* **53**, 71.

Rabinowitz, M. (1941). Studies on the cytology and early embryology of the egg of *Drosophila melanogaster*. *J. Morphol.* **69**, 1.

Rachkus, Yu. A., Kupriyanova, N. S., Timofeeva, M. Ya., and Kafiani, K. A. (1969). Homologies in RNA populations at different stages of embryogenesis. *Mol. Biol. (Engl. Transl.)* **3**, 486.

Rachkus, Yu. A., Kafiani, K. A., and Timofeeva, M. Ya. (1971). Some characteristics of ribonucleic acids synthesized in loach embryogenesis. *Ontogenez (Sov. J. Dev. Biol.) (Engl. Transl.)* **2**, 222.

Raff, R. A., and Kaumeyer, J. F. (1973). Soluble microtubule proteins of the sea urchin embryo: Partial characterization of the proteins and behavior of the pool in early development. *Dev. Biol.* **32**, 309.

Raff, R. A., Greenhouse, G., Gross, K. W., and Gross, P. R. (1971). Synthesis and storage of microtubule proteins by sea urchin embryos. *J. Cell Biol.* **50**, 516.

Raff, R. A., Colot, H. V., Selvig, S. E., and Gross, P. R. (1972). Oogenetic origin of messenger RNA for embryonic synthesis of microtubule proteins. *Nature (London)* **235**, 211.

Raff, R. A., Newrock, K. M., and Secrist, R. D. (1975). Regulation of microtubule protein synthesis in embryos of the marine snail, *Ilyanassa obsoleta*. *Dev. Biol.* **44**, 369.

Rattenbury, J. C., and Berg, W. E. (1954). Embryonic segregation during early development of *Mytilus edulis*. *J. Morphol.* **95**, 393.

Rau, D. C., and Klotz, L. C. (1976). On the theory and practice of DNA renaturation kinetics followed by optical density. *J. Mol. Biol.*, in press.

Raven, C. P. (1958). "Morphogenesis: The Analysis of Molluscan Development." Pergamon, Oxford.

Raven, C. P. (1961). "Oogenesis: The Storage of Developmental Information." Pergamon, Oxford.

Reeder, R. H. (1973). Transcription of chromatin by bacterial RNA polymerase. *J. Mol. Biol.* **80**, 229.

Reiger, J. C., and Kafatos, F. C. (1976). Absolute rates of protein synthesis in sea urchins, with specific activity measurements of radioactive leucine and leucyl-tRNA. *Dev. Biol.*, in press.

Renkawitz, R., and Kunz, W. (1975). Independent replication of the ribosomal RNA genes in the polytrophic-meroistic ovaries of *Calliphora erythrocephala*, *Drosophila hydei*, and *Sarcophaga barbata*. *Chromosoma* **53**, 131.

Reverberi, G. (1970). The ultrastructure of *Dentalium* egg at the trefoil stage. *Acta Embryol. Exp.* p. 31.

Reverberi, G. (1971a). Ctenophores. *In* "Experimental Embryology of Marine and Fresh-Water Invertebrates" (G. Reverberi, ed.), p. 83. North-Holland Publ., Amsterdam.

Reverberi, G. (1971b). Annelids. *In* "Experimental Embryology of Marine and Fresh-Water Invertebrates" (G. Reverberi, ed.), p. 126. North-Holland Publ., Amsterdam.

Reverberi, G. (1971c). Ascidians. *In* "Experimental Embryology of Marine and Fresh-Water Invertebrates" (G. Reverberi, ed.), p. 507. North-Holland Publ., Amsterdam.

Reverberi, G., and Minganti, A. (1947). La distribuzione delle potenze nel germe di Ascidie allo stadio di otto blastomeri, analizzata mediante le combinazione e i trapianti di blastomeri. *Pubbl. Stn. Zool. Napoli* **21**, 1.

Rhoads, R. E., McKnight, G. S., and Schimke, R. T. (1973). Quantitative measurement of ovalbumin messenger ribonucleic acid activity. Localization in polysomes, induction by estrogen and effect of actinomycin D. *J. Biol. Chem.* **248**, 2031.

Ribbert, D., and Bier, K. (1969). Multiple nucleoli and enhanced nucleolar activity in the nurse cells of the insect ovary. *Chromosoma* **27**, 178.

Ribbert, D., and Kunz, W. (1969). Lampenbürsten-chromosomen in den Oocytenkernen von *Sepia officinalis*. *Chromosoma* **28**, 93.

Ricard, B., and Salser, W. (1975). The structure of bacteriophage T4-specific messenger RNAs. I. Tightly folded conformation revealed by sedimentation in denaturing solvents. *J. Mol. Biol.* **94**, 163.

Rice, T. B., and Garen, A. (1975). Localized defects of blastoderm formation in maternal effect mutants of *Drosophila*. *Dev. Biol.* **43**, 277.

Rinaldi, A. M., and Parente, A. (1976). Rate of protein synthesis in oocytes of *Paracentrotus lividus*. *Dev. Biol.* **49**, 260.

Rinaldi, A. M., Sconzo, G., Albanese, I., Ramirez, F., Bavister, B. D., and Giudice, G. (1974). Cytoplasmic giant RNA in sea urchin embryos. III. Polysomal localization. *Cell Differ.* **3**, 305.

Ritossa, F. M., Atwood, K. C., Lindsley, D. L., and Spiegelman, S. (1966). On the chromosomal distribution of DNA complementary to ribosomal and soluble RNA. *Natl. Cancer Inst., Monogr.* **23**, 449.

Roberts, R. B., Abelson, P. H., Cowie, D. B., Bolton, E. J., and Britten, R. J. (1955). Studies of biosynthesis in *Escherichia coli*. *Carnegie Inst. Washington Publ.* **607**, 455.

Rochaix, J.-D., Bird, A., and Bakken, A. (1974). Ribosomal RNA gene amplification by rolling circles. *J. Mol. Biol.* **87**, 473.

Roeder, R. G. (1974a). Multiple forms of deoxyribonucleic acid-dependent ribonucleic acid polymerase in *Xenopus laevis*. Isolation and partial characterization. *J. Biol. Chem.* **249**, 241.

Roeder, R. G. (1974b). Multiple forms of deoxyribonucleic acid-dependent ribonucleic acid polymerase in *Xenopus laevis*. Levels of activity during oocyte and embryonic development. *J. Biol. Chem.* **249**, 249.

Roeder, R. G., and Rutter, W. J. (1970). Multiple RNA polymerases and RNA synthesis during sea urchin development. *Biochemistry* **9**, 2543.

Roeder, R. G., Reeder, R. H., and Brown, D. D. (1970). Multiple forms of RNA polymerase in *Xenopus laevis*: Their relationship to RNA synthesis *in vivo* and their fidelity to transcription *in vitro*. *Cold Spring Harbor Symp. Quant. Biol.* **35**, 727.

Romanoff, A. L. (1960). "The Avian Embryo," p. 13. Macmillan, New York.

Rosbash, M., and Ford, P. J. (1974). Polyadenylic acid-containing RNA in *Xenopus laevis* oocytes. *J. Mol. Biol.* **85**, 87.

Rosbash, M., Ford, P. J., and Bishop, J. O. (1974). Analysis of the C-value paradox by molecular hybridization. *Proc. Natl. Acad. Sci. U.S.A.* **71**, 3746.

Ross, J., Gielen, J., Packman, S., Ikawa, Y., and Leder, P. (1974). Globin gene expression in cultured erythroleukemic cells. *J. Mol. Biol.* **87**, 697.

Rossant, J. (1975a). Investigation of the determinative state of the mouse inner cell mass. I. Aggregation of isolated inner cell masses with morulae. *J. Embryol. Exp. Morphol.* **33**, 979.

Rossant, J. (1975b). Investigation of the determinative state of the mouse inner cell mass. II. The fate of isolated inner cell masses transferred to the oviduct. *J. Embryol. Exp. Morphol.* **33**, 991.

Roux, W. (1883). "Ueber die Bedentung der Kerntheilungsfiguren. Eine hypothetische Erörterung." Engelmann, Leipzig.

Rückert, J. (1892). Zur Entwicklungsgeschichte des Ovarialeies bei Selachiern. *Anat. Anz.* **7**, 107.

Ruderman, J. V., and Gross, P. R. (1974). Histones and histone synthesis in sea urchin development. *Dev. Biol.* **36**, 286.

Ruderman, J. V., and Pardue, M. L. (1976). *In vitro* translation analysis of messenger RNA in echinoderm and amphibian early development. *Dev. Biol.*, in press.

Ruderman, J. V., Baglioni, C., and Gross, P. R. (1974). Histone mRNA and histone synthesis during embryogenesis. *Nature (London)* **247**, 36.

Runnström, J., and Markman, B. (1966). Gene dependency of vegetalization in sea urchin embryos treated with lithium. *Biol. Bull.* **130**, 402.

Ruud, G. (1925). Die Entwicklung isolierter Keimfragmente frühester Stadien von *Triton taeniatus*. *Wilhelm Roux' Arch. Entwicklungsmech. Org.* **105**, 1.

Ryffel, G. U., and McCarthy, B. J. (1975). Complexity of cytoplasmic RNA in different mouse tissues measured by hybridization of polyadenylated RNA to complementary DNA. *Biochemistry* **14**, 1379.

Samarina, O. P., Molnar, J., Lukaniden, E. M., Bruskov, V. I., Krichevskaya, A. A., and Georgiev, G. P. (1967). Reversible dissociation of nuclear ribonucleoprotein particles containing mRNA into RNA and protein. *J. Mol. Biol.* **27**, 187.

Sander, K. (1971). Pattern formation in longitudinal halves of leaf hopper eggs (Homoptera) and some remarks on the definition of "embryonic regulation." *Wilhelm Roux' Arch. Entwicklungsmech. Org.* **167**, 336.

Sargent, T. D., and Raff, R. A. (1976). Protein synthesis and messenger RNA stability in activated, enucleate sea urchin eggs are not affected by actinomycin D. *Dev. Biol.* **48**, 327.

Schaffner, W., Gross, K., Telford, J., and Birnstiel, M. (1976). Molecular analysis of the histone gene cluster of *Psammechinus miliaris*. II. The arrangement of the five histone-coding and spacer sequences. *Cell* **8**, 471.

Scheer, U. (1973). Nuclear pore flow rate of ribosomal RNA and chain growth rate of its precursor during oogenesis of Xenopus laevis. Dev. Biol. 30, 13.

Scheer, U., Trendelenburg, M. F., and Franke, W. W. (1976). Regulation of transcription of genes of ribosomal RNA during amphibian oogenesis: A biochemical and morphological study. J. Cell Biol. 69, 465.

Schmid, C. W., and Deininger, P. L. (1975). Sequence organization of the human genome. Cell 6, 345.

Schmidt, O., Zissler, D., Sander, K., and Kalthoff, K. (1975). Switch in pattern formation after puncturing the anterior pole of Smittia eggs (Chironomidae, Diptera). Dev. Biol. 46, 216.

Schönmann, W. (1938). Der diploide bastard Triton palmatus ♀ × Salamandra ♂. Wilhelm Roux' Arch. Entwicklungsmech. Org. 138, 345.

Schubinger, G. (1976). Adult differentiation from partial Drosophila embryos after egg ligation during stages of nuclear multiplication and cellular blastoderm. Dev. Biol. 50, 476.

Schultz, G. A. (1973). Characterization of polyribosomes containing newly synthesized messenger RNA in preimplantation rabbit embryos. Exp. Cell Res. 82, 168.

Schultz, G. A. (1974). The stability of messenger RNA containing polyadenylic acid sequences in rabbit blastocysts. Exp. Cell Res. 86, 190.

Schultz, G. A. (1975). Polyadenylic acid-containing RNA in unfertilized and fertilized eggs of the rabbit. Dev. Biol. 44, 270.

Schultz, G. A., Manes, C., and Hahn, W. E. (1973a). Estimation of the diversity of transcription in early rabbit embryos. Biochem. Genet. 9, 247.

Schultz, G. A., Manes, C., and Hahn, W. E. (1973b). Synthesis of RNA containing polyadenylic acid sequences in preimplantation rabbit embryos. Dev. Biol. 30, 418.

Schwartz, M. C. (1970). Nucleic acid metabolism in oocytes and embryos of Urechis caupo. Dev. Biol. 23, 241.

Schwartz, R. J., and Wilde, C. E., Jr. (1973). Changes in protein synthesis in the morphogenesis of Fundulus heteroclitus. Nature (London) 245, 376.

Sconzo, G., Pirrone, A. M., Mutolo, V., and Giudice, G. (1970a). Synthesis of ribosomal RNA during sea urchin development. III. Evidence for an activation of transcription. Biochim. Biophys. Acta 199, 435.

Sconzo, G., Pirrone, A. M., Mutolo, V., and Giudice, G. (1970b). Synthesis of ribosomal RNA in disaggregated cells of sea urchin embryos. Biochim. Biophys. Acta 199, 441.

Sconzo, G., Bono, A., Albanese, I., and Giudice, G. (1972). Studies on sea urchin oocytes. II. Synthesis of RNA during oogenesis. Exp. Cell Res. 72, 95.

Sconzo, G., Albanese, I., Rinaldi, A. M., Lo Presti, G., and Giudice, G. (1974). Cytoplasmic giant RNA in sea urchin embryos. II. Physicochemical characterization. Cell Differ. 3, 297.

Scott, S. E. M., and Sommerville, J. (1974). Location of nuclear proteins on the chromosomes of newt oocytes. Nature (London) 250, 680.

Seale, R. L., and Aronson, A. I. (1973a). Chromatin-associated proteins of the developing sea urchin embryo. I. Kinetics of synthesis and characterization of non-histone proteins. J. Mol. Biol. 75, 633.

Seale, R. L., and Aronson, A. I. (1973b). Chromatin-associated proteins of the developing sea urchin embryos. II. Acid-soluble proteins. J. Mol. Biol. 75, 647.

Seidel, F. (1960). Die Entwicklungsfähigkeiten isolierter Furchungszellen aus dem Ei des Kaninchens Oryctolagus cuniculus. Wilhelm Roux' Arch. Entwicklungsmech. Org. 152, 43.

Selman, K., and Kafatos, F. C. (1974). Transdifferentiation in the labial gland of silk moths: Is DNA required for cellular metamorphosis? Cell Differ. 3, 81.

Selman, K., and Kafatos, F. C. (1975). Differentiation in the cocoonase producing silkmoth galea: Ultrastructural studies. *Dev. Biol.* **46**, 132.

Shearer, R. W., and McCarthy, B. J. (1967). Evidence for ribonucleic acid molecules restricted to the cell nucleus. *Biochemistry* **6**, 283.

Shih, R. J. (1975). Analysis of amino acid pools, rates of protein synthesis, and nuclear proteins of *Rana pipiens* oocytes and embryos. Ph.D. Thesis, Purdue University, Lafayette, Indiana.

Shiokawa, K., and Yamana, K. (1967a). Pattern of RNA synthesis in isolated cells of *Xenopus laevis* embryos. *Dev. Biol.* **16**, 368.

Shiokawa, K., and Yamana, K. (1967b). Inhibitor of ribosomal RNA synthesis in *Xenopus laevis* embryos. *Dev. Biol.* **16**, 389.

Shiokawa, K., and Yamana, K. (1968). Ribonucleic acid (RNA) synthesis in dissociated embryonic cells of *Xenopus laevis*. IV. Synthesis of messenger RNA in the presence of an inhibitory factor of ribosomal RNA synthesis. *Proc. Jpn. Acad.* **44**, 379.

Shutt, R. H., and Kedes, L. H. (1974). Synthesis of histone mRNA sequences in isolated nuclei of cleavage stage sea urchin embryos. *Cell* **3**, 283.

Singer, R. H., and Penman, S. (1972). Stability of HeLa cell mRNA in actinomycin. *Nature (London)* **240**, 100.

Skalko, R. G., and Morse, J. M. D. (1969). The differential response of the early mouse embryo to actinomycin D treatment in vitro. *Teratology* **2**, 47.

Skoultchi, A., and Gross, P. R. (1973). Maternal histone messenger RNA: Detection by molecular hybridization. *Proc. Natl. Acad. Sci. U.S.A.* **70**, 2840.

Slater, D. W., and Spiegelman, S. (1966). An estimation of genetic messages in the unfertilized echinoid egg. *Proc. Natl. Acad. Sci. U.S.A.* **56**, 164.

Slater, D. W., Slater, I., and Gillespie, D. (1972). Post-fertilization synthesis of polyadenylic acid in sea urchin embryos. *Nature (London)* **240**, 333.

Slater, I., and Slater, D. W. (1974). Polyadenylation and transcription following fertilization. *Proc. Natl. Acad. Sci. U.S.A.* **71**, 1103.

Slater, I., Gillespie, D., and Slater, D. W. (1973). Cytoplasmic adenylation and processing of maternal RNA. *Proc. Natl. Acad. Sci. U.S.A.* **70**, 406.

Smith, L. D. (1966). Role of a "germinal plasm" in the formation of primordial germ cells in *Rana pipiens*. *Dev. Biol.* **14**, 330.

Smith, L. D. (1975). Molecular events during oocyte maturation. *In* "Biochemistry of Animal Development" (R. Weber, ed.), Vol. 3, p. 1. Academic Press, New York.

Smith, L. D., and Ecker, R. E. (1965). Protein synthesis in enucleated eggs of *Rana pipiens*. *Science* **150**, 777.

Smith, L. D., and Ecker, R. E. (1969a). Role of the oocyte nucleus in physiological maturation in *Rana pipiens*. *Dev. Biol.* **19**, 281.

Smith, L. D., and Ecker, R. E. (1969b). Cytoplasmic regulation in early events of amphibian development. *In* "Canadian Cancer Conference," Proceedings of the Eighth Canadian Cancer Research Conference, Honey Harbour, Ontario, 1968 (J. F. Morgan, ed.), p. 103. Pergamon, Oxford.

Smith, L. D., and Ecker, R. E. (1970). Regulatory processes in the maturation and early cleavage of amphibian eggs. *Curr. Top. Dev. Biol.* **5**, 1.

Smith, L. D., and Williams, M. A. (1975). Germinal plasm and determination of the primordial germ cells. *In* "The Developmental Biology of Reproduction" (C. L. Markert and J. Papaconstantinou, eds.), p. 3. Academic Press, New York.

Smith, L. D., Ecker, R. E., and Subtelny, S. (1966). The initiation of protein synthesis in eggs of *Rana pipiens*. *Proc. Natl. Acad. Sci. U.S.A.* **56**, 1724.

Smith, L. J. (1956). A morphological and histochemical investigation of a preimplantation lethal (t^{12}) in the house mouse. *J. Exp. Zool.* **132**, 51.

Smith, M., Stavnezer, J., Huang, R. C., Gurdon, J. B., and Lane, C. D. (1973). Translation of messenger RNA for mouse immunoglobulin light chains in living frog oocytes. *J. Mol. Biol.* **80**, 553.

Smith, M. J., Hough, B. R., Chamberlin, M. E., and Davidson, E. H. (1974). Repetitive and non-repetitive sequence in sea urchin heterogeneous nuclear RNA. *J. Mol. Biol.* **85**, 103.

Smith, M. J., Britten, R. J., and Davidson, E. H. (1975). Studies on nucleic acid reassociation kinetics: Reactivity of single stranded tails in DNA–DNA renaturation. *Proc. Natl. Acad. Sci. U.S.A.* **72**, 4805.

Smith, M. J., Galau, G. A., Klein, W. H., Britten, R. J., and Davidson, E. H. (1976). Studies on nucleic acid reassociation kinetics. V. Kinetics of RNA-DNA hybridization in DNA excess. *Proc. Natl. Acad. Sci. U.S.A.*, in press.

Smith-Gill, S. J., Richards, C. M., and Nace, G. W. (1972). Genetic and metabolic bases of two "albino" phenotypes in the leopard frog, *Rana pipiens. J. Exp. Zool.* **180**, 157.

Solomon, J. (1957). Nucleic acid content of early chick embryos and hen's egg. *Biochim. Biophys. Acta* **24**, 584.

Solov'eva, I. A., and Timofeeva, M. Ya. (1974). Low-molecular-weight RNAs in loach embryogeny. *Mol. Biol. (Engl. Transl.)* **7**, 748.

Solov'eva, I. A., Timofeeva, M. Ya., and Sosinsakaya, I. E. (1974). Transfer RNAs in early embryogenesis of fish. Communication I. Content of tRNA in the unfertilized egg and developing embryo of the loach (*Misgurnus fossilis* L.). *Ontogenez (Sov. J. Dev. Biol.) (Engl. Transl.)* **4**, 315.

Sommerville, J. (1973). Ribonucleoprotein particles derived from the lampbrush chromosomes of newt oocytes. *J. Mol. Biol.* **78**, 487.

Sommerville, J. (1974). Informational ribonucleoprotein particles of newt oocytes: Polyribosome-associated ribonucleoproteins. *Biochim. Biophys. Acta* **349**, 96.

Sommerville, J., and Malcolm, D. B. (1976). Transcription of genetic information in amphibian oocytes. *Chromosoma* **55**, 183.

Sonneborn, T. M. (1950). The cytoplasm in heredity. *Heredity* **4**, 11.

Spemann, H. (1903). Entwicklungsphysiologische Studien am *Triton*-Ei. II. *Arch. Entwicklungsmech. Org.* **15**, 448.

Spemann, H. (1914). Uber verzögerte Kernversorgung von Keimteilen. *Verh. Dtsch. Zool. Ges.* **24**, 216.

Spiegel, M., Ozaki, H., and Tyler, A. (1965). Electrophoretic examination of soluble proteins synthesized in early sea urchin development. *Biochem. Biophys. Res. Commun.* **21**, 135.

Spirin, A. S. (1966). On "masked" forms of messenger RNA in early embryogenesis and in other differentiating systems. *Curr. Top. Dev. Biol.* **1**, 1.

Spirin, A. S., and Nemer, M. (1965). Messenger RNA in early sea-urchin embryos: Cytoplasmic particles. *Science* **150**, 214.

Spradling, A., Penman, S., Campo, M. S., and Bishop, J. O. (1974). Repetitious and unique sequences in the heterogeneous nuclear and cytoplasmic messenger RNA of mammalian and insect cells. *Cell* **3**, 23.

Stafford, D. W., and Iverson, R. M. (1964). Radioautographic evidence for the incorporation of leucine-carbon-14 into the mitotic apparatus. *Science* **143**, 580.

Stafford, D. W., Sofer, W. H., and Iverson, R. M. (1964). Demonstration of polyribosomes after fertilization of the sea urchin egg. *Proc. Natl. Acad. Sci. U.S.A.* **52**, 313.

Stavy, L., and Gross, P. R. (1969). Availability of mRNA for translation during normal and transcription-blocked development. *Biochim. Biophys. Acta* **182**, 203.

Stedman, E., and Stedman, E. (1950). Cell specificity of histones. *Nature (London)* **166**, 780.

Steinert, G., and Van Gansen, P. (1971). Binding of ³H-actinomycin to vitelline platelets of amphibian oocytes. *Exp. Cell Res.* **64**, 355.

Stephens, R. E. (1972). Studies on the development of the sea urchin *Strongylocentrotus droebachiensis*. III. Embryonic synthesis of ciliary proteins. *Biol. Bull.* **142**, 489.

Stern, M. S. (1972). Experimental studies on the organization of the preimplantation mouse embryo. II. Reaggregation of disaggregated embryos. *J. Embryol. Exp. Morphol.* **28**, 255.

Stern, M. S. (1973). Development of cleaving mouse embryos under pressure. *Differentiation* **1**, 407.

Stern, M. S., and Wilson, I. B. (1972). Experimental studies on the organization of the preimplantation mouse embryo. I. Fusion of asynchronously cleaving eggs. *J. Embryol. Exp. Morphol.* **28**, 247.

Stern, R., and Friedman, R. M. (1970). Double-stranded RNA synthesis in animal cells in the presence of actinomycin D. *Nature (London)* **226**, 612.

Stevens, N. M. (1909). The effect of ultra-violet light upon the developing eggs of *Ascaris megalocephala*. *Arch. Entwicklungsmech. Org.* **27**, 622.

Stone, L. S. (1950). Neural retina degeneration followed by regeneration from surviving retinal pigment cells in grafted adult salamander eyes. *Anat. Rec.* **106**, 89.

Strasburger, E. (1877). Ueber Befruchtung und Zelltheilung. *Jena. Z. Naturwiss.* **11**, 437.

Strasburger, E. (1884). New investigations on the course of fertilization in the phanerograms as basis for a theory of heredity. *In* "The Chromosome Theory of Inheritance: Classic Papers in Development and Heredity" (B. R. Voeller, ed.), p. 22. Appleton, New York, 1968 (transl.).

Straus, N. A. (1971). Comparative DNA renaturation kinetics in amphibians. *Proc. Natl. Acad. Sci. U.S.A.* **68**, 799.

Sullivan, D., Palacios, R., Stavnezer, J., Taylor, J. M., Faras, A. J., Kiely, M. L., Summers, N. M., Bishop, J. M., and Schimke, R. T. (1973). Synthesis of a deoxyribonucleic acid sequence complementary to ovalbumin messenger ribonucleic acid and quantification of ovalbumin genes. *J. Biol. Chem.* **248**, 7530.

Summers, R. G. (1970). The effect of actinomycin D on demembranated *Lytechinus variegatus* embryos. *Exp. Cell Res.* **59**, 170.

Suzuki, Y., Gage, L. P., and Brown, D. D. (1972). The genes for silk fibroin in *Bombyx mori*. *J. Mol. Biol.* **70**, 637.

Swift, H., and Kleinfeld, R. (1953). DNA in grasshopper spermatogenesis, oogenesis and cleavage. *Physiol. Zool.* **26**, 301.

Tanabe, K., and Kotani, M. (1974). Relationship between the amount of the "germinal plasm" and the number of primordial germ cells in *Xenopus laevis*. *J. Embryol. Exp. Morphol.* **31**, 89.

Tarkowski, A. K. (1959a). Experiments on the development of isolated blastomeres of mouse eggs. *Nature (London)* **184**, 1286.

Tarkowski, A. K. (1959b). Experimental studies on regulation in the development of isolated blastomeres of mouse eggs. *Acta Theriol.* **3**, 191.

Tarkowski, A. K. (1961). Mouse chimaeras developed from fused eggs. *Nature (London)* **190**, 857.

Tarkowski, A. K. (1963). Studies on mouse chimaeras developed from eggs fused *in vitro*. *Natl. Cancer Inst., Monogr.* **11**, 37.

Tarkowski, A. K., and Wróblewska, J. (1967). Development of blastomeres of mouse eggs isolated at the 4- and 8-cell stages. *J. Embryol. Exp. Morphol.* **18**, 155.

Tasca, R. J., and Hillman, N. (1970). Effects of actinomycin D and cycloheximide on RNA and protein synthesis in cleavage stage mouse embryos. *Nature (London)* **225**, 1022.

Teitelman, G. (1973). Protein synthesis during *Ilyanassa* development: Effect of the polar lobe. *J. Embryol. Exp. Morphol.* **29**, 267.

Telfer, W. H. (1965). The mechanism and control of yolk formation. *Ann. Rev. Entomol.* **10**, 161.

Tennent, D. H. (1914). The early influence of the spermatozoan upon the characters of echinoid larvae. *Carnegie Inst. Washington Publ.* **182**, 129.

Tennent, D. H. (1922). Studies on the hybridization of echinoids, *Cidaris tribuloides*. *Carnegie Inst. Washington Publ.* **312**, 1.

Tennent, D. H., and Ito, T. (1941). A study of the oogenesis of *Mespilia globulus* (Linné). *J. Morphol.* **69**, 347.

Terman, S. A. (1970). Relative effect of transcription-level and translation-level control of protein synthesis during early development of the sea urchin. *Proc. Natl. Acad. Sci. U.S.A.* **65**, 985.

Terman, S. A. (1972). Extent of post-transcriptional level control of protein synthesis in the absence of cell division. *Exp. Cell Res.* **72**, 576.

Terman, S. A., and Gross, P. R. (1965). Translation level control of protein synthesis during early development. *Biochem. Biophys. Res. Commun.* **21**, 595.

Thaler, M. M., Cox, M. C. L., and Villee, C. A. (1970). Histones in early embryogenesis. Developmental aspects of composition and synthesis. *J. Biol. Chem.* **245**, 1479.

Thomas, C. (1974). RNA metabolism in previtellogenic oocytes of *Xenopus laevis*. *Dev. Biol.* **39**, 191.

Tilney, L. G., and Gibbins, J. R. (1969). Microtubules in the formation and development of the primary mesenchyme in *Arbacia punctulata*. II. An experimental analysis of their role in development and maintenance of cell shape. *J. Cell Biol.* **41**, 227.

Timourian, H., and Watchmaker, G. (1970). Protein synthesis in sea urchin eggs. II. Changes in amino acid uptake and incorporation at fertilization. *Dev. Biol.* **23**, 478.

Tocco, G., Orengo, A., and Scarano, E. (1963). Ribonucleic acids in early embryonic development of the sea urchin. I. Quantitative variations and ^{32}P orthophosphate incorporation studies of the RNA of subcellular fractions. *Exp. Cell Res.* **31**, 52.

Tompkins, R., and Rodman, W. P. (1971). The cortex of *Xenopus laevis* embryos: Regional differences in composition and biological activity. *Proc. Natl. Acad. Sci. U.S.A.* **68**, 2921.

Trelstad, R. L., Hay, E. D., and Revel, J.-P. (1967). Cell contact during early morphogenesis in the chick embryo. *Dev. Biol.* **16**, 78.

Trinkaus, J. P. (1966). Morphogenetic cell movements. *In* "Major Problems of Developmental Biology" 25th Symp. Soc. Dev. Biol. (M. Locke, ed.), p. 125. Academic Press, New York.

Trinkaus, J. P. (1973). Surface activity and locomotion of *Fundulus* deep cells during blastula and gastrula stages. *Dev. Biol.* **30**, 68.

Trinkaus, J. P., and Lentz, T. L. (1967). Surface specialization of *Fundulus* cells and their relation to cell movements during gastrulation. *J. Cell Biol.* **32**, 139.

Tyler, A. (1963). The manipulations of macromolecular substances during fertilization and early development of animal eggs. *Am. Zool.* **3**, 109.

Tyler, A. (1965). The biology and chemistry of fertilization. *Am. Nat.* **99**, 309.

Tyler, A. (1967). Masked messenger RNA and cytoplasmic DNA in relation to protein synthesis and processes of fertilization and determination in embryonic development. *Dev. Biol. Suppl.* **1**, 170.

Ullerich, F.-H. (1970). DNS-gehalt und chromosomenstruktur bei amphibien. *Chromosoma* **30**, 1.

Vacquier, V. D. (1975). The isolation of intact cortical granules from sea urchin eggs: Calcium ions trigger granule discharge. *Dev. Biol.* **43**, 62.

Van Beneden, E. (1883). Recherches sur la maturation de l'oeuf et la fécondation et la division cellulaire. *Arch. Biol.* **4**, 265.

Van Blerkom, J., and Brockway, G. O. (1975). Qualitative patterns of protein synthesis in the preimplantation mouse embryo. I. Normal pregnancy. *Dev. Biol.* **44**, 148.

Van Blerkom, J., and Manes, C. (1974). Development of preimplantation rabbit embryos *in vivo* and *in vitro*. *Dev. Biol.* **40**, 40.

Van Blerkom, J., Barton, S. C., and Johnson, M. H. (1976). Molecular differentiation in the preimplantation mouse embryo. *Nature (London)* **259**, 319.

Van de Kerckhove, D. (1959). Content of deoxyribonucleic acid of the germinal vesicle of the primary oocyte in the rabbit. *Nature (London)* **183**, 329.

Vanderslice, R., and Hirsh, D. (1976). Temperature-sensitive zygote defective mutants of *Caenorhabditis elegans*. *Dev. Biol.* **49**, 236.

Van Dongen, C. A. M., and Geilenkirchen, W. L. M. (1974). The development of *Dentalium* with special reference to the significance of the polar lobe. I. Division chronology and development of the cell pattern in *Dentalium dentale* (Scaphopoda). *Proc. K. Ned. Akad. Wet., Ser. C* **77**, 57.

Van Gansen, P., and Schram, A. (1969). Etude des ribosomes et du glycogéne des gastrules de *Xenopus laevis* par cytochimie ultrastructurale. *J. Embryol. Exp. Morphol.* **22**, 69.

Van Gansen, P., and Schram, A. (1974). Incorporation of [³H]uridine and [³H]thymidine during the phase of nucleolar multiplication in *Xenopus laevis* oögenesis: A high-resolution autoradiographic study. *J. Cell Sci.* **14**, 85.

van Holde, K. E., Sahasrabuddhe, C. G., and Shaw, B. R. (1974). Electron microscopy of chromatin subunit particles. *Biochem. Biophys. Res. Commun.* **60**, 1365.

Vassart, G., Brocas, H., Nokin, P., and Dumont, J. E. (1973). Translation in *Xenopus* oocytes of thyroglobulin mRNA isolated by poly(U)-Sepharose affinity chromatography. *Biochim. Biophys. Acta* **324**, 575.

Vassart, G., Brocas, H., Lecocq, R., and Dumont, J. E. (1975). Thyroglobulin messenger RNA: Translation of a 33-S mRNA into a peptide immunologically related to thyroglobulin. *Eur. J. Biochem.* **55**, 15.

Verdonk, N. H. (1968). The effect of removing the polar lobe in centrifuged eggs of *Dentalium*. *J. Embryol. Exp. Morphol.* **19**, 33.

Verdonk, N. H., and Cather, J. N. (1973). The development of isolated blastomeres in *Bithynia tentaculata* (Prosobranchia, Gastropoda). *J. Exp. Zool.* **186**, 47.

Verdonk, N. H., Geilenkirchen, W. L. M., and Timmermans, L. P. M. (1971). The localization of morphogenetic factors in uncleaved eggs of *Dentalium*. *J. Embryol. Exp. Morphol.* **25**, 57.

Versteegh, L. R., Hearn, T. F., and Warner, C. M. (1975). Variations in the amounts of RNA polymerase forms I, II and III during preimplantation development in the mouse. *Dev. Biol.* **46**, 430.

Vlad, M., and MacGregor, H. C. (1975). Chromomere number and its genetic significance in lampbrush chromosomes. *Chromosoma* **50**, 327.

von Ubisch, L. (1954). Ueber Seeigelmerogone. *Pubbl. Stn. Zool. Napoli* **25**, 246.

Vorobyev, V. I., Gineitis, A. A., and Vinogradova, I. A. (1969). Histones in early embryogenesis. *Exp. Cell Res.* **57**, 1.

Wada, K., Shiokawa, K., and Yamana, K. (1968). Inhibitor of ribosomal RNA synthesis in *Xenopus laevis* embryos. I. Changes in activity of the inhibitor during development and its distribution in early gastrulae. *Exp. Cell Res.* **52**, 252.

Waldeyer, W. (1870). "Eirstock und Ei." Engelmann, Leipzig.

Walker, P. M. B. (1971). Origin of satellite DNA. *Nature (London)* **229**, 306.

Wallace, H. (1966). Autoradiographic studies of RNA and protein synthesis in anucleolate *Xenopus laevis*. *Natl. Cancer Inst., Monogr.* **23**, 425.

Wallace, H., and Birnstiel, M. L. (1966). Ribosomal cistrons and the nucleolar organizer. *Biochim. Biophys. Acta* **114**, 296.

Wallace, H., and Elsdale, T. R. (1963). Effects of actinomycin D on amphibian development. *Acta Embryol. Morphol. Exp.* **6**, 275.

Wallace, R. A., Nickol, J. M., Ho, T., and Jared, D. W. (1972). Studies on amphibian yolk. X. The relative roles of autosynthetic and heterosynthetic processes during yolk protein assembly by isolated oocytes. *Dev. Biol.* **29**, 255.

Waring, M., and Britten, R. J. (1966). Nucleotide sequence repetition: A rapidly reassociating fraction of mouse DNA. *Science* **154**, 791.

Warn, R. (1975). Restoration of the capacity to form pole cells in UV-irradiated *Drosophila* embryos. *J. Embryol. Exp. Morphol.* **33**, 1003.

Warneck, N. A. (1850). Ueber die Bildung und Entwickelung des Embryos bei Gasteropoden. *Moskovskoe Obshchestvo Ispytatelel Prirody* **23**, No. 1, 90.

Wasserman, P. M., Hollinger, T. G., and Smith, L. D. (1972). RNA polymerases in the germinal vesicle contents of *Rana pipiens* oocytes. *Nature (London), New Biol.* **240**, 208.

Watterson, R. L. (1956). Selected invertebrates. *In* "Analysis of Development" (B. H. Willier, P. A. Weiss, and V. Hamburger, eds.), p. 315. Saunders, Philadelphia, Pennsylvania.

Webb, A. C., LaMarca, M. J., and Smith, L. D. (1975). Synthesis of mitochondrial RNA by full-grown and maturing oocytes of *Rana pipiens* and *Xenopus laevis*. *Dev. Biol.* **45**, 44.

Wegnez, M., and Denis, H. (1972). Recherches biochimiques sur l'oogénèse. 4. Absence d'amplification des gènes organisateurs du RNA 5 S et du tRNA dans les petits oocytes de *Xenopus laevis*. *Biochimie* **54**, 1069.

Wegnez, M., Monier, R., and Denis, H. (1972). Sequence heterogeneity of 5 S RNA in *Xenopus laevis*. *FEBS Lett.* **25**, 13.

Weinberg, E. S., Birnstiel, M. L., Purdom, I. F., and Williamson, R. (1972). Genes coding for polysomal 9S RNA of sea urchins: Conservation and divergence. *Nature (London)* **240**, 225.

Weismann, A. (1885). The continuity of the germ-plasm as the foundation of a theory of heredity. Translated in "Essays upon Heredity and Kindred Biological Problems" (E. B. Poulton, S. Schönland, and A. E. Shipley, eds.), Vol. 1, p. 167. Oxford Univ. Press (Clarendon), London and New York, 1891.

Wellauer, P. K., Reeder, R. H., Carroll, D., Brown, D. D., Deutch, A., Higashinakagawa, T., and Dawid, I. B. (1975). Amplified ribosomal DNA from *Xenopus laevis* has heterogeneous spacer lengths. *Proc. Natl. Acad. Sci. U.S.A.* **71**, 2823.

Wells, R., Royer, H. D., and Hollenberg, C. P. (1976). Non *Xenopus*-like DNA sequence organization in the *Chironomus tentans* genome. *Molec. Gen. Genet.* **147**, 45.

Westin, M., Perlmann, H., and Perlmann, P. (1967). Immunological studies of protein synthesis during sea urchin development. *J. Exp. Zool.* **166**, 331.

Wetmur, J. G. (1971). Excluded volume effects on the rate of renaturation of DNA. *Biopolymers* **10**, 601.

Wetmur, J. G. (1976). Hybridization and renaturation kinetics of nucleic acids. *Annu. Rev. Biophys. Bioeng.* **5**, 337.

Wetmur, J. G., and Davidson, N. (1968). Kinetics of renaturation of DNA. *J. Mol. Biol.* **31**, 349.

Whiteley, A. H. (1949). The phosphorus compounds of sea urchin eggs and the uptake of radio-phosphate upon fertilization. *Am. Nat.* **83**, 249.

Whiteley, A. H., and Baltzer, F. (1958). Development, respiratory rate and content of desoxyribonucleic acid in the hybrid *Paracentrotus* ♀ × *Arbacia* ♂. *Pubbl. Stn. Zool. Napoli* **30**, 402.

Whiteley, A. H., and Whiteley, H. R. (1972). The replication and expression of maternal and paternal genomes in a blocked echinoid hybrid. *Dev. Biol.* **29**, 183.

Whiteley, A. H., McCarthy, B. J., and Whiteley, H. R. (1966). Changing populations of messenger RNA during sea urchin development. *Proc. Natl. Acad. Sci. U.S.A.* **55**, 285.

Whiteley, H. R., McCarthy, B. J., and Whiteley, A. H. (1970). Conservatism of base sequences in RNA for early development of echinoderms. *Dev. Biol.* **21**, 216.

Whitman, C. O. (1878). The embryology of *Clepsine*. *Q. J. Microsc. Sci.* [N.S.] **18**, 215.

Whitman, C. O. (1895a). Evolution or epigenesis. *In* "Biological Lectures Delivered at the Marine Biological Laboratory of Woods Holl," 10th Lecture, p. 205. Ginn. Boston, Massachusetts.

Whitman, C. O. (1895b). Bonnet's theory of evolution. A system of negations. *In* "Biological Lectures Delivered at the Marine Biological Laboratory of Woods Holl," 12th Lecture, p. 225. Ginn, Boston, Massachusetts.

Whittaker, J. R. (1973a). Segregation during ascidian embryogenesis of egg cytoplasmic information for tissue-specific enzyme development. *Proc. Natl. Acad. Sci. U.S.A.* **70**, 2096.

Whittaker, J. R. (1973b). Tyrosinase in the presumptive pigment cells of ascidian embryos: Tyrosine accessibility may initiate melanin synthesis. *Dev. Biol.* **30**, 441.

Whittaker, J. R. (1973c). Evidence for the localization of an RNA template for endodermal alkaline phosphatase in an ascidian egg. *Biol. Bull.* **145**, 459.

Wieschaus, E., and Gehring, W. (1976). Clonal analysis of primordial disc cells in the early embryo of *Drosophila melanogaster*. *Dev. Biol.* **50**, 249.

Wilde, C. E., Jr., and Crawford, R. B. (1966). Cellular differentiation in the anamniota. III. Effects of actinomycin D and cyanide on the morphogenesis of *Fundulus*. *Exp. Cell Res.* **44**, 471.

Williams, M. A., and Smith, L. D. (1971). Ultrastructure of the "germinal plasm" during maturation and early cleavage in *Rana pipiens*. *Dev. Biol.* **25**, 568.

Wilson, E. B. (1893). *Amphioxus*, and the mosaic theory of development. *J. Morphol.* **18**, 579.

Wilson, E. B. (1896). On cleavage and mosaic work. An Appendix to Crampton (1896), p. 19.

Wilson, E. B. (1903). Experiments on cleavage and localization in the nemertine egg. *Arch. Entwicklungsmech. Org.* **16**, 411.

Wilson, E. B. (1904). Experimental studies in germinal localization. II. Experiments on the cleavage-mosaic in *Patella* and *Dentalium*. *J. Exp. Zool.* **1**, 197.

Wilson, E. B. (1925). "The Cell in Development and Heredity," 3rd ed. Macmillan, New York.

Wilson, I. B., and Stern, M. S. (1975). Organization in the preimplantation embryo. *In* "Early Development of Mammals" (M. Balls and A. E. Wild, eds.), p. 81. Cambridge Univ. Press, London and New York.

Wilson, I. B., Bolton, E., and Cuttler, R. H. (1972). Preimplantation differentiation in the mouse egg as revealed by microinjection of vital markers. *J. Embryol. Exp. Morphol.* **27**, 467.

Wilson, M. C., Melli, M., and Birnstiel, M. L. (1974). Reiteration frequency of histone coding sequences in man. *Biochem. Biophys. Res. Commun.* **61**, 354.

Wilt, F. H. (1963). The synthesis of ribonucleic acid in sea urchin embryos. *Biochem. Biophys. Res. Commun.* **11**, 447.

Wilt, F. H. (1964). Ribonucleic acid synthesis during sea urchin embryogenesis. *Dev. Biol.* **9**, 299.

Wilt, F. H. (1965). Regulation of the initiation of chick embryo hemoglobin synthesis. *J. Mol. Biol.* **12**, 331.

Wilt, F. H. (1970). The acceleration of ribonucleic acid synthesis in cleaving sea urchin embryos. *Dev. Biol.* **23**, 444.

Wilt, F. H. (1973). Polyadenylation of maternal RNA of sea urchin eggs after fertilization. *Proc. Natl. Acad. Sci. U.S.A.* **70**, 2345.

Wilt, F. H., and Hultin, T. (1962). Stimulation of phenylalanine incorporation by polyuridylic acid in homogenates of sea urchin eggs. *Biochem. Biophys. Res. Commun.* **9**, 313.

Wilt, F. H., Sakai, H., and Mazia, D. (1967). Old and new protein in the formation of the mitotic apparatus in cleaving sea urchin eggs. *J. Mol. Biol.* **27**, 1.

Wilt, F. H., Aronson, A. I., and Wartiovaara, J. (1969). Function of the nuclear RNA of sea urchin embryos. *In* "Problems in Biology: RNA in Development" (E. W. Hanly, ed.), p. 331. Univ. of Utah Press, Salt Lake City.

Wilt, F. H., Anderson, M., and Ekenberg, E. (1973). Centrifugation of nuclear ribonucleoprotein particles of sea urchin embryos in cesium sulfate. *Biochemistry* **12**, 959.

Wolpert, L. (1969). Positional information and the spatial pattern of cellular differentiation. *J. Theor. Biol.* **25**, 1.

Wolpert, L., and Gustafson, T. (1961). Studies on the cellular basis of morphogenesis of the sea urchin embryo. Development of the skeletal pattern. *Exp. Cell Res.* **25**, 311.

Wolstenholme, D. R. (1973). Replicating DNA molecules from eggs of *Drosophila melanogaster. Chromosoma* **43**, 1.

Woodland, H. R. (1974). Changes in the polysome content of developing *Xenopus laevis* embryos. *Dev. Biol.* **40**, 90.

Woodland, H. R., and Adamson, E. D. (1976). The synthesis and storage of histones during the oogenesis of *Xenopus laevis. Dev. Biol.*, in press.

Woodland, H. R., and Graham, C. F. (1969). RNA synthesis during early development of the mouse. *Nature (London)* **221**, 327.

Woodland, H. R., and Gurdon, J. B. (1968). The relative rates of synthesis of DNA, sRNA and rRNA in the endodermal region and other parts of *Xenopus laevis* embryos. *J. Embryol. Exp. Morphol.* **19**, 363.

Woodland, H. R., and Pestell, R. Q. W. (1972). Determination of the nucleoside triphosphate contents of eggs and oocytes of *Xenopus laevis. Biochem. J.* **127**, 597.

Wright, D. A., and Shaw, C. R. (1970). Time of expression of genes controlling specific enzymes in *Drosophila* embryos. *Biochem. Genet.* **4**, 385.

Wright, D. A., and Subtelny, S. (1971). Nuclear and cytoplasmic contributions to dehydrogenase phenotypes in hybrid frog embryos. *Dev. Biol.* **24**, 119.

Wu, R. S., and Wilt, F. H. (1973). Poly A metabolism in sea urchin embryos. *Biochem. Biophys. Res. Commun.* **54**, 704.

Wu, R. S., and Wilt, F. H. (1974). The synthesis and degradation of RNA containing polyriboadenylate during sea urchin embryogeny. *Dev. Biol.* **41**, 352.

Wylie, C. C. (1972). The appearance and quantitation of cytoplasmic ribonucleic acid in the early chick embryo. *J. Embryol. Exp. Morphol.* **28**, 367.

Yajima, H. (1960). Studies on embryonic determination of the harlequin-fly, *Chironomus dorsalis* I. Effects of centrifugation and its combination with constriction and puncturing. *J. Embryol. Exp. Morphol.* **8**, 198.

Yamada, T. (1967). Cellular and subcellular events in Wolffian lens regeneration. *Curr. Top. Dev. Biol.* **2**, 247.

Yamana, K., and Shiokawa, K. (1966). Occurrence of an inhibitory factor of ribosomal RNA synthesis in embryonic cells of *Xenopus laevis. Exp. Cell Res.* **44**, 283.

Yatsu, N. (1903). Experiments on the development of egg fragments in *Cerebratulus. Biol. Bull.* **6**, 123.

Yatsu, N. (1910a). Experiments on cleavage in the egg of *Cerebratulus*. *J. Coll. Sci., Imp Univ. Tokyo* **27**, No. 10.

Yatsu, N. (1910b). Experiments on germinal localization in the egg of *Cerebratulus*. *J. Coll. Sci., Imp. Univ. Tokyo* **27**, No. 17.

Yatsu, N. (1912). Observations and experiments on the Ctenophore egg. III. Experiments on germinal localization of the egg of *Beroe ovata*. *Ann. Zool. Jpn.* **8**, 5.

Yeoman, L. C., Olson, M. O. J., Sugano, N., Jordan, J. J., Taylor, C. W., Starbuck, W. C., and Busch, H. (1972). Amino acid sequence of the center of the arginine-lysine-rich histone from calf thymus. *J. Biol. Chem.* **247**, 6018.

Young, P. G., and Zimmerman, A. M. (1973). Synthesis of mitochondrial RNA in disaggregated embryos of *Xenopus laevis*. *Dev. Biol.* **33**, 196.

Zalokar, M. (1973). Transplantation of nuclei into the polar plasm of *Drosophila* eggs. *Dev. Biol.* **32**, 189.

Zalokar, M., Audit, C., and Erk, I. (1975). Developmental defects of female-sterile mutants of *Drosophila melanogaster*. *Dev. Biol.* **47**, 419.

Zamboni, L., and Gondos, B. (1968). Intercellular bridges and synchronization of germ cell differentiation during oogenesis in the rabbit. *J. Cell Biol.* **36**, 276.

Zamboni, L., Mishell, D. R., Jr., Bell, J. H., and Baca, M. (1966). Fine structure of the human ovum in the pronuclear stage. *J. Cell Biol.* **30**, 579.

Zeleny, C. (1904). Experiments on the localization of developmental factors in the nemertine egg. *J. Exp. Zool.* **1**, 293.

Zissler, D., and Sander, K. (1973). The cytoplasmic architecture of the egg cell of *Smittia* spec. (Diptera, Chironomidae). *Wilhelm Roux' Arch. Entwicklungsmech. Org.* **172**, 175.

Index

A

Aboral pole, 261–262
Acetylcholinesterase, 78, 299–300
Acheta domestica, duration of oogenesis, 355
Acmaea scutum, complexity of repetitive sequence transcripts, 211
Actin, electrophoretic identification of, 60–61
Actinomycin D, 45–51, 54, 67–68, 76, 121, 124, 134–135, 278, 281, 300
  analog of, 47
  chemical enucleation, 45–51
  delayed effect on morphogenesis, 51
  effects
    on chordate embryos, 49–51
    on early morphogenesis, 51–55
    on echinoderm embryos, 46–49, 78
    on enzyme synthesis, 300
    on histone synthesis, 121, 124
    on lithium chloride treatment, 54, 316
    on mesenchyme cell differentiation, 78
    on microtubule protein synthesis, 134–135
    on protein synthesis, 46, 50, 67–68, 278, 281
    on ribonucleoprotein particles, 368, 382
    on RNA synthesis, 46–50
    on tyrosine-DOPA oxidase accumulation, 76
  experiments, limitation of data from, 241
  model of action on transcription, 54
Alkaline phosphatase, 276–277, 300
  inner cell mass marker, 306
Allocentrotus fragilis, species hybrid experiments in, 37
Altered cell fate experiments, 4, *see also*
Transdifferentiation
α-Amanitin, effect on embryonic RNA synthesis, 50, 102, 176
Ambystoma mexicanum, *see* Axolotl
Amino acid, free, measurement of pool of, 87–88
Aminoacyl-tRNA synthetase, 94–95
Amphibians eggs, *see also* Axolotl; *Engystomops*; *Plethodon*; *Rana*; *Salamandra*; *Triton*; *Triturus*; *Xenopus*
  axial determinants in, 300–304
    localization of, 302–304, 315–316
  cortical localization in, 300–304
  enucleated, maternal messenger RNA in, 99
  external permeability barrier, 139
  germ cell determinants in, 286–287
  nuclear growth stages, 341
  polar granules in, 287–289
Amphibian embryos
  cell disaggregation experiments, 150–151
  electrophoretic protein pattern, 62, 64–65
  genome control in hybrids, 38–40
  heterogeneous nuclear RNA sequence organization in, 218
  lobopodia in, 74
  protein synthesis rate in, 91–92
  regulative development, 310
  RNA synthesis in, 139–154
Amphibian oogenesis
  lampbrush phase duration, 340–343
  stages of, 340–343
Amphioxus, blastomere isolation experiment, 310
Androgenetic haploid, development in, 53, *see also* Merogone
Animal pole, 287
Animalization, 316

433